Upgrading Urban Power Grids

Other related titles:

You may also like

- PBPO086 | Federico Milano | Advances in Power System Modelling, Control and Stability Analysis | 2016
- PBPO147 | Juan Manuel Gers | Distribution Systems Analysis and Automation, 2nd Edition | 2020
- PBPO161 | Marcelo Godoy Simões | Artificial Intelligence for Smarter Power Systems | 2021
- PBRN106 | Salvador Acha | Modelling Distributed Energy Resources in Energy Service Networks | 2013

We also publish a wide range of books on the following topics:
Computing and Networks
Control, Robotics and Sensors
Electrical Regulations
Electromagnetics and Radar
Energy Engineering
Healthcare Technologies
History and Management of Technology
IET Codes and Guidance
Materials, Circuits and Devices
Model Forms
Nanomaterials and Nanotechnologies
Optics, Photonics and Lasers
Production, Design and Manufacturing
Security
Telecommunications
Transportation

All books are available in print via https://shop.theiet.org or as eBooks via our Digital Library https://digital-library.theiet.org.

IET ENERGY ENGINEERING SERIES 244

Upgrading Urban Power Grids

Juan M. Gers

The Institution of Engineering and Technology

About the IET

This book is published by the Institution of Engineering and Technology (The IET).

We inspire, inform and influence the global engineering community to engineer a better world. As a diverse home across engineering and technology, we share knowledge that helps make better sense of the world, to accelerate innovation and solve the global challenges that matter.

The IET is a not-for-profit organisation. The surplus we make from our books is used to support activities and products for the engineering community and promote the positive role of science, engineering and technology in the world. This includes education resources and outreach, scholarships and awards, events and courses, publications, professional development and mentoring and advocacy to governments.

To discover more about the IET please visit https://www.theiet.org/.

About IET books

The IET publishes books across many engineering and technology disciplines. Our authors and editors offer fresh perspectives from universities and industry. Within our subject areas, we have several book series steered by editorial boards made up of leading subject experts.

We peer review each book at the proposal stage to ensure the quality and relevance of our publications.

Get involved

If you are interested in becoming an author, editor, series advisor, or peer reviewer please visit https://www.theiet.org/publishing/publishing-with-iet-books/ or contact author_support@theiet.org.

Discovering our electronic content

All of our books are available online via the IET's Digital Library. Our Digital Library is the home of technical documents, eBooks, conference publications, real-life case studies and journal articles. To find out more, please visit https://digital-library.theiet.org.

In collaboration with the United Nations and the International Publishers Association, the IET is a Signatory member of the SDG Publishers Compact. The Compact aims to accelerate progress to achieve the Sustainable Development Goals (SDGs) by 2030. Signatories aspire to develop sustainable practices and act as champions of the SDGs during the Decade of Action (2020–30), publishing books and journals that will help inform, develop, and inspire action in that direction.

In line with our sustainable goals, our UK printing partner has FSC accreditation, which is reducing our environmental impact to the planet. We use a print-on-demand model to further reduce our carbon footprint.

Published by The Institution of Engineering and Technology, London, United Kingdom

The Institution of Engineering and Technology (the "**Publisher**") is registered as a Charity in England & Wales (no. 211014) and Scotland (no. SC038698).

First published 2024

The Institution of Engineering and Technology
Futures Place
Kings Way, Stevenage
Herts, SG1 2UA, United Kingdom
www.theiet.org

British Library Cataloguing in Publication Data

A catalogue record for this product is available from the British Library

ISBN 978-1-83953-784-4 (hardback)
ISBN 978-1-83953-785-1 (PDF)

Typeset in India by MPS Limited

Cover image credit: Juan Gers

Contents

Preface

Distribution systems evolved notoriously from the last decades of the 20th century. Undoubtedly more emphasis and investment were given to transmission and generation systems throughout the century. A main factor that came to change definitively this perspective was the notorious importance given to the efficiency, quality and reliability of the electrical service whose standards have been greatly risen by regulatory bodies all over the world. A second factor was the great developments in technology applied to distribution systems in equipment, hardware and software platforms. A third factor undoubtedly has been the development of green energy sources and in particular of solar facilities that are connected to distribution systems that gave the origin to what is called Distribute Energy Resources-DER including roof-type installations. These sources are changing the operation of the entire electrical system worldwide.

National grids feed distribution systems through electrical substations from where feeders start, to serve either big urban and/or rural feeders. This book concentrates mainly in urban areas that have special characteristics that differentiate them from rural areas. In urban areas, the load is highly interconnected and normally the standards of service are more stringent.

The book has been divided into eight chapters. Chapter 1 presents the introduction to urban power grids and to basic concepts applied in distribution systems. Chapter 2 makes reference to benefits, concepts and modelling of maturity models to implement smart grids. Chapter 3 is devoted to basic concepts of load management and demand response which are essential to improve the operating condition and efficiency of the distribution systems. Essential components of distribution automation related to service restoration are discussed in Chapter 4. These are essential to increase the reliability of energy service. In Chapter 5, reference is done to renewable sources and microgrids which are some of the most important topics of distribution systems. The high penetration of generation adjacent to the loads has changed dramatically how distribution systems are operated. A topic which receives every day more importance, not only in distribution systems but also in generation and transmission systems, is that of protection and automation, which is discussed in Chapter 6. Here, some important concepts of communications and protocols are also discussed due to the need to have fast clearing when events happen. Voltage control became a very demanding aspect to keep appropriate quality of service, which is the objective of Chapter 7. This includes considerations not only on capacitor application but also on voltage regulators. Optimal power flow is included as a way to improve performance of systems operation.

The technology associated with urban systems and in general with distribution systems is changing at fast rate and therefore some discussion points are suggested in the last chapter of the book. Important developments like real-time simulation are discussed. Surely new concepts and techniques will be developed and that oblige our industry to keep pace with more challenges to come.

The approach to develop the book has been to cover fundamentals on each section and then develop concepts from there in a clear and updated way. The technical concepts presented in the book are well complemented with case studies. Some of them have been taken from real experiences and others are more theoretical.

Thanks are given to colleagues of my consulting engineers practice and to students of the universities where I teach currently as Adjunct Professor. Both experiences have enriched my professional career and provided valuable information that is somehow shared in this book. Very special thanks are given to my colleagues Carlo Viggiano and Juan Martin Guardiola for providing ideas and exercises to prepare the book. Long discussions with them helped to improve the content. Thanks are giving also to Leinyker Palacios for the good discussions too and for providing illustrative case examples. I am indebted to Dr Adam Lupinski and other colleagues of NEPLAN for authorizing the use of references material. As a matter of fact, most cases in the book use NEPLAN in the illustrations, although the course data is provided for readers to use any platform that they have available. Finally special thanks are given to many officials from IET for the encouragement, guidance and support to complete this new effort.

I certainly hope that the book provides useful insight and information to readers on this important and everyday changing, evolving knowledge associated with urban networks of distribution systems. Comments and questions will be gladly received and attended to my contact information.

Juan M. Gers
Juan.gers@gers.com
Weston, FL, USA
May 2024

About the author

Juan M. Gers is president and founder of GERS Consulting Engineers which has completed projects in more than 45 countries. He has more than 40 years of experience designing and analysing transmission and distribution systems. His books, *Protection of Electricity Distribution Networks* and *Distribution System Analysis and Automation*, published by the IET, are used by utilities, consulting engineers and universities worldwide. Dr Gers is a Chartered member of the IET and Senior Member of the IEEE.

Chapter 1

Urban power grids

Nowadays, a growing trend in renewable generation is the planning and operation strategies of transmission and distribution networks. Inverter-based resources (IBRs) are typical technologies in renewable generation that allow the implementation of control strategies to maximise the use of available resources. However, the transformation of conventional networks becomes a challenge considering that power systems were based on unidirectional flows and rotative machines for generation.

Different technologies and information services are being used for the modernisation of networks and the introduction of smart grids (SG). The latter includes aspects of energy generation, transmission and distribution and aims for a more reliable service, higher efficiency, more security, two-way utility-user communications and promotion of green energy, among other goals.

1.1 Introduction to urban power grids

A power distribution system is a set of electrical equipment that allows the energy to safely and reliably service loads at different voltage levels distributed throughout a region. Depending on the load characteristics and the size of the power system, the distribution systems are classified as industrial, commercial, urban and rural.

Urban power grids are systems that provide the supply of energy to towns and high-consumption urban centres. The system is composed of a high number of transformers in a meshed configuration (but operated radially), since topologies with different interconnection possibilities are suitable for this type of network. The type of load in urban distribution networks is mainly domestic and commercial.

However, rural or semi-urban systems are usually characterised by long lines with radial configuration and a reduced number of transformers, which support the electricity supply to areas with a low load density. In some cases, it is required to generate energy locally due to the long distances and the size of the loads depending on the feasibility for interconnection to the bulk power system.

Figure 1.1(a) illustrates a geographic reference of an urban power grid and Figure 1.1(b) the same for a semi-urban power grid.

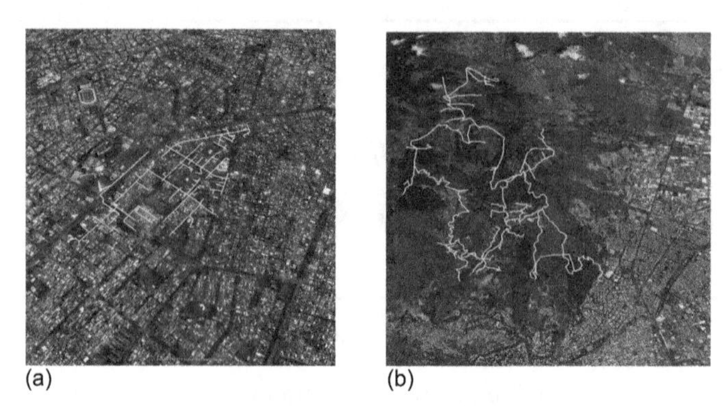

Figure 1.1 (a) Urban and (b) semi-urban power grids

1.2 Power systems evolution

Power systems are electrical networks composed of different elements installed to supply, transfer and use the electrical energy [1]. Its development began in the last two decades of the 19th century [2]. Initially, electrical systems fulfilled the function to supply energy for street lighting like the Pearl Street station in New York, which began operation on 4 September 1882. This energy was generated from direct current (DC) machines moved by steam turbines. The energy was also used for heating homes and the operation of motors.

At that time, the alternating current (AC) system had several advantages over the DC system [3]. Even so, DC seemed to prevail [4]. One of the advantages of AC was its easier transmission for long distances. This is due to the inherent variability of voltage and current (sinusoidal in nature). Transformers made the use of different voltage levels for generation, transmission and distribution (GTD) of electricity possible, which allowed electrical power systems to transmit electricity to remote places.

Some of the key stages for the electricity grid development in the 19th century are shown in Figure 1.2 and can be summarised as follows:

- 1886: George Westinghouse and William Stanley demonstrated the transmission in AC to remote loads for Great Barrington, Massachusetts, USA. The same year, the Westinghouse Electric Company successfully tested a four-mile transmission line in Lawrenceville, Pennsylvania.
- 1889: Thomas Edison buys Thomson-Houston and merges it with Edison General Electric to form the General Electric Company that starts a fierce competition with Westinghouse Electric Company for AC power projects.
- 1890: Westinghouse Electric had about 300 generating stations to supply AC electricity to lighting loads.
- 1893: George Westinghouse won the contract for the lighting of the Chicago Columbian Exposition and takes the opportunity to show his two-phase system with 12-1000 HP alternators.

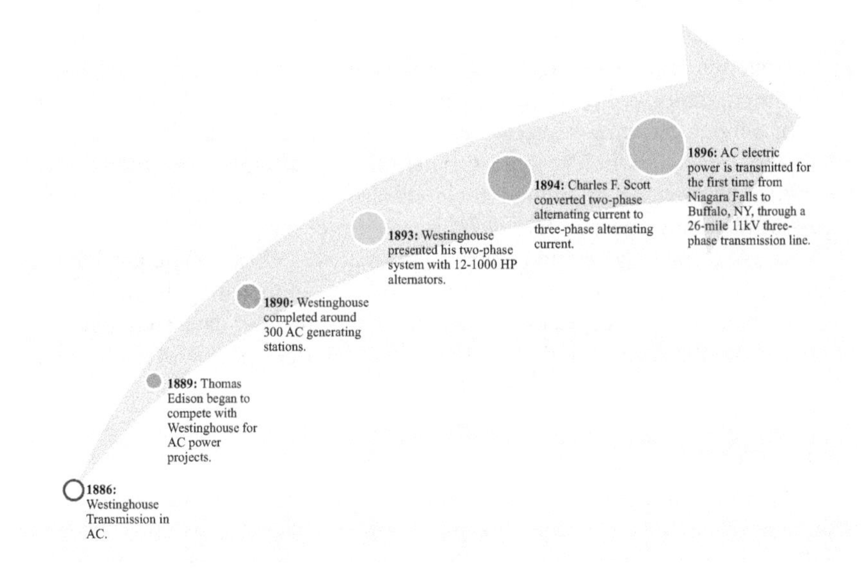

Figure 1.2 Power systems evolution

- 1894: Charles F. Scott from England invented the ‘T’ connection for transformers to convert two-phase AC to three-phase AC.
- 1896: AC electric power is transmitted for the first time from Niagara Falls to Buffalo, NY, through a 26-mile 11 kV three-phase transmission line.

Thus, the conception of a three-phase AC electrical power system was accepted as a standard, allowing electricity to increase its participation in the industrial, commercial and residential segments around the world [5]. However, power systems have also experienced problems such as voltage regulation as a result of the transmission of electrical energy over long distances, among others [6].

1.2.1 Vertical integration

Due to the operation of today’s electrical systems, electricity companies have emerged to supply electricity to final consumers. Initially, the companies were responsible for generation, transmission and energy distribution (vertical integration model).

Economic theory understands vertical integration as a response to pre-existing market power problems or as a strategic action by a company to increase its market power. This phenomenon implies that a company integrates new complementary activities to provide goods or services, which can occur with the purpose of reducing the production and transaction costs. However, a powerful incentive is having control of the market for the electricity sector.

There are different reasons to justify the traditional vertical integration model and the monopoly for electricity generation. It is possible to highlight the existence of scale economies for generation and the development of interconnections, where the sector has been very intensive in the use of capital. Thus, it favours the use of large production units, which still are maintained in current electrical systems.

However, traditional power companies, which were previously able to produce power by relying on economies of scale, have seen their era begin to come to an end. In the 1960s and 1970s, large generation thermal units were found to have operational problems that limited their efficiency. This limitation stopped the process of obtaining cheaper energy from ever larger power units [7].

Moreover, small generation units have improved their efficiency with new technologies such as mini-hydraulics, gas turbines, combined cycle and fuel cells [8,9]. In addition, the integration of computer systems and data communications has helped in the supervision and control of electrical power systems, in turn reducing the costs of their operation [10].

1.2.2 Present

Currently, electrical systems are undergoing different changes due to the penetration of new technologies in generation and demand. This transformation, which can be called the 'energy transition', mainly refers to two major changes: the integration of renewable energy sources and the use of electric vehicles (EVs).

Conventional power grids were not planned with these technological innovations, so they must change significantly to support them [1]. In addition, it is expected that these technologies can be acquired and managed by final consumers, who would become owners of the resource and an active part of the electrical system.

The end-user will become a key player in the energy market, and its influence on the development of electricity networks will gradually grow. Based on this, electric companies and governments must fundamentally change their ways for planning and operation of the grid [11]. The modernisation of the electrical network must consider:

- Integration of intermittent energy sources such as wind and solar.
- Bi-directional flow of energy from the grid to consumers and from consumers to the grid.
- Bi-directional flow of information in real time between producers and consumers' smart devices and meters.
- Maximum quality, efficiency, reliability and safety.
- Optimal cost for producers and consumers.

Therefore, new generation, information, communication and control technologies will play a fundamental role due to the new trends in renewable sources and the active participation of demand, where a clean environment is also promoted. In addition, it considers decentralised platforms of generation, distribution and consumption on a small scale. Hence, the questions about how to facilitate the modernisation of the electrical system and what requirements must be considered for its planning are important considerations for those who directly face the new energy challenges [12]. Figure 1.3 illustrates a conventional distribution system and Figure 1.4 illustrates how the same conventional distribution system can be upgraded with new technologies, including but not limited to non-conventional renewable generation and EVs.

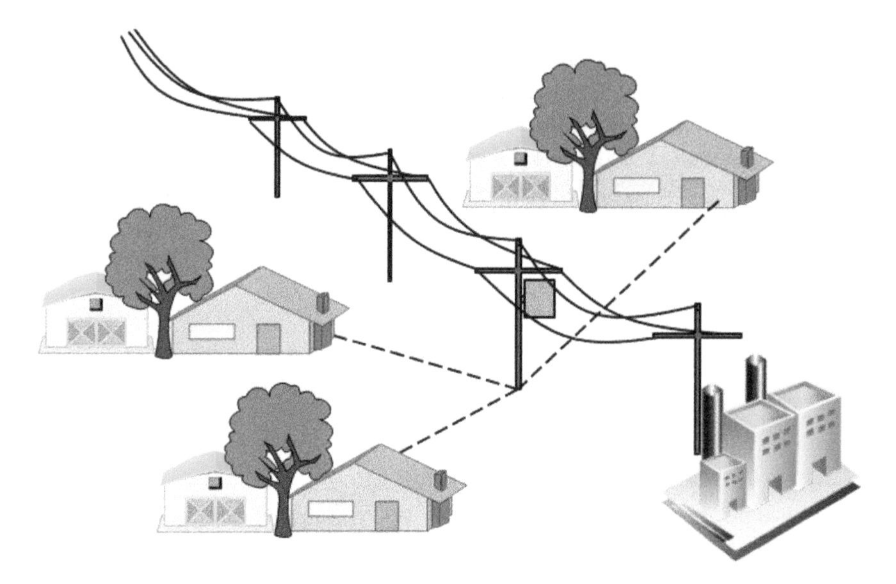

Figure 1.3 Illustration of a conventional distribution power system

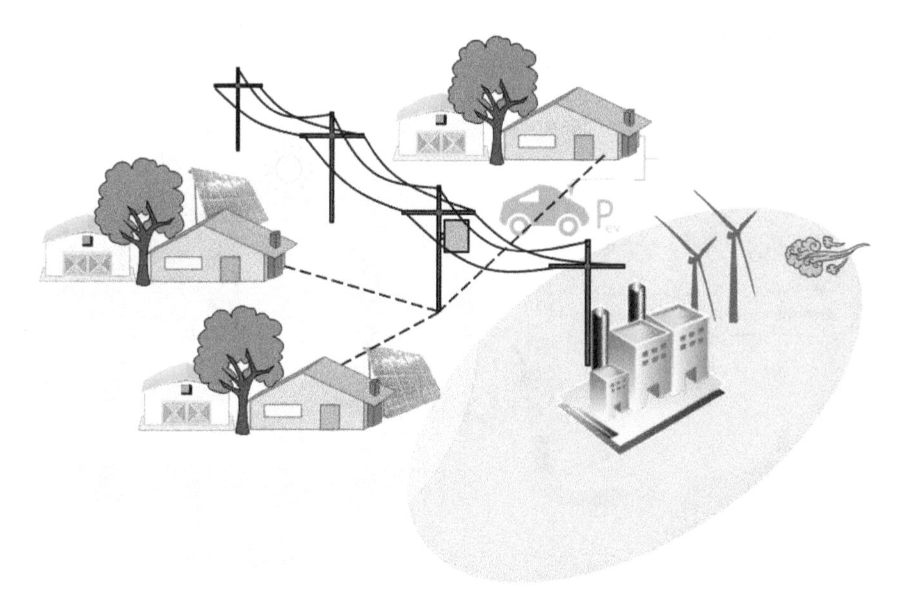

Figure 1.4 Illustration of the upgraded distribution power system

1.3 Conventional and non-conventional energy sources

Traditional schemes for GTD face challenges related to sustainability, efficiency and scalability [13]. Since most of these schemes have centralised generation models with large-scale power generation plants, there are some risks for operational and environmental problems that could make the electrical network unsustainable and inefficient [14]. On the other hand, conventional energy source reserves, such as coal, oil and gas, are experiencing high restrictions worldwide due to environmental concerns.

Countries around the world are implementing pilot cases for networks with distributed generation and non-conventional energy resources, to evaluate their strengths and weaknesses. These energy solutions would allow the decentralisation of GTD operations, savings in fossil fuel reserves, expansion costs, reduction of energy losses in transmission and distribution and reduction of emissions [15].

1.3.1 Conventional generation technologies

Conventional energy sources have been used for decades [16]. Figure 1.5 presents a summary for conventional generation and some of their characteristics are described in the following sections.

1.3.1.1 Natural gas thermal technology

Natural gas power plants use high-pressure steam from natural gas combustion to drive a turbine. These types of thermal power plants are generally located near the coast or near water sources due to the feasibility to obtain water for the cooling process. Gas turbines take advantage of the Brayton and Rankine thermal cycles to generate electricity [17]. There are two gas-based power generation technologies; these are open cycle gas turbines and combined cycle gas turbines.

In combined cycle power plants, the goal is to recover part of the energy lost in open cycle gas turbines, to increase the total efficiency of the system. Therefore, the hot air that leaves the gas turbine is captured and directed as the main heat

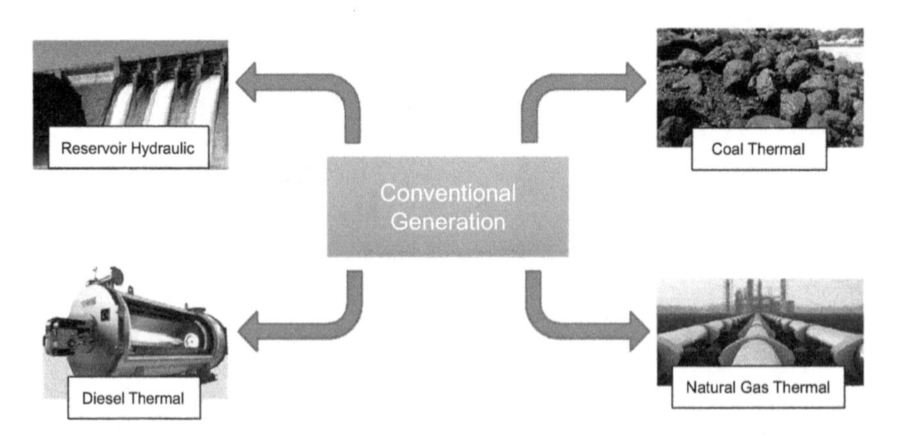

Figure 1.5 Conventional generation technologies

source for the heating water process and steam generation in heat exchangers that supply the steam turbine [18,19].

1.3.1.2 Diesel thermal technology

Thermal technology from diesel consists of an internal combustion engine operated with diesel for a synchronous electric generator. This technology presents good electrical performance in open cycle gas turbines and the possibility of intermittent operation (cycle of 4 hours running 1 hour pause). On the other hand, the main equipment for this type of generation technology are motor-generator groups, with lubrication systems [20].

1.3.1.3 Coal thermal technology

Conventional coal-fired power plants burn coal to produce the steam that drives the turbines to generate electricity [21]. The efficiency of these plants can be increased by improving combustion. This is achieved by pulverising the coal to increase the effective area over which combustion occurs in the boiler [22].

The hot gases are directed towards tubes through which the water passes, converting it into steam at a high pressure. The steam moves a turbine that is connected (mechanically) to an electric generator. Once the steam passes through the turbine, it is condensed using water from the cooling towers.

1.3.1.4 Reservoir hydraulic technology

Reservoir hydroelectric plants consider the use of water resources by storing them through a dam. The stored water is used to move a hydraulic turbine coupled to an electric power generator [23]. The most representative parts of a hydroelectric plant are reservoir, dam, intake, conduction tunnel, equilibrium chimney, pressure tunnel, valve cavern, powerhouse, gates, evacuation tunnel and final channel.

Reservoir hydroelectric projects depend on the hydrological, geological and construction conditions [24]. Therefore, technical studies are necessary and include the gathering of relevant information, simulations and field measurements.

1.3.1.5 Water's edge hydraulic technology

It is also known as run-of-the-river hydraulic power plants where the powerhouse is located outside and not in caverns. As is the case with hydraulic reservoir technology, in run-of-the-river hydroelectric plants, energy is obtained from the transformation of the potential energy of water into electrical energy through the movement of hydraulic turbines coupled to electrical generators. However, while reservoir power plants store water through a dam, run-of-river hydroelectric plants divert only a portion of the water to drive hydraulic turbines and generate electricity; then the water returns to the river [12].

1.3.2 Non-conventional renewable energy sources

Non-conventional renewable energy sources are those that take advantage of a natural resource that is cyclical, conserved and renewed. They are sources of energy considered as a clean alternative. Its integration into the electrical system has reduced

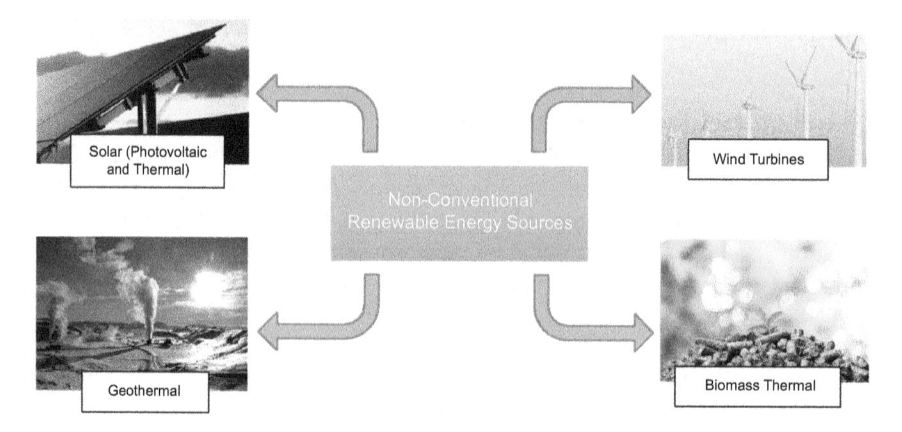

Figure 1.6 Non-conventional generation technologies

the generation from fossil fuels and facilitate the fulfilment of the increasing demand [12]. Figure 1.6 illustrates some non-conventional renewable energy sources and a description for them will be presented in the following sections.

1.3.2.1 Photovoltaic solar technology

The photovoltaic solar system is an intermittent renewable energy source. Its technology is based on the cells of semiconductor materials, which convert the energy of sunlight into DC electrical energy through a photoelectric effect.

Solar power plants are mainly made up of photovoltaic modules or panels that are connected to power inverters and transformers. Likewise, other devices are required in photovoltaic plants such as protection and grounding systems, measurement systems, instrumentation, control, automation and communications.

1.3.2.2 Wind generation technology

These power plants are based on the transformation of wind speed, which is variable, into electrical energy, through wind turbines coupled to synchronous or asynchronous electrical generators [25]. The typical power of wind turbines varies between 1 MW and 8 MW, and they typically operate between 3 and 25 m/s. However, the modularity of the technology makes it possible to install wind power plants totalling hundreds of MW, depending on the location.

1.3.2.3 Solar thermal technology

Solar thermal power plants concentrate solar energy using mirrors to heat a fluid to produce steam, driving a conventional steam turbine. Among the technologies of solar thermal power plants, the most outstanding are the solar concentration towers, which consist of hundreds of mirrors that follow the position of the sun to guide and concentrate the solar rays reflected in a central receiver.

The main components of this type of technology are mirrors, concentration tower, receiver, storage, heat exchangers, turbogenerator and cooling system.

1.3.2.4 Geothermal technology

Geothermal energy takes advantage of the thermal potential stored in reservoirs near volcanic areas. Geothermal power plants produce electrical energy from the heat contained in the Earth, through systems that obtain water, steam or hot air from the lower layers of the earth's crust.

In general, geothermal fluid is extracted from the underground deposit and flows through pipes to separate the vapour from the volcanic liquid. The steam is directed towards a turbine and an electrical generator coupled to it. In the final stage of the process, the steam is cooled in cooling towers and is condensed and reinjected into the geothermal system.

1.3.2.5 Biomass and biogas thermal technology

Biomass, understood as organic matter of vegetable or animal origin, is a renewable type that is considered carbon-neutral since its emissions are balanced by the carbon dioxide absorbed by plants. There are different processes that give rise to fuels derived from biomass, called biofuels, and that can be used to replace fossil fuels (coal, diesel) for electricity generation.

Biofuels generated from biomass can be in a solid, liquid or gas state. The use of these goes through generation technologies based on steam turbines, motor-generator groups and gas turbines.

1.4 Demand response and electrical vehicles

1.4.1 Demand response

Demand response (DR) corresponds to the changes in the use of electricity by end-consumers, in response to changes in electricity prices over time, or to incentivise less use of electricity [26,27].

Incentive-based programs offer payments to end-users to reduce their electricity use or to incentivise its use during certain periods. By shifting the load to off-peak periods, customers help reduce the level of demand. Consequently, the customers who participate in incentive-based programs receive rate discounts or direct payments [28].

1.4.1.1 Demand response benefits

In addition to some market efficiency improvements, DR creates multiple benefits for the entire electrical system. Consumers can use their DR potential with existing assets and see savings on their bills. This is achieved by using less energy when prices are high or shifting its consumption to hours with lower prices [29]. Transmission and distribution system operators have an additional market agent to deal with power imbalance and congestion problems [30].

Moreover, DR can be used as a planning alternative for the electrical system. It is considered a means that mitigates the intermittency of renewable generation and improves the efficiency of the system. In areas with high density of electrical demand (such as urban areas), DR can be especially effective in alleviating network contingencies (or blackouts) during peak load periods or emergency network reconfigurations.

Currently, DR has mainly helped to reduce the maximum load and postpone investments in GTD systems [31]. The applicability and potential of DR in new smart generation, distribution and consumption environments, such as microgrids, have become an area of interest, especially as it relates to end-user value of DR, energy cost, pollution reduction, peak load reduction and reliability improvement.

1.5 Electrical vehicles

In most countries, the transport sector represents one of the main consumers of energy, predominantly from fossil fuels such as gas and oil [32,33]. From an economic perspective, it is observed how the price of these fossil fuels changes constantly due to economies of scale and the depletion of fossil fuel sources. From the environmental point of view, it can be seen how the use of fossil resources in the transport sector negatively affects the environment with the production of carbon dioxide. The transition towards electric mobility contributes to the achievement of energy efficiency objectives and reduction of polluting emissions in the transport sector, and at the same time allows the use of renewable energies [34].

The EV is a car whose mobility is produced by the release of electrical energy stored in a battery and transmitted to an electric motor. There are other storage devices that can be used as an auxiliary source of energy such as supercapacitors or fuel cells [35].

Within the past couple of decades, EVs have seen rapid development in the automotive industry. An important factor that has promoted the growth of electric mobility technology has been climate change, caused by the exhaustive use of fossil fuels, both for transportation and for the generation of electrical energy. However, the EV must overcome some barriers such as its initial cost and its autonomy compared to that of a conventional vehicle [36].

The EV is classified into pure EV, hybrid EV, plug-in and fuel cells. Some characteristics of each type of EV are described below.

1.5.1 Pure electric vehicle

It is a vehicle that uses the electrical energy accumulated in a rechargeable battery bank as the only energy resource to move the vehicle. The basic components of an EV include the electric motor, the battery and the electric motor controller. Under normal operating conditions, the controller is powered by the battery, which transmits energy to the electric motor to generate vehicle movement. On the other hand, the autonomy of these vehicles is defined by the capacity and weight of the batteries.

1.5.2 Hybrid electric vehicle

It is a vehicle that uses two sources of energy to move the vehicle, and therefore, uses two motors, an internal combustion engine and an electric motor. The battery and electric motor improve fuel economy or performance of the conventionally operated vehicle. An important aspect of this type of EV is that, during deceleration, the regenerative brakes cause the electric motor to operate in reverse mode, which means that it behaves like an electricity generator. In this way, during deceleration processes, the electric motor generates electrical energy, charging the vehicle's batteries.

1.5.3 Plug-in electric vehicle

This vehicle has a similar operating mechanism to the hybrid EV. The difference is that it does not have a regenerative braking system, but instead its batteries are recharged in an external source, either domestic or in public points adapted for this purpose. These vehicles can run on fossil fuels, electricity or a combination of both, which leads to a wide variety of advantages such as reduced dependence on oil, economic savings from reduced fuel use, increased energy efficiency and, in general, the reduction of greenhouse gases.

1.5.4 Fuel cell electric vehicle

These vehicles use hydrogen stored in a pressurised tank and a fuel cell for power generation. They are hybrid vehicles since braking energy is recovered and stored in a battery through the regenerative braking system. Therefore, the electrical energy from the battery is used to reduce the peak demand of the fuel cell during acceleration and optimise its operating efficiency.

1.6 Control and protection challenges

Distribution networks are receiving more and more generation from renewable sources, which not only changes the topology and power flow but also introduces a high number of harmonics [37,38]. This has created a critical condition in the protection of distribution systems which has started to change relays technologies and filters implementation to avoid the impact of harmonics [39].

Moreover, distributed generation (DG) can affect the power flows and require changes in an electric power system's operation, commercial and regulatory arrangements [31]. The fault response of IBR is different than the synchronous generators response [32]. Based on this, protection settings calculated with methodologies in traditional power systems may cause mis-operation of the protection devices [40].

For example, IBRs do not have a rotating mechanical mass that is synchronously coupled with the electrical network. Therefore, it does not provide inertia [41]. It can cause unintended operation of DG protection due to the higher rate of change of frequency (RoCoF) during a power imbalance [42]. In addition, in a low inertia system, protection devices can be overly sensitive and can cause an undesired trip in response to faults on the transmission network [43,44].

1.6.1 Protection challenges due to renewable generation

Fault contributions from DG change the short-circuit levels in the system [45]. Miscoordination problems occur due to the unplanned fault currents contribution from DG to the faulted point [46].

Renewable generation helps the grid reliability and resilience [47]. However, the growing integration of these systems could impose a major challenge to the protection strategies due to some issues described below:

1.6.1.1 Low/zero inertia

The inertial response of the power system is an inherent physical response of synchronous generators [48]. IBRs do not provide inertia because they do not have a rotating mechanical mass synchronously coupled with the electrical network [41].

1.6.1.2 Short-circuit level

Synchronous generation provides a high asymmetrical fault current that decays over the transient period of the fault [49]. In contrast, IBR contributions are in the range of up to 2 or 3 per unit in the first 1–2 cycles and between 1 and 1.5 per unit thereafter [50].

1.6.1.3 Protection blinding

The contribution of fault current by the grid varies with respect to the location and size of the DG [51]. An effect is produced when the grid protection doesn't detect fault current owing to current division and change in Thevenin impedance due to the DG contribution [52].

1.6.1.4 False tripping

An undesirable operation can be caused by the tripping of a healthy feeder relay when the fault occurs at the adjacent feeder, due to the participation of DG [51].

1.6.1.5 Unintentional islanding

It occurs when DG continues generating power to the load after a fault in the grid that trips the associated protection. It may cause power imbalance, voltage and frequency issues to the network if such unintentional islanding remains undetected [51].

1.7 Connection request in distribution systems

The high penetration of variable renewable energies in power systems, as well as the technological evolution of distributed energy resources (DERs) and their accelerated incorporation into the electrical infrastructure, especially the predominantly photovoltaic distributed generation systems, makes it pertinent to define a procedure for the safe penetration capacity of variable renewable energies in the distribution systems.

DER hosting capacity varies throughout the circuit. The capacity does not remain constant over time due to possible topological changes, as well as changes in load and generation, connection of new DERs, among others, so the penetration capacity must be periodically estimated.

The determination of DER penetration capacity per circuit must consider at least the following analyses:

- Loading: thermal overloads in conductors and transformers due to an increase in the power flow. Figure 1.7 presents a heat map for a georeferenced distribution system with overloads illustrated in red.
- Voltage: voltage variations, high and low voltage conditions in MV and LV must be considered, as well as the impact on the voltage regulation equipment in the circuit. Figure 1.8 presents a heat map for a georeferenced distribution

Figure 1.7 Line loading in a distribution network

Figure 1.8 Voltage profile in a distribution network

system with voltage profiles along the network, where under voltage conditions are identified in the white areas.

- Protection: selectivity problems and coordination of the existing protections due to changes in the fault currents from the contributions of the DER. Figure 1.9 illustrates an increase in the short-circuit level on the same faulted node, but also a reduction for the contribution from the main substation due to the new DER on the system.

Studies to estimate the capacity of circuits to accommodate DERs are based on computer simulations of future scenarios that model the behaviour of the network for different levels of penetration in DERs, thus evaluating the effects of these resources on the network.

Among the main factors that define the amount of DER that can be installed in a distribution circuit are:

- Location and behaviour of the circuit demand.
- Location and behaviour of power injections in the circuit.
- The topology and characteristics of the circuit.

1.7.1 Criteria for connection requests in distribution systems

In order to estimate the capacity of DER, the distribution companies and the OS must consider voltage evaluation criteria, control, thermal and protection actions. Some of them are described below.

1.7.1.1 Increase of voltage

Electricity distribution companies must maintain customer voltage at ±5% of nominal voltage for the majority of the time, as stipulated in ANSI C84.1. To apply this criterion, power flow simulations are performed with DER futures at the different evaluated penetration levels and it is checked that the voltage in all medium and low voltage nodes does not exceed 1.05 p.u.

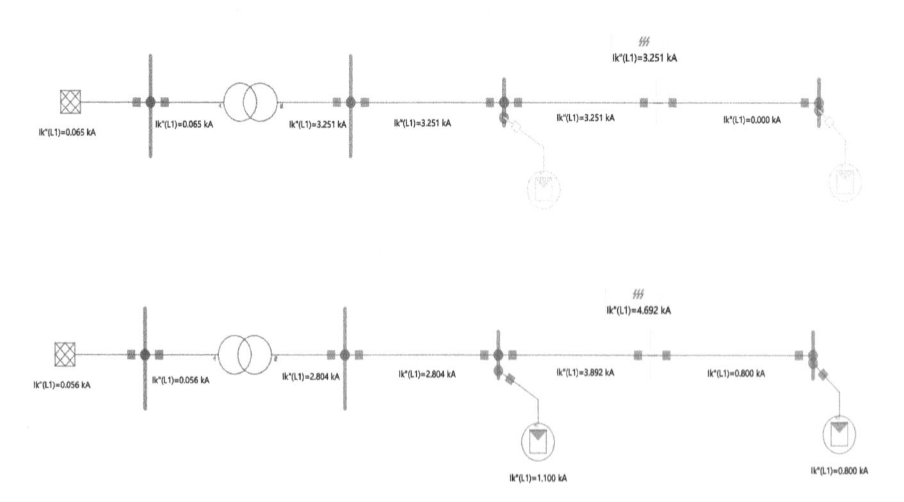

Figure 1.9 Short-circuit changes due to DER contributions

1.7.1.2 Voltage variation

Variations in DER power produces rapid voltage fluctuations that can cause the interruption of sensitive equipment or more frequent operation of voltage regulating equipment for some customers. To apply this criterion, the difference between the voltage magnitude with and without DER is calculated (base study for different scenarios of power flows) for all nodes of the circuit.

1.7.1.3 Voltage imbalance

The imbalance of loads between phases becomes a power quality problem when the imbalance of voltage is greater than a certain threshold. Voltage imbalance leads to additional losses, heating and the premature failure of induction motors and transformers, affecting the utility and its customers. To apply this criterion, the voltage unbalance is calculated in all MV and LV three-phase nodes of the circuit for the simulations with DER futures.

1.7.1.4 Overload of conductors and transformers

The evaluation criterion consists of determining the installed capacity of DER that implies currents greater than 100% of the ampacity of the conductors or the capacity of the distribution transformers due to reverse flows. To apply the criterion, it is checked whether the future DER penetration level does not lead to currents in conductors that exceed the respective ampacities. In such a case, the DER penetration level is allowed.

Analogously to the conductor overload review, the load level of the transformers is reviewed for future DER simulations. If the through power in the transformers does not exceed its nominal value, the penetration level of DER is allowed.

Example 1.1

Consider the power system of Figure 1.10 that has one PV facility. The system has been modelled for load flow analysis. Figure 1.11(a) illustrates the load flow result when the distance of the PV facility (L1-2 length) is about 1,640.42 ft and Figure 1.11(b) illustrates the impact on the voltage result when the distance is increased 10 times. Both figures consider the same generation of 100 kW, PF = 1.

Based on the methodology of the previous results, Table 1.1 indicates that when PV facilities are modelled as PC nodes (active power constant and power

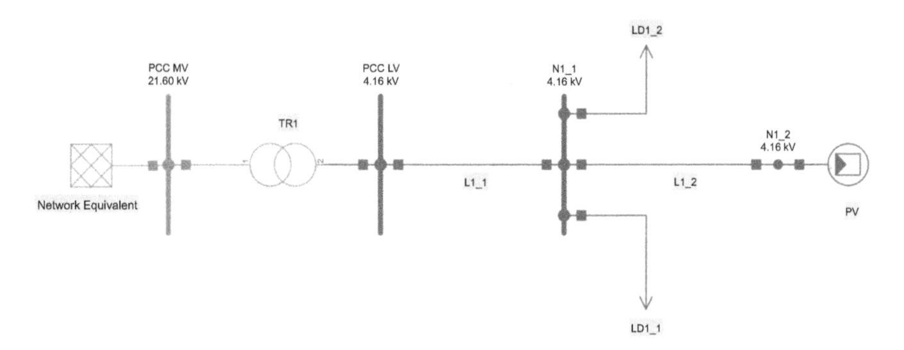

Figure 1.10 Single line diagram of a distribution system

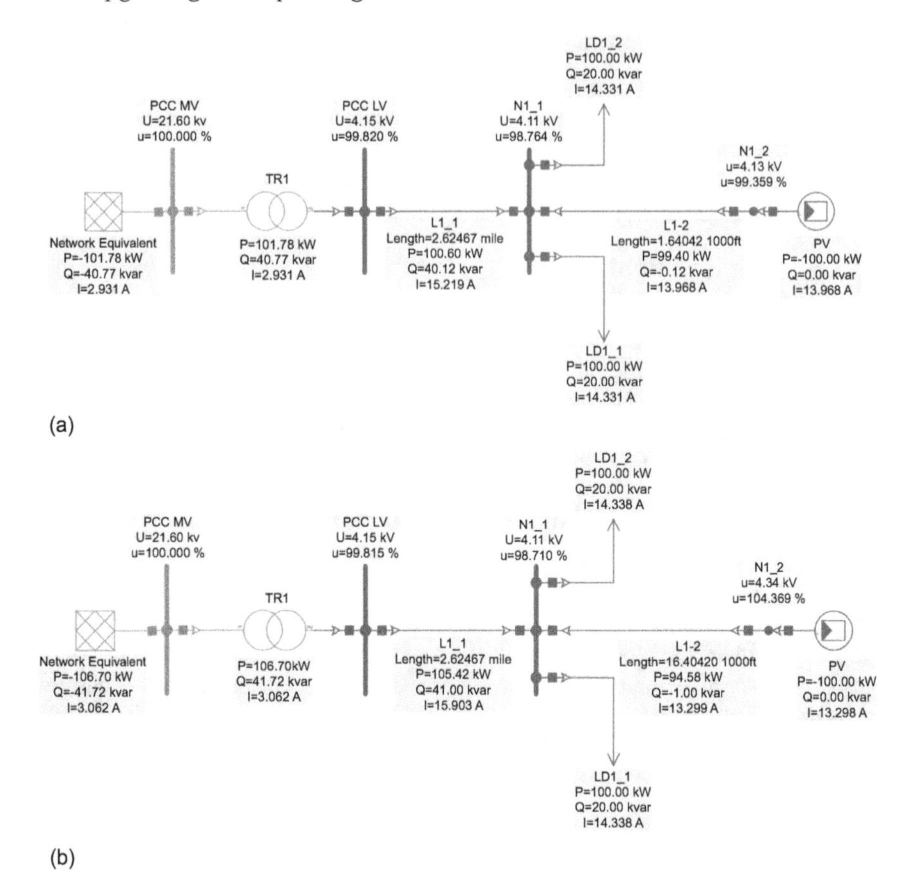

Figure 1.11 Impact on the voltage result for different line lengths: (a) Load Flow results for line length 1,640.42 ft and (b) Load Flow results for line length 16,404.2 ft

factor constant), the voltage at the corresponding node tends to rise whenever the length of the line is increased. This is illustrated in Figure 1.12.

These types of situations are very common, especially with the large number of photovoltaic facilities installed around the world either in solar farms or in roof-type projects. Care must be exercised to ensure that the voltage on the user end does not go beyond the current limits specified by the corresponding standards.

Negative values in Table 1.1 represent the current and power flow in a reverse direction, from the PV unit to the load or to the grid.

Example 1.2

Figure 1.13 presents a section of a typical distribution network that indicates the total short-circuit current values for a three-phase fault at each node.

Table 1.1 Load flow results for different line lengths

Line distance		Voltage (%)			Current (A)		Power flow (kW)			Ploss (kW)	
PV distance (km)	PV distance (kft)	PCC_LV	N1_1	N1_2	L1_1	L1_2	L1_1	L1_2	PV	L1_1	L1_2
0.5	1.64042	99.82	98.76	99.36	15	−14	102	−99	−100	1	1
0.8	2.62467	99.82	98.76	99.71	15	−14	102	−99	−100	1	1
1	3.28084	99.82	98.76	99.94	15	−14	102	−99	−100	1	1
1.5	4.92126	99.82	98.75	100.52	15	14	103	−98	−100	1	2
2	6.56168	99.82	98.75	101.08	15	14	103	−98	−100	1	2
3	9.84252	99.82	98.73	102.2	16	−14	105	−97	−100	1	3
5	16.4042	99.82	98.71	104.37	16	−13	107	−95	−100	1	5
10	32.808	99.81	98.66	109.44	17	−13	111	−90	−100	1	10
15	49.212	99.81	98.62	114.11	17	−12	115	−86	−100	1	14

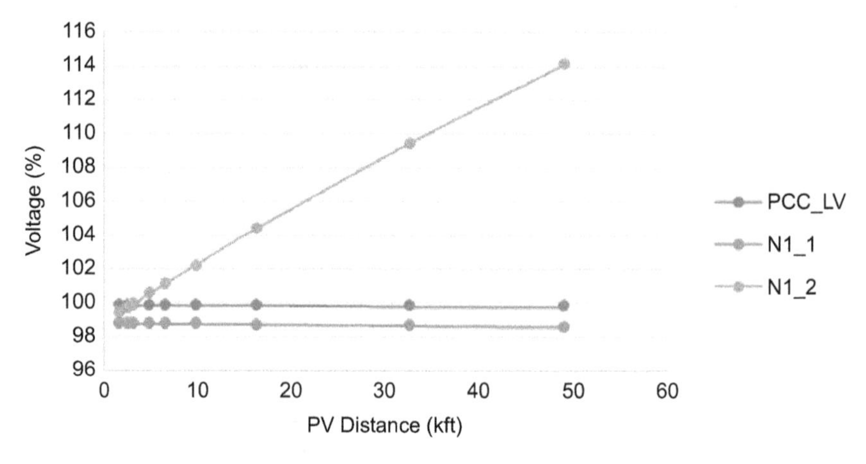

Figure 1.12 Voltage results for the different line lengths

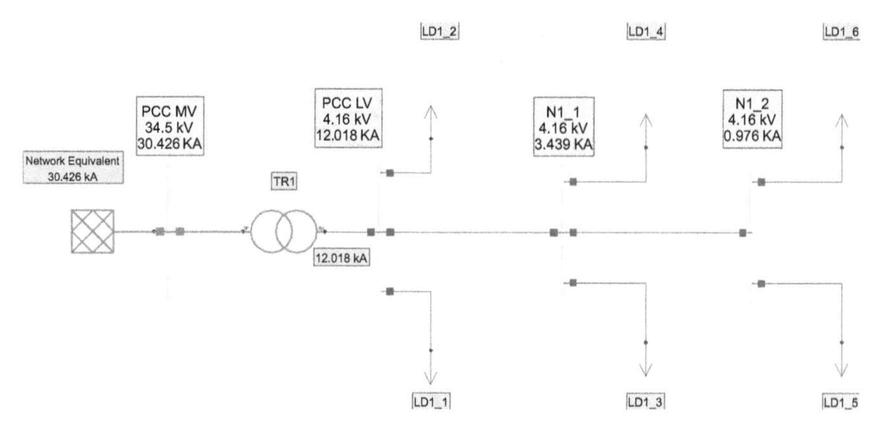

Figure 1.13 Short-circuit calculations for a three-phase fault

The following generation units are connected to the node N1_2 as shown in Figure 1.14.

- A PV system, which consists of three (3) PV modules.
- A type 1 wind generation unit.
- A type 2 wind generation unit.
- A type 3 wind generation unit.
- A type 4 wind generation unit.

Considering the short-circuit contributions, the impact of the different generation sources can be seen as illustrated in Figure 1.15.

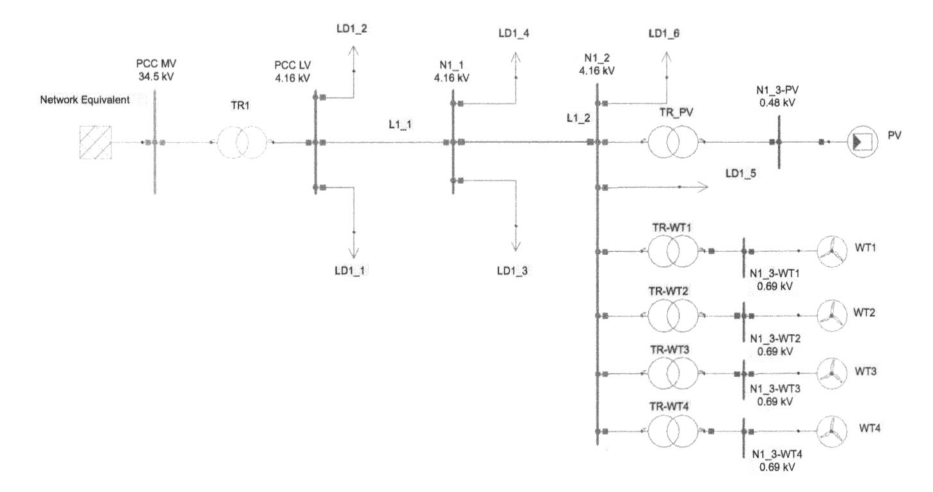

Figure 1.14 New generation systems in the network

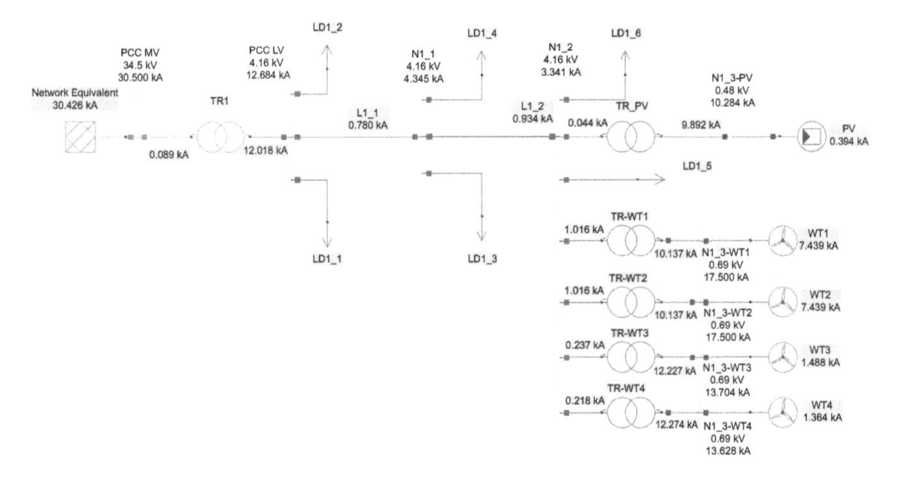

Figure 1.15 New generation systems connected to the distribution network

According to the results, short-circuit currents are increased due to the integration of generation units into the distribution system. Hence, the implemented technology (synchronous/asynchronous or inverter-based generators) defines the contributions to the short-circuit currents.

As a consequence of the short-circuit level variations, the integration of several generation units can lead to protection miss-coordination, affecting the reliability of the distribution system. On the other hand, upstream the generation unit, voltage can be increased significantly due to the change on the net power flow.

1.8 Expansion of transmission systems

The transmission network is a natural monopoly (i.e., an infrastructure that cannot be duplicated and in which competition makes no sense) and is used by generation operators to supply electricity to end-customers. The electrical infrastructure is generally designed to operate correctly under the worst conditions. However, the worst future condition is a priori unknown, and therefore this uncertainty must be faced, which leads to both economic and reliability consequences. To face this challenge, it is necessary to carefully describe the uncertain parameters involved in the transmission scheduling problem.

Transmission expansion planning (TEP) refers to the decision-making process faced by a transmission system operator (TSO) to find the best way to expand or reinforce an existing electricity transmission network. The TSO is the public control entity in charge of operating, maintaining, reinforcing and expanding the electric power transmission network within a given jurisdiction and with the objective of maximising social well-being. The problem of transmission expansion is motivated by several reasons, including aged infrastructure, the expected growth in demand and the construction of new generation plants, mainly renewable and normally far from consumption centres.

These issues make it essential to reinforce and expand the existing transmission network to facilitate the supply of energy from the generation centres to the consumption centres, as well as to guarantee the balance between generation and demand.

1.8.1 Frameworks of the TEP problem

The TEP problem is usually approached in two different frameworks: centralised and competitive. In a centralised framework, one entity controls both generation and transmission plants and jointly oversees planning generation and transmission (as is the case in most countries around the world). In a competitive environment, an independent and regulated entity is usually in charge of operating and expanding the transmission network to maximise energy trading opportunities between generators and consumers. There is a third option that consists of considering the possibility of a for-profit commercial investor, who carries out the task of expanding the transmission network.

In general, the TSO decides the best way to reinforce and expand the existing transmission network with the objective of facilitating energy trading opportunities for market agents, for example, by minimising generation costs or reducing non-energy costs. In this way, by expanding the transmission network, generation and load interruption costs are reduced (economic objective), but at the same time the reliability of demand supply is improved (technical objective).

1.8.2 Solution methods for the TEP problem

TEP is a complex decision problem, as it usually involves an objective with several nonlinear constraints and a non-convex feasible region. As a result, different approaches have been proposed to deal with this complexity. These approaches are based on the application of non-linear programming techniques or the use of heuristics.

An important observation is that the TEP problem is usually considered for a long-term planning horizon (e.g., 20 years). When the TSO decides on the transmission expansion plan to be carried out, it must consider future demand growth, the availability of existing generation units and the construction of new generation facilities. Transmission expansion decisions are made in an uncertain environment, and such uncertainties must be adequately accounted for to make informed expansion decisions. For this, techniques such as stochastic programming and robust optimisation have been used.

Stochastic programming is based on the generation of scenarios that describe the uncertain parameters. However, this scenario generation usually requires knowing the probability distribution functions of the uncertain parameters, which is often a complex task. In addition to this, a sufficiently large number of scenarios must be generated to accurately represent the uncertainty, which increases the computational complexity of the problem.

Robust optimisation focuses on finding optimal solutions that are resistant to disturbances or uncertainties in the data or model. It seeks to minimise the impact of the variations in the model parameters, to obtain a solution that is good in a wide variety of situations. In more technical terms, robust optimisation uses uncertainty theory tools to model uncertainties in the data or model, and then finds the optimal solution that minimises the worst possible outcome in all possible situations within a set of uncertainties. A general disadvantage of robust optimisation is that the results are often very conservative.

Finally, there are two expansion strategies for transmission networks within the framework of an optimisation problem. The first consists of building new transmission lines in a single moment (usually at the beginning of the planning horizon). The model resulting from this strategy is known as a static or single-stage model. The second consists of making transmission expansion decisions at different moments in the planning horizon. In this case, the model is called dynamic or multistage. The dynamic approach usually provides more precise solutions since it allows the transmission planner to adapt to future changes in the system. However, it further increases the complexity of the TEP problem.

References

[1] L. F. N. Delboni, D. Marujo, P. P. Balestrassi, and D. Q. Oliveira, 'Electrical power systems: evolution from traditional configuration to distributed generation and microgrids', in A. Zambroni de Souza and M. Castilla (eds.), *Microgrids Design and Implementation*, (pp. 1–25). Cham: Springer, 2019.

[2] M. Safiuddin, 'History of electric grid', in D. Apple, R. Elmoudi, I. Grinberg, S. M. Macho, and M. Safiuddin (eds), *Foundations of Smart Grid*, 1st edn (pp. 6–11). Hampton, NH: Pacific Crest, 2013.

[3] F. Charaabi, M. Dali, A. Ben Rhouma, and J. Belhadj, 'Technical and economic evaluation of remote DC and AC microgrids', in *IEEE International Conference on Electrical Sciences and Technologies in Maghreb (CISTEM)*, IEEE, 2022, pp. 1–6, doi:10.1109/CISTEM55808.2022.10044042.
[4] P. Fairley, 'DC versus AC: the second war of currents has already begun', *IEEE Power Energy Mag.*, vol. 10, no. 6, pp. 104–103, 2012, doi:10.1109/MPE.2012.2212617.
[5] D. I. Stern, P. J. Burkes, and S. B. Bruns, 'The impact of electricity on economic development: a macroeconomic perspective', *Int. Rev. Environ. Resour. Econ.*, vol. 12, no. 1, pp. 85–127, 2018, [Online]. Available: https://escholarship.org/uc/item/7jb0015.
[6] H. Sun, Q. Guo, J. Qi, *et al.*, 'Review of challenges and research opportunities for voltage control in smart grids', *IEEE Trans. Power Syst.*, vol. 34, no. 4, pp. 2790–2801, 2019, doi:10.1109/TPWRS.2019.2897948.
[7] W. P. Schill, M. Pahle, and C. Gambardella, 'Start-up costs of thermal power plants in markets with increasing shares of variable renewable generation', *Nat. Energy*, vol. 2, no. 6, pp. 1–6, 2017, doi:10.1038/nenergy.2017.50.
[8] T. S. Kishore, E. R. Patro, V. S. K. V Harish, and A. T. Haghighi, 'A comprehensive study on the recent progress and trends in development of small hydropower projects', *Energies (Basel)*, vol. 14, no. 10, 2882, 2021, doi:10.3390/en14102882.
[9] A. Arsalis, 'A comprehensive review of fuel cell-based micro-combined-heat-and-power systems', *Renew. Sustain. Energy Rev.*, vol. 10, no. 1016, pp. 391–414, 2019.
[10] L. S. Vedantham, Y. Zhou, and J. Wu, 'Information and communications technology (ICT) infrastructure supporting smart local energy systems: a review', *IET Energy Syst. Integr.*, vol. 4, no. 4, pp. 460–472, 2022, doi:10.1049/esi2.12063.
[11] L. Thomas, Y. Zhou, C. Long, J. Wu, and N. Jenkins, 'A general form of smart contract for decentralized energy systems management', *Nat. Energy*, vol. 4, no. 2, pp. 140–149, 2019, doi:10.1038/s41560-018-0317-7.
[12] J. D. Mina-Casaran, D. F. Echeverry, and C. A. Lozano, 'Demand response integration in microgrid planning as a strategy for energy transition in power systems', *IET Renew. Power Gener.*, vol. 15, no. 4, pp. 889–902, 2021.
[13] IRENA and IEA-ETSAP, *Renewable Energy Integration in Power Grids*. Abu Dhabi, United Arab Emirates, 2015.
[14] X. Xia and J. Xia, 'Evaluation of potential for developing renewable sources of energy to facilitate development in developing countries', in *Asia-Pacific Power and Energy Engineering Conference*, pp. 1–3, 2010, doi:10.1109/APPEEC.2010.5449477.
[15] REN21, 'Renewables 2023 Global Status Report', 2023 [Online]. Available from: https://www.ren21.net/gsr-2023/modules/energy_demand [accessed 23 March 2024].
[16] Y. Khan and F. Liu, 'Consumption of energy from conventional sources a challenge to the green environment: evaluating the role of energy imports,

and energy intensity in Australia', *Environ. Sci. Pollut. Res.*, vol. 30, no. 9, pp. 22712–22727, 2023.

[17] L. D. Smoot and L. L. Baxter, 'Fossil fuel power stations—coal utilization', in R.A. Meyers (ed), *Encyclopedia of Physical Science and Technology*, 3rd edn (pp. 121–144). New York: Academic Press, 2003.

[18] M. J. B. Kabeyi and O. A. Olanrewaju, 'Performance analysis of an open cycle gas turbine power plant in grid electricity generation', in *IEEE International Conference on Industrial Engineering and Engineering Management (IEEM)*, IEEE, pp. 524–529, 2020, doi:10.1109/IEEM45057.2020.9309840.

[19] A. L. Polyzakis, C. Koroneos, and G. Xydis, 'Optimum gas turbine cycle for combined cycle power plant', *Energy Convers. Manag.*, vol. 49, no. 4, pp. 551–563, 2007.

[20] D. R. Karana and R. R. Sahoo, 'Thermal, environmental and economic analysis of a new thermoelectric cogeneration system coupled with a diesel electricity generator', *Sustain. Energy Technol. Assessments*, vol. 10, p. 1, 2020.

[21] J. W. Butler and P. Basu, 'Clean power generation from coal', in M. Kutz (ed), *Environmentally Conscious Alternative Energy Production* (pp. 207–265). Hoboken, NJ: John Wiley & Sons, Inc., 2011, doi:10.1002/9780470209738.ch8.

[22] C. Fu, R. Anantharaman, K. Jordal, and T. Gundersen, 'Thermal efficiency of coal-fired power plants: from theoretical to practical assessments', *Energy Convers. Manag.*, vol. 10, no. 1016, pp. 530–544, 2015.

[23] V. K. Singh and S. K. Singal, 'Operation of hydro power plants: a review', *Renew. Sustain. Energy Rev.*, vol. 10, no. 1016, pp. 610–619, 2016.

[24] A. A. A. Abuelnuor, K. Ahmed, K. M. Saqr, Y. A. M. Nogoud, and M. E. A. M. Babiker, 'Exergy analysis of large and impounded hydropower plants: case study El Roseires Dam (280 MW)', *Environ. Prog. Sustain. Energy*, vol. 39, no. 3, pp. 1–11, 2020.

[25] P. Sorensen, B. Andresen, J. Fortmann, and P. Pourbeik, 'Modular structure of wind turbine models in IEC 61400-27-1', in 2013 *IEEE Power & Energy Society General Meeting*, pp. 1–5, 2013, doi:10.1109/PESMG.2013.6672279.

[26] Federal Energy Regulatory Commission, 'Assessment of demand response and advanced metering', pp. 1–240, 2006.

[27] M. Asensio, G. Munoz-Delgado, and J. Contreras, 'Bi-level approach to distribution network and renewable energy expansion planning considering demand response', *IEEE Trans. Power Syst.*, vol. 32, no. 6, pp. 4298–4309, 2017.

[28] M. Hussain and Y. Gao, 'A review of demand response in an efficient smart grid environment', *Electr. J.*, vol. 31, no. 5, pp. 55–63, 2018.

[29] C. Roldán-Blay, G. Escrivá-Escrivá, and C. Roldán-Porta, 'Improving the benefits of demand response participation in facilities with distributed energy resources', *Energy*, vol. 169, pp. 710–718, 2019.

[30] C. Kok, J. Kazempour, and P. Pinson, 'A DSO-level contract market for conditional demand response', in *IEEE Milan PowerTech*, pp. 1–6, 2019.

[31] A. T. Davda and B. R. Parekh, 'System impact analysis of renewable distributed generation on an existing radial distribution network', in *IEEE Electrical Power and Energy Conference*, pp. 128–132, 2012. doi:10.1109/EPEC.2012.6474936.

[32] J. Holbach, R. Zhang, T. Charton, D. Kerven, and S. Ward, 'Protection functionality and performance with declining system fault levels and inertia within national grid electricity transmission system in the United Kingdom', in *15th International Conference on Developments in Power System Protection (DPSP 2020),* Liverpool, UK, 2020, pp. 1–6, doi:10.1049/cp.2020.0065.

[33] B. L. Salvi, K. A. Subramanian, and N. L. Panwar, 'Alternative fuels for transportation vehicles: a technical review', *Renew. Sustain. Energy Rev.*, vol. 10, no. 1016, pp. 404–419, 2013.

[34] M. Adhikari, L. P. Ghimire, Y. Kim, P. Aryal, and S. B. Khadka, 'Identification and analysis of barriers against electric vehicle use', *Sustainability*, vol. 12, no. 12, 2020, doi:10.3390/su12124850.

[35] M. A. Hannan, M. M. Hoque, A. Mohamed, and A. Ayob, 'Review of energy storage systems for electric vehicle applications: issues and challenges', *Renew. Sustain. Energy Rev.*, vol. 10, no. 1016, pp. 771–789, 2016.

[36] S. Goel, R. Sharma, and A. K. Rathore, 'A review on barrier and challenges of electric vehicle in India and vehicle to grid optimisation', *Transp. Eng.*, vol. 57, p. 2021, 2021.

[37] A. Olatoke and M. Darwish, 'Relay coordination and harmonic analysis in a distribution network with over 20% renewable sources', in *48th International Universities' Power Engineering Conference (UPEC)*, Dublin, Ireland, 2013, pp. 1–6, doi:10.1109/UPEC.2013.6714871.

[38] J. Gers, *Distribution System Analysis and Automation*, 2nd ed. London: The Institution of Engineering and Technology, 2020.

[39] H. Al-Nasseri and M. A. Redfern, 'Harmonics content based protection scheme for micro-grids dominated by solid state converters', in *12th International Middle-East Power System Conference*, IEEE, pp. 50–56, 2008, doi:10.1109/MEPCON.2008.4562361.

[40] A. Haddadi, E. Farantatos, I. Kocar, and U. Karaagac, 'Impact of inverter based resources on system protection', *Energies (Basel)*, vol. 14, p. 1050, 2021, doi:10.3390/en14041050.

[41] F. Milano, F. Dörfler, G. Hug, D. J. Hill, and G. Verbič, 'Foundations and challenges of low-inertia systems (Invited Paper)', *Power Systems Computation Conference (PSCC)*, Dublin, Ireland, 2018, pp. 1–25, doi:10.23919/PSCC.2018.8450880.

[42] National Grid ESO: System Operability Framework, 'Operating a Low Inertia System' [Online]. Available: https://tinyurl.com/ytk42der [accessed 20 March 2024].

[43] A. Dýsko, D. Tzelepis, and C. Booth, 'Practical risk assessment of the relaxation of LOM protection settings in NIE networks', *IET Gener. Transm. Distrib.*, vol. 15, pp. 1335–1339, 2018.
[44] J. Gers and E. Holmes, *Protection of Electricity Distribution Networks*, 4th ed. London: Institution of Engineering and Technology, 2021.
[45] F. M. Nuroglu and A. B. Arsoy, 'Voltage profile and short circuit analysis in distribution systems with DG', in *IEEE Canada Electric Power Conference*, pp. 1–5, 2008, doi:10.1109/EPC.2008.4763309.
[46] B. Bhalja and P. H. Shah, 'Miscoordination of relay in radial distribution network containing distributed generation', in *2011 IEEE Recent Advances in Intelligent Computational Systems*, pp. 72–75, 2011, doi:10.1109/RAICS.2011.6069275.
[47] Summer Dean at Renewable Northwest, 'The Contribution of Distributed Solar to Reliability, Grid Resilience, and Community Resilience', Washington, 2018.
[48] IEEE Power System Dynamic Performance Committee, 'Stability definitions and characterization of dynamic behaviour in systems with high penetration of power electronic interfaced technologies', in *IEEE Power and Energy Society*, pp. 1–42, 2020.
[49] N. D. Tleis, *Power Systems Modelling and Fault Analysis: Theory and Practice*, 2nd ed. Cambridge, MA: Elsevier Academic Press, 2019.
[50] R. A. Walling, E. Gursoy, and B. English, 'Current contributions from Type 3 and Type 4 wind turbine generators during faults', PES T&D 2012, Orlando, FL, USA, 2012, pp. 1–6, doi:10.1109/TDC.2012.6281623.
[51] M. Usama, H. Mokhlis, and M. Moghavvemi, 'A comprehensive review on protection strategies to mitigate the impact of renewable energy sources on interconnected distribution networks', *IEEE Access*, vol. 9, pp. 35740–35765, 2021.
[52] V. A. Papaspiliotopoulos, G. N. Korres, V. A. Kleftakis, and N. D. Hatziargyriou, 'Hardware-in-the-loop design and optimal setting of adaptive protection schemes for distribution systems with distributed generation', *IEEE Trans. Power Del.*, vol. 32, no. 1, pp. 393–400, 2017.

Chapter 2

Modelling and trends of urban smart grids

Smart grids are a rather new concept that includes aspects of energy generation, transmission and distribution and aims for a more reliable service, higher efficiency, more security, two-way utility-user communications and the promotion of green energy, among other goals.

When the term 'smart grid' was first used, it was mainly associated with remote metering, which was later called automatic meter reading (AMR) [1]. The activities of AMR were encompassed within those of a broader field that was eventually called advanced metering infrastructure (AMI). The metering system is one of the major elements of the smart grid, but certainly is not the only one.

Distribution systems have been autonomously operated for many years, with only occasional manual setting changes and a rather primitive automation that is known today as local intelligence. In fact, automation was first implemented on generation and transmission systems and gradually also became popular on distribution systems.

A good example of local intelligence is used in reclosers operating in coordination with sectionalisers. After a local fault, reclosers start a set of reclosing operations before locking out. Another good example is the operation of capacitor bank switches which rely on local signals such as voltage level, power factor or even time.

In the 1980s, distribution systems gained greater attention, prompting the evolution of the ideas behind smart grids. Prior to this, most attention had been directed towards generation and transmission systems. During that time, contributions in the area included immediate detection and isolation of faulted feeders and reducing the amount of time required by line crews to locate and repair faults. There was also an anticipated increase in distributed generation, then referred to as dispersed storage and generation (DSG) systems [1].

Governments and utilities funding the development and modernisation of grids have defined the functions required for smart grids. According to the United States Department of Energy's Modern Grid Initiative Report, a modern smart grid must satisfy the following requirements:

- Motivate consumers to actively participate in operations of the grid
- Be able to heal itself
- Resist attack
- Provide a higher quality power that will save money wasted during outages
- Accommodate all generation and storage options
- Enable electricity markets to flourish

- Run more efficiently
- Enable higher penetration of intermittent power generation sources

In order to achieve the goals of smart grids mentioned above and in particular the improvement in reliability, security and efficiency, it is essential to have a well-developed digital technology. Among the significant challenges facing development of a smart grid is the cost of implementing it and the new standards that regulatory bodies have to enact. Interoperability standards certainly will allow the operation of highly interconnected systems that include distributed generation plants.

Another challenge that the implementation of smart grids and distribution automation faces is the huge variety of technologies produced by multiple vendors. Establishing a proper development path is highly recommended to any utility before embarking on a comprehensive project. Maturity models that are discussed in a latter chapter help to establish this plan.

Smart grids have the great advantage of allowing two-way communication, i.e. utility-user and user-utility. This will allow a better and more effective relationship between the user and the utility. The latter will be able to monitor and control small appliances of each user. In turn, the user will have the great advantage of receiving information regarding the consumption level, new rates available and load management schemes. This requires powerful communication systems that need to be flexible and reliable.

2.1 Definitions of smart grid

Many definitions have been written to describe smart grids [2]. Every utility might have its own definition. In general, a smart grid refers to a sustainable modernisation of the electricity grid, integrating information and communication technologies (ICTs) to intelligently manage and operate generation, transmission, distribution, consumption or even the electric energy market [3,4]. This concept is illustrated in Figure 2.1. The smart grid comprises many or

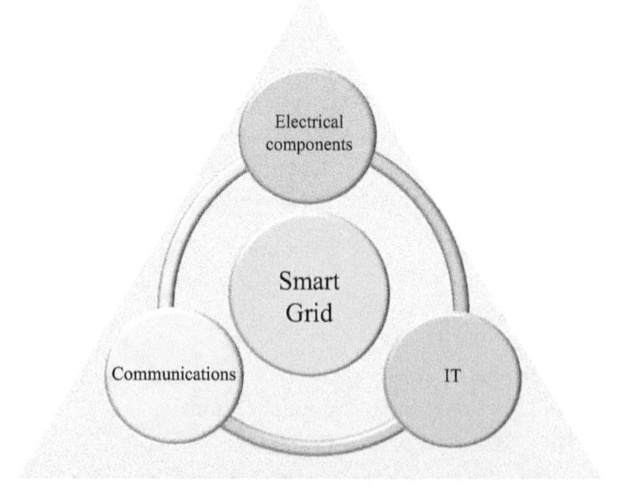

Figure 2.1 Smart grid concept

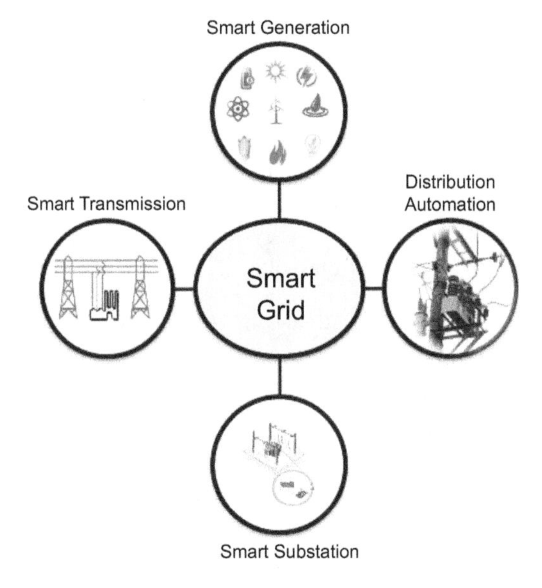

Figure 2.2 Smart grid components

almost all the elements of the utility and their relationships. Figure 2.2 includes some of these components such as distribution automation, generation, transmission and the substation, which is the connection element of the different components.

2.2 Benefits of smart grids on distribution systems

The benefits of implementing smart grids are many [5–7]. They can be summarised in the following categories.

2.2.1 Enhancing reliability

Smart grids dramatically reduces the cost of power disturbances. This can be achieved by means of system reconfiguration using switches placed along the feeders. Communications and control technologies greatly help to isolate faults and allow a faster restoration of service.

2.2.2 Improving system efficiency

The technical and non-technical reduction of losses in electrical systems is one of the goals of every utility in the world. It not only reduces the demanded power but also benefits the environment. The reduction of system losses results in capital deferral which gives an attractive payback of investments. Power capacitors, voltage regulators and proper design criteria are required to achieve this.

2.2.3 Distributed energy resources

Construction of generation plants at the user level is becoming more frequent every day. These sources are referred to as distributed generation or distributed energy resources (DERs) and are getting much attention from government authorities and environmental institutions as they reduce the pollution levels that some plants have, in particular those burning coal and oil. DERs also contribute to better operating conditions of distribution systems as they are sources connected directly to the users' loads, increasing control of the voltage.

2.2.4 Optimising asset utilisation and efficient operation

Real-time data makes it possible to utilise assets more effectively during both normal and adverse conditions and to reduce the costs of outages. This results in a longer service life of the assets.

2.3 Prioritisation in smart grid projects

Defining the priorities of which projects have to go first is a hard decision process inside any utility [8,9]. Cost-benefits analysis is the most popular method for making these decisions [10]. Nonetheless, this method is difficult to use when social and environmental impacts play an important role while evaluating projects [11,12].

The structure of the considered smart grid decision problem consists of eight criteria and four sub-criteria. The smart grid evaluation system for the modernisation of the electrical systems explained in this section considers the structure of the smart grid maturity model (SGMM). The number of alternatives can vary, though it is recommended not to be greater than 10 due to the number of comparisons needed.

The list of criteria for a smart grid decision problem is based on the domains of the SGMM. As reference, the model developed by Carnegie Mellon University is presented here (this model and domains will be discussed later in this chapter) as shown in Table 2.1.

The list of sub-criteria for a smart grid decision problem is based on the type of benefits related to the smart grid, presented by the US Department of Energy (DOE) and it is shown in Table 2.2.

Table 2.1 Criteria for prioritising smart grid projects

Item	Criterion
C1	*Strategy, management and regulatory (SMR)*
C2	*Organisation and structure (OS)*
C3	*Grid operations (GO)*
C4	*Work and asset management (WAM)*
C5	*Technology (TECH)*
C6	*Customer (CUST)*
C7	*Value chain integration (VCI)*
C8	*Societal and environmental (SE)*

Table 2.2 Sub-criteria for prioritising smart grid projects

Item	Sub-criterion
C11	Economic benefits
C21	Power quality and reliability benefits
C31	Environmental benefits
C41	Security benefits

Table 2.3 Alternatives proposed for reaching the objectives of a smart grid project

Item	Alternatives
A1	Vision and strategic planning
A2	Business architecture
A3	Assets and workforce management
A4	Physical and cyber security
A5	AMI
A6	Two-way communication
A7	Energy efficiency
A8	Advance distribution automation
A9	Microgrids
A10	Distributed generation
A11	Energy storing
A12	Demand site management
A13	Transmission automation
A14	Smart mobility
A15	Smart lighting

Table 2.3 illustrates some of the alternatives needed to develop an integral smart grid into any utility. These alternatives are an example of some projects that must be evaluated in accordance with the SGMM results.

2.4 Maturity models for smart grid applications

Utilities are aware of the need to implement distribution automation and smart grid programs. It is easy to appreciate the various individual efforts being made, given the number of existing applications. However, there is a lack of definite procedures and recommended practices to help utility organisations establish an order of priority in their implementation and development of these technologies.

Developing a maturity model is important for modern grid management and for selecting the best way of creating a sustainable path towards integration. This involves tools such as having a common strategy and vision, which not only helps in the development of an organised work plan but also allows for the implementation of profitable projects as needed.

The maturity model helps utilities to implement smart grid applications by prioritising tasks and measuring the progress achieved. It also helps to identify the characteristics of the organisation by designing a roadmap and by promoting the exchange of common terms among internal and external actors. All share experiences with the community and prepare the organisation to undertake the required changes.

There are maturity models focused on smart grids analysis available, which can help to develop a profile for energy companies and understand the challenges and required targets to be achieved according to short-term, medium-term or long-term goals [13–15]. Some of them are listed below:

- Electricity Subsector Cybersecurity Capability Maturity Model (ES-C2M2): This tool was developed by the US Department of Energy and the Department of Homeland Security [16]. It is tailored to assess the cybersecurity of utilities, specifically aiming to enhance cybersecurity within the energy sector by evaluating current capabilities, sharing knowledge and best practices to promote cybersecurity and facilitating planning and coordination among distribution system operators.

  This maturity model was conceived as a self-assessment framework with a comprehensive toolkit. Like other maturity models, it is designed to accommodate the diverse needs of various sectors with a specific focus. Therefore, industry subsectors can interpret the model according to their distinct requirements. To ensure adaptability, the model draws upon existing cybersecurity standards and frameworks and considers the efforts of various programmes and initiatives.

  The model comprises ten domains, each containing different objectives (such as approach objectives and management objectives). Approach objectives pertain to domain-specific goals, whereas management objectives encompass common objectives shared across different domains. Specific practices or practices for management are consolidated within each objective. In essence, four distinct maturity (indicator) levels (MILs, which go from MIL0 to MIL3) are employed in the model, with MILs applied to individual domains. Achieving a level requires fulfilling all practices assigned to a domain. For instance, if an organisation has a risk management MIL0, it signifies that all practices are at least at MIL0. At the moment to define specific goals, each organisation should set a target MIL for each domain to improve the organisational state. It is expected that the achieved MILs align with the business strategy and cybersecurity strategy of the company. Therefore, achieving higher levels of maturity would not be necessarily requested or beneficial. This will depend only on the required profile.
- Smart grid interoperability maturity model (SG-IMM): The GridWise Architecture Council (GWAC) developed the SG-IMM, a conceptual

framework accompanied by a corresponding tool, aimed at organisations involved in smart grid development with a focus on interoperability [17]. The rationale behind creating a maturity model for interoperability within the smart grid stems from the recognition that the interoperability of components and systems is pivotal for its advancement. The objective of SG-IMM is to furnish a maturity model for assessing and delineating the process of achieving interoperability. A key objective of GWAC is to formulate a model that encompasses multiple organisations yet remains applicable for individual companies for self-assessment purposes. Additionally, SG-IMM aims to aid in the development of new tools and provide recommendations to enhance interoperability. The model should facilitate methods and processes conducive to integrating and maintaining automated components within the smart grid context. Despite the complexity of interoperability, the development of the model intends to ensure that it fosters understanding and delivers benefits at both a high level of abstraction and in detail [18,19]. The design of maturity levels comprises six maturity levels ranging from level 0 (lacking processes to support cross-organisational interoperability) to level 5 (pioneering continuous interoperability innovations, fostering community/social goals).

Figure 2.3 summarises the different domains and levels of this maturity model. The structure of the SG-IMM is founded upon the GridWise Interoperability Context-Setting Framework, with additional areas identified that traverse interoperability categories, including configuration and evolution, operation and performance, and security and safety.

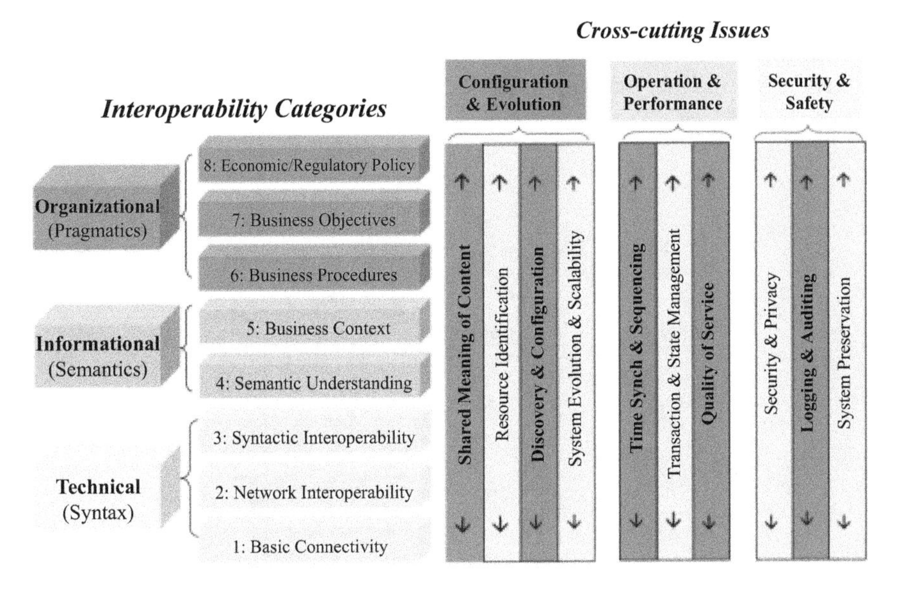

Figure 2.3 Interoperability context-setting framework [19]

Figure 2.4 shows a simplified interoperability landscape serving as a starting point for the application of SG-IMM. Detailed levels on both the categories axis and the issues axis are employed to channel user communities towards specific interoperability areas. Each issue area encompasses related cross-cutting objectives. For instance, within the 'technical model' category, parties agree on vocabularies, concepts and definitions. Moreover, each interoperability category features subsidiary interoperability objectives. Over 70 metric statements were formulated for the nine cells of the framework. SG-IMM is tailored for various applications within the electrical power system. The common thread among these applications is that they involve at least two information systems or devices. Examples of SG-IMM applications include retail service providers, enterprise smart grid applications and energy market operations [19].

- The SGMM of the Software Engineering Institute (SGMM-SEI): this model has been developed for the energy domain and specifically for the smart grid. The SGMM was developed by the Software Engineering Institute (SEI) of Carnegie Mellon University [20]. The aim of the SGMM-SEI is to provide a common framework for the management to define elements of smart grid transformation. Utilities should use the SGMM-SEI to develop a programmatic approach and track their overall processes. This helps utilities to plan their future smart grid endeavours, to prioritise their option and support decision-making processes. The characteristics and capabilities of an organisation can be determined with the help of the definition matrix of SGMM as proposed in a document entitled 'SGMM Model Definition. A framework for smart grid transformation' Version 1.2 of the Software Engineering Institute of the Carnegie Mellon University. The definition matrix of SGMM is composed of six levels in eight maturity domains.

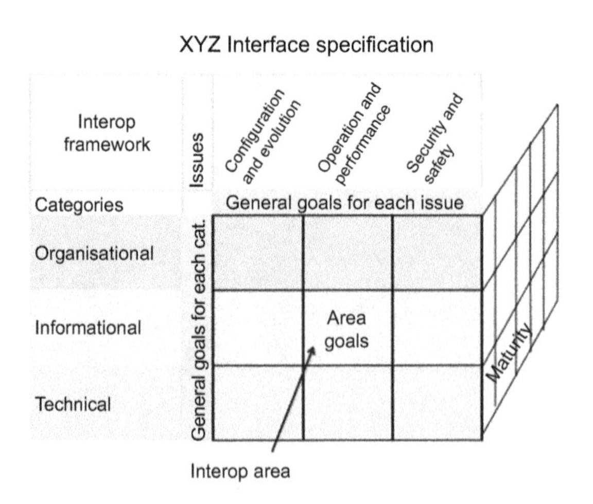

Figure 2.4 Simplified interoperability framework for SG-IMM [19]

The model encompasses eight different domains. In this context, a domain means a logical group of smart grid-related characteristics: Strategy – management – regulatory (vision planning, governance, etc.), organisation – structure (training, knowledge management, etc.), grid operations (reliability, security, safety, etc.), technology (IT architecture, standards, infrastructure, etc.), work – asset management (monitoring, tracking, maintenance), customer (pricing, customer participation, etc.), value chain integration (demand and supply management, etc.) and societal – environmental (sustainability, efficiency, critical infrastructure) [20].

The SGMM-SEI has six maturity levels that represent well-defined states. Each one describes the capabilities and the characteristics of the organisation to achieve the smart grid vision regarding efficiency, automation, reliability, energy savings, interaction with the user, integration of distribution energy resources and access to new opportunities of the business. To describe the organisation in terms of the domain, an overall of 175 characteristics is used.

The lowest level (level 0) represents the default position of the organisation when the study starts. An organisation operating in a traditional way without modernisation will be at this level. It is important for an organisation to evaluate its condition to establish the future vision in a predefined time interval. Since level 0 represents the starting condition, the model does not have precise characteristics for this level.

When the organisation achieves a level 1, it is in the starting process of exploring smart grid technologies. At this level, the organisation has a vision but does not have a clear strategy and is capable of communicating its vision to the community and industry. When it goes to level 2, the organisation has a defined strategy and is already investing to attain the modernisation of the electrical network. At this level, tests are performed with the business case already implemented to assess the changes in the organisation. A company with a level 3 integrates its smart grid program with the operating departments. The procedures should be repeatable and the information should be shared within the entire organisation. When the company achieves a level 4, the functionality and benefits of the smart grid can be assessed. The organisation performs an analysis and makes corrections in real time. Finally, a company with a level 5 is in a permanent innovation state, develops standards and improves procedures. The organisation becomes a leader in the industry. The vision and strategies of the organisation fulfil national, regional and local interests.

The organisation will be at a higher maturity level when the implementation of changes starts. Each organisation must establish its own target maturity levels based on its own operating system, strategy and timeline. It is obvious that higher levels of maturity in the model indicate a successful adoption from its grid modernisation efforts. The target level is not the same for each organisation. Therefore, any level could be appropriate to an organisation but not necessarily to others.

All the SGMM tools provide the basis to help an organisation guide, evaluate and improve its efforts to best select applications of smart grid in order to achieve a proper transformation and modernisation. From a methodological standpoint, the model allows the creation of a map defining the task and technologies to identify gaps in the strategy and execution, to support business opportunities that promote smart grid projects, to delineate the organisation vision and strategy, to evaluate alternative solutions and future goals that will help guide the future of the electric network.

All models presented and others available for similar purposes share similar characteristics. However, each one pays attention to specific requirements. Various criteria can be considered for comparing maturity models. The developer of the model can start by defining the kind of question they want to answer. For instance, the main things to define after applying the model are the target audience, the application drivers and areas, the number of maturity levels to incorporate, etc. According to the answers, the model to be used as a reference to evaluate the organisation can be selected. Table 2.4 summarises the differences between the presented models [15].

2.4.1 Benefits of using a smart grid maturity model

In today's competitive world, the industry demands that organisations strive to achieve repeatable and scalable procedures to promote sustainable improvements within an organisation. Maturity models were initially intended to be applied in the software development industry, driven by the necessity of evaluating various organisations under identical parameters. These models gave the possibility of developing improved planning, engineering and governing practices to guarantee higher quality levels in both processes and results.

Many organisations approved the use of the maturity models in electrical utilities, considering the successful experience in the software industry. It allowed them to determine the development level of networks and to visualise the gap between the current and the future situation. From this, the best solutions can be proposed. Maturity models must be combined with an entire work methodology to identify the standards and technical solutions that will be considered in developing the smart grid roadmap. Maturity models also support the implementation of applications.

A robust model needs to recognise not only management activities being carried out at the individual project level but also those activities within an

Table 2.4 Comparison between maturity models

	ES-C2M2	**SG-IMM**	**SGMM-SEI**
Objective	Improvement of cybersecurity	Smart grid interoperability development	Smart grid development
Target group	Organisations in the electricity market	Energy domain	Energy domain
Maturity scale	4	5	5
Origin	Politics	Consortium	Research institute
Application area/focus	Energy sector and cybersecurity	Smart grid	Smart grid

organisation that build and maintain a framework of effective project approaches and management practices.

By undertaking a maturity assessment against an industry standard model, an organisation will be able to verify what they have achieved, where their strengths and weaknesses are and then to identify a prioritised action plan to take them to an improved level of capability.

2.4.2 Development process of a SGMM

The maturity model development process can be carried out with the following steps:

1. *Information gathering in the eight domains of the related organisation:* in this stage, the consulting engineer must explore all organisation characteristics from a strategic point of view. Since the navigator assists the organisation in comprehending the questions included in the evaluation, it is vital that they have a full understanding of the model.
2. *Smart grid and SGMM concept awareness:* the navigator must be familiar with the organisation; likewise, the organisation itself must understand the aspects to be evaluated. The navigator must prepare the organisation to understand the concepts of the SGMM. Next, the criteria must be adjusted and the evaluation process must be established.
3. *SGMM application:* in this stage, the Navigator researches the questions included in the maturity model. The idea is to evaluate the current condition of the organisation as well as the desired future condition. It is vitally important that the organisation responds to questions in a critical, objective and honest manner. Also, it is important to involve individuals from various areas of the organisation, since the model covers all aspects of the organisation and requires that answers are given with an exhaustive view of the problem. If this is not the case, key points will be missed and efforts may be focused on irrelevant aspects. It is recommended to re-evaluate after the first model attempt.
4. *Final results:* results are obtained regarding the actual condition of the organisation and the future condition. The results help identify gaps that must be filled in order to reach the desired condition. The cost and time required to do so should be carefully analysed to assure the viability of the project. If the gap is large, a viable economic solution may be to extend the time or reduce the scope of the future vision.

2.4.3 Results and analysis obtained by SGMM

Once the results of the maturity model are obtained (the actual state and the desired future), a gap analysis is needed to determine what actions to take. Figure 2.5 helps to illustrate an example of results obtained by applying the SGMM, particularly the model from SEI.

It is important to mention that establishing a high future condition is not precluded by a low present rating. On the other hand, not aiming for a level of 5 in a particular

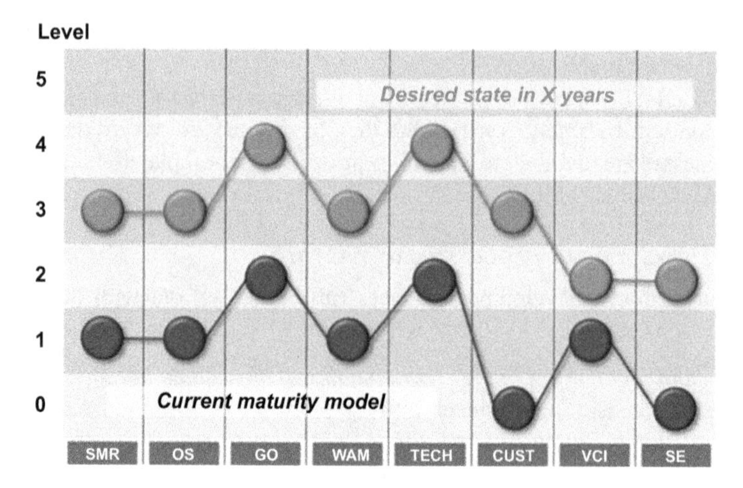

Figure 2.5 Example of results obtained from the SGMM

domain does not mean that the organisation is not focused on the smart grid technology. It is possible for the business model not to involve all aspects in the domain or for the organisation not to consider it profitable to improve their level in that domain. Similarly, in the software industry, where ideas are born from Maturity Models, some organisations consider an optimisation condition acceptable in the processes in a domain.

2.4.4 Example case

The methodology presented was applied to the evaluation of a real utility that has around 500,000 users. The utility has a vertical integration as it handles generation, transmission, distribution and energy marketing. The results presented here refer to the SGMM-SEI model domain and the answers given by the utility.

A score is obtained once all questions are evaluated indicating the maturity level of the organisation under study. Figure 2.6 shows the results from this organisation. The red line indicates the required limits to satisfy the requirements of each level. For this example, the values are established by a certification institution using particular standards.

For the example shown in the figure, it is assumed that the criterion of the certification institution to pass the first three levels establishes a score minimum of 0.6 for each one. After the third level, a higher score is required to approve (level 4 specifically requires at least 0.7 and level 5 requires 0.8). Once the score gets below one of the defined limits at a level, the total is then calculated with the summation of the scores obtained up to that level, including the score of the last one.

The GO domain for this organisation indicates a maturity around 2.25 which indicates that the organisation manages higher levels of operations automation and network optimisation processes. The example shows that the current condition (blue bars) only reaches the red line in the first level with a 0.70 score (minimum is 0.6), a

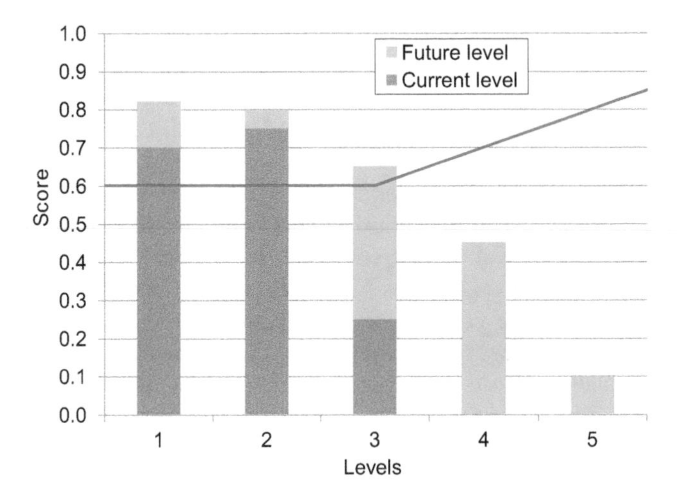

Figure 2.6 Example of results in grid operation (GO) domain

second level of 0.75 and in the third level only 0.25 is achieved. Since the third level is the first level that does not approve the objective, the equivalent level will be the score summation up to this point (1.5). According to the results shown in Figure 2.4, for the GO domain, this organisation could reach the third level in three years at the most. That could be acceptable for that company. In that case, the total score would be 2.8 which is calculated from the results of the table (0.85 points for the first level, 0.80 for the second level, 0.65 for the third level and 0.45 for the fourth level). The organisation aims to improve its current condition by 1.3 points in a three-year term for a total of 2.8.

The profile of this organisation assessed in this example indicates that it has been starting an evaluation of potential opportunities for grid automation and is investigating the capabilities of optimisation processes. In pursuit of this objective, approval is sought for various business cases pertaining to new equipment and systems associated with a smart grid within at least one line of business. Additionally, the organisation is anticipated to explore various switching, sensor and communication systems aimed at monitoring the network and its assets. It is preferred that aspects related to security (both physical and cyber) are already integrated into network operation initiatives. This demonstrates that the organisation currently has instituted smart grid initiatives where technologies and equipment have been elected for pilot testing, some of which may have concluded, and implementations are in process. For example, the distribution automation and the interoperable data system employ tools like CIM.

Once current and future state results have been obtained for the company under review, an analysis is conducted to evaluate what efforts must be made to reach such a condition. It can also be defined based on the conditions the model expects for the level. For this example, according to desired future expectations, considerable emphasis should be placed on prioritising customer satisfaction. To achieve this goal, the organisation needs to possess the necessary technological tools that enhance its decision-making capabilities through the utilisation of measurements and actual network data. This information aims to optimise various operational aspects, including enhancing network reliability and stability, as well as minimising technical losses.

To attain these objectives, the organisation must have measurement tools that enable the collection of real network data and facilitate decision-making for both the network operations team and crews. This should be complemented by tools equipped with analytical capabilities for processing the obtained information. The acquisition of compatible communication equipment to gather such information plays a crucial role in most smart grid-focused developments.

It is particularly advised to address the aspects of the model that pertain to the approval of business cases for new equipment and systems related to smart grids. Similarly, proof-of-concept projects and component testing for grid monitoring and control should be underway within three years. Lastly, exploration and evaluation of outage and distribution management systems linked to substation automation are essential. Potential obstacles, such as interoperability issues with technologies used, restrictions on IT acquisition and the intricacies of the electrical grid, must be carefully considered to overcome challenges and achieve aspirations in this domain.

Table 2.5 summarises the maturity level in each of the domains discussed in the model for this example, and Figure 2.7 shows a graphical representation of this score. Based on the results obtained, an analysis will identify the gaps to be improved. The distance from actual to future in the evaluated conditions represents the efforts required to reach the desired future condition. The larger the gap between actual and future conditions (radar chart area differences), the more aspects need to be improved. Based on the estimated time and budget, necessary efforts must be defined and the probability of success evaluated.

Table 2.5 Example of results after applying the SGMM

Domains	**Current maturity level**	**Future maturity level (5 years)**
SMR	2.0	3.0
OS	2.0	3.0
GO	1.5	2.7
WAM	1.5	2.0
TECH	1.5	2.0
CUST	1.0	2.0
VCI	1.0	2.0
SE	0.0	1.0
Mean maturity level	1.31	2.21

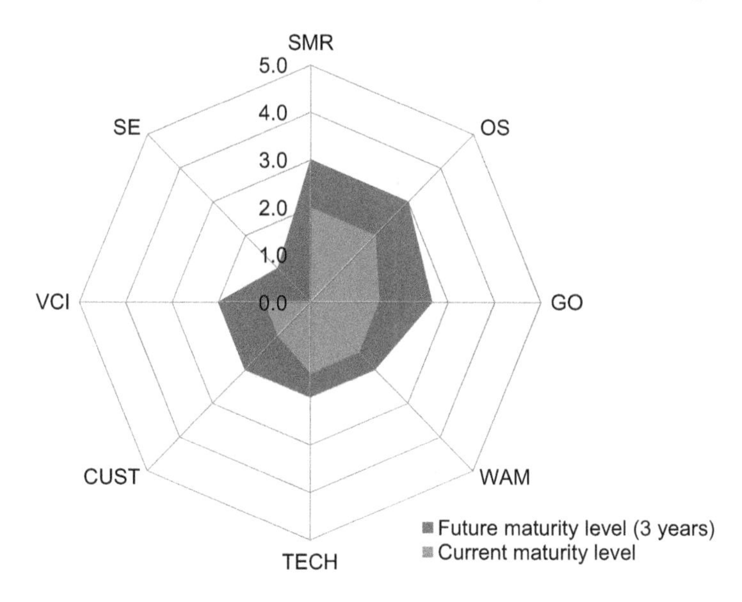

Figure 2.7 Graphical example of results after applying the SGMM

2.5 Simulation requirements for smart grids

Nowadays, various components and stakeholders of power systems are interconnected across diverse technological and conceptual levels [21]. Consequently, examining the effects and implications of novel components or altered parameters presents an increasingly challenging task. As a result, the significance of simulation has become progressively more pronounced.

Several research projects and initiatives have demonstrated that the shift from conventional power systems to smart grids requires the expertise of interdisciplinary professionals to tackle diverse challenges. One such challenge is effectively managing an ever-growing level of complexity, resulting from interactions between various subsystems. These primarily comprise the power system, the ICT, customers, energy markets, environmental factors affecting renewable sources, as well as legal and regulatory considerations.

The use of simulation-based assessments for complex systems has witnessed an upward trend since the advent of more advanced computational solutions. Typically, varying factors can prompt the adoption of simulation tools and environments in lieu of studying actual systems. For example, limited monitoring may impede the measurement of critical internal states, particularly in distribution systems. Moreover, executing certain scenarios may require a substantial investment, and in certain instances, replicating minor pilot projects can prove to be costly and unfeasible without undertaking further technical analyses.

Despite being a new discipline, simulations of smart grids have seen the emergence of several methods for simulating ICT-powered system concepts. This underscores the requirement for new smart grid-specific simulation concepts that can determine the behaviour and impact of technologies such as DER, energy storage solutions and varied control mechanisms.

2.6 Modelling and simulation

Simulation entails depicting the functions or attributes of a system via the operation of an alternate system [22]. For the digital simulation varieties examined in this chapter, it is presumed that a discrete-time simulation with constant step duration is conducted. In discrete-time simulation, time progresses through regular intervals of uniform duration, commonly termed fixed time-step simulation. It is noteworthy that other solving methods exist that employ variable time-steps, which are used for resolving high-frequency dynamics and nonlinear systems. Nevertheless, such techniques are unsuitable for real-time simulation.

2.6.1 Off-line simulations

To resolve mathematical equations and functions during a particular time-step, each variable or system state is solved in turn as a function of the variables and states at the conclusion of the preceding time-step [22]. In discrete-time simulation, the real time taken to compute all equations and functions representing a system during a given time-step can be shorter or longer than the duration of the simulation time-step. Figure 2.8 exhibits these two scenarios, where in part (a), the computing

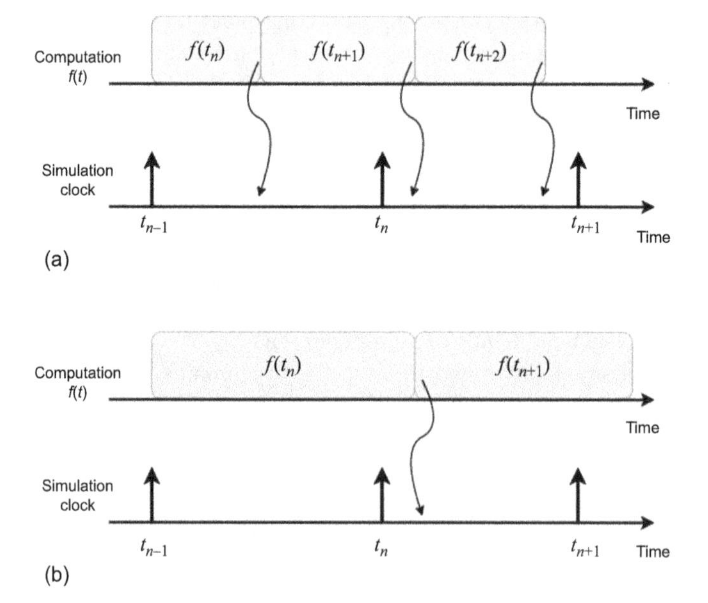

Figure 2.8 Real-time simulation requisites and other simulation techniques. (a) Offline simulation: faster than real-time and (b) offline simulation: lower than real-time.

time is shorter than the fixed time-step (also known as accelerated simulation) while in part (b), the computing time is longer. These two situations are categorised as offline simulation. In both scenarios, the timing of result availability is unimportant. Typically, when conducting offline simulation, the aim is to attain outcomes as fast as possible. The velocity of system solving is dependent on available computation power and the complexity of the system's mathematical model.

2.6.2 Real-time simulations

In real-time simulation, the precision of computations relies not only on an exact dynamic representation of the system but also on the time duration taken to produce results. As depicted in Figure 2.9, the chronological principle of real-time simulation requires the real-time simulator to accurately generate the internal variables and outputs of the simulation within the same duration as its physical counterpart. Specifically, the time required to compute the solution at a given time-step must be shorter than the wall clock duration of the time-step. This enables the real-time simulator to execute all necessary operations such as driving inputs and outputs (I/O) to and from connected externals. Any idle time preceding or following simulator operations for a given time-step is lost. This is different from accelerated simulation, where idle time is utilised to compute equations at the next time-step, and the simulator waits until the clock ticks to the next time-step. However, if simulator operations are not completed within the fixed time-step, the real-time simulation is deemed erroneous. This is commonly referred to as an 'overrun'.

Based on these fundamental definitions, it can be deduced that a real-time simulator is operating effectively if the equations and states of the simulated system are resolved accurately, with an acceptable resemblance to its physical counterpart, and without any overruns.

2.6.3 Hardware in the loop

There are drawbacks regarding off-line simulations: simulation speed will be very slow when the model is complicated, and limits of computer memory do not allow simulation to run a long period of time [23]. In contrast with off-line simulation, the real-time simulation platforms have the advantages of real-time simulation speed,

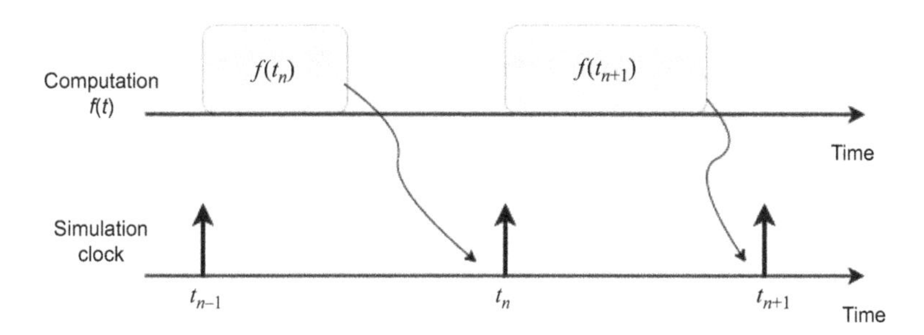

Figure 2.9 Synchronised real-time simulation techniques

ability of hardware in the loop (HIL) simulations and online parameter adjustment. These features can not only improve the simulation speed but also bring efficiency and precision to models as well as results.

HIL simulation is an emerging method for advanced experiments in power system analysis. None of the analytical and experimental models can have the same accuracy as a real system. However, a test bed consisting of only physical systems is very expensive. Therefore, physical devices will be used only for the important modules. The other modules of the testbed will be modelled in software. The physical system and the software simulation must be interconnected. In the literature, research papers have been dedicated to HIL studies in the power system field such as big portions of a power system, microgrids or a hybrid energy system consisting of a fuel cell and a battery storage system.

The HIL techniques can be broadly categorised into two methods [24]: controller hardware-in-the-loop (CHIL) and power-hardware-in-the-loop (PHIL). In CHIL, the real controller hardware is used as the HIL system, while the other power systems are simulated inside the simulator. PHIL, on the other hand, is an extension of CHIL simulation, where a part of the power hardware under test is simulated internally and connected to other real parts through input/output (I/O) interfaces. CHIL focuses on the internal emulation of the power system, while PHIL includes the rest of the system as an externally connected real hardware power device. A representation of both methods is shown in Figure 2.10.

During CHIL testing, a specific device, such as a power converter controller, is subjected to a closed-loop environment in which signals are exchanged between a real-time simulator and the hardware-under-test via its information ports. The interface used in CHIL includes analogue-to-digital and digital-to-analogue

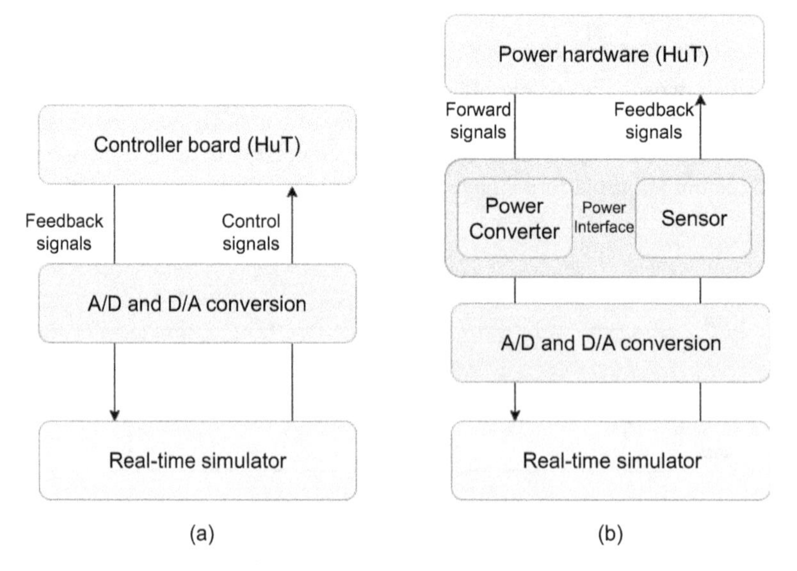

Figure 2.10 Basic HIL simulation systems: (a) CHIL and (b) PHIL

converters and/or digital communication interfaces, and the signals exchanged are low-power control signals, typically ranging from −10 V to 10 V or 0–20 mA. In addition to control devices, real-time simulations coupled to other units like DERS controllers, relays, Phasor Measurement Units (PMU) or monitoring components can be classified as CHIL. Such devices are validated under various dynamic and fault conditions, which enhances the validation of control and protection systems for power systems and energy components. In contrast, PHIL testing involves devices that generate or absorb power to be tested in realistic environments such as converter devices, electric vehicles, charging equipment or entire microgrids. A high precision power interface amplifier is required to facilitate the testing.

2.6.4 Applications of RT-HIL

Real-time HIL simulation has been extensively employed in various applications such as protection systems, microgrids, power plants, power converters and renewable energy sources. PHIL simulations provide a means of testing electric components and their interactions with complex systems. PHIL simulations combine the advantages of pure software simulation and hardware testing. However, prior to starting a PHIL experiment, several crucial considerations must be noted since PHIL simulations are not currently 'plug and play'. Improving the connection between the hardware part of a PHIL simulation and the real-time computing system involves adding a current filter into the feedback path. This method employs an inverter-based interface connected to a low voltage AC grid with linear and non-linear loads. The inverter is connected as real hardware to the simulation environment, while the low voltage AC grid and the loads are simulated. The introduction of a feedback current filter assures stable PHIL experiments that offer insight into the interaction between the inverter and the nonlinear load.

The growing adoption of high-voltage DC (HVDC) and flexible AC transmission systems in AC networks has significant impacts on grid performance and reliability. The presence of HVDC links in close proximity may lead to undesired interactions and affect the performance of the AC network. To evaluate such risks, vendors, utilities and third parties carry out off-line and real-time simulations. For instance, Réseau de Transport d'Électricité, the electricity transmission system operator of France, has undertaken three pioneering projects aimed at studying these interactions, utilising physical control and protection (C&P) cubicles in an industrial setting through HIL simulation. Each case presents distinct challenges and experiences that contribute to the development of HIL simulation techniques for interaction studies.

In wind turbine development, manufacturers employ HIL systems to perform realistic ground-level testing. An example is depicted in Figure 2.11, where the test rig is equipped with HIL functionality, consisting of both software parts such as electric motors and hardware parts such as the wind-load-unit, which serve as the test bench's drive unit and provide mechanical power to the mounted device under test. This HIL system permits ground-based testing of full-scale wind turbines in the multi-megawatt range. The HIL simulation system can be experimentally validated at a test rig with a mounted full-scale wind turbine. The proposed system

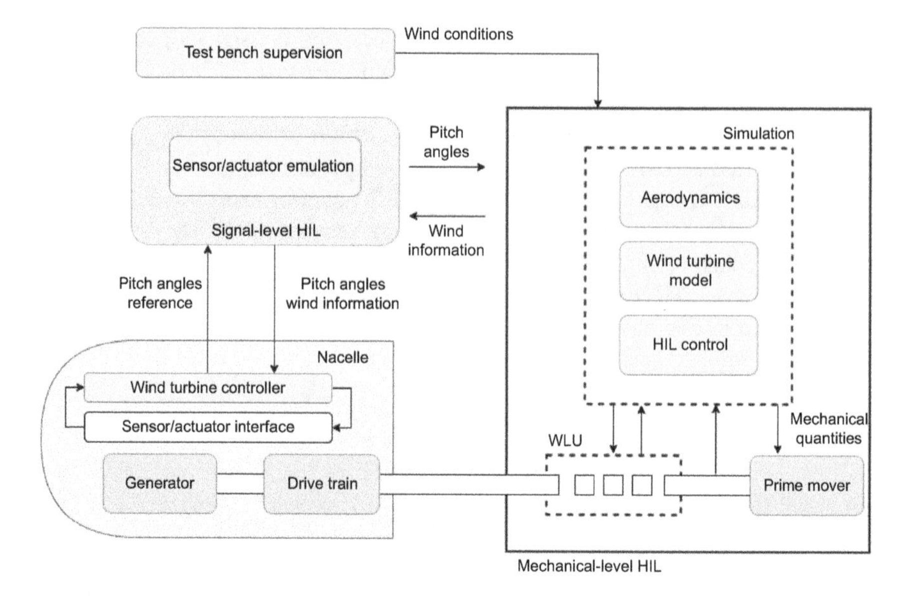

Figure 2.11 Relevant time steps for real-time simulation in state-of-the-art simulation methods

includes a state estimator and an internal wind turbine model that features an aeroelastic rotor model for reference generation, both of which are integral parts of the control loops for setpoint tracking and active drivetrain damping.

2.6.5 Limitations of RT-HIL techniques

The present HIL technology provides opportunities for validation and testing of smart grid systems, but there remain some constraints:

- Integrating HIL technology with the communication layer can be difficult, particularly regarding the synchronisation of real-time and offline simulation, as well as continuous and discrete timelines. Communication networks in simulations or labs do not always reflect the real scenario where long geographic distances between different equipment may cause unexpected delay, signal loss and failure for timely control. Until now, the communication network has been separately simulated with dedicated software to study the effect of realistic latency, packet loss or failure in the ICT system on the reliability and performance of monitoring and control applications. Communication simulators also facilitate cyber-security related experiments, such as denial-of-service protection, confidentiality and integrity testing, which are important but not always easy for the electrical community. A holistic consideration of the cyber-physical energy system, as well as the impact of ICT issues on the power system, requires a seamless integration of HIL techniques into the co-simulation framework.

- There is lack of a general framework to facilitate the reusability of models and information exchange among different proprietary interfaces or different partners of a joint HIL experiment.
- The limited capacity of HIL simulations for complex systems such as those with scale effects, synchronicity, diversification, nonlinearity, high harmonics and transient phases, can be a major limitation. In some cases, the fidelity to represent the dynamics of complex power components, such as power electronics converters, must be compromised due to the fixed time-step of real-time simulations. This limits the size of simulations and transient performance.
- Remote HIL and geographically distributed HIL experiments have limited capacity due to synchronisation issues in CHIL and PHIL experiments. Loop delay can affect the stability and accuracy of the PHIL power interface.

References

[1] J. Gers, *Distribution Systems Analysis and Automation*, 2nd ed., London: IET, 2020.

[2] XcelEnergy, ‘Xcel Energy smart grid: a white paper’, *Denver Bus. J.*, p. 13, 2007.

[3] J. Momoh, *Smart Grid: Fundamentals of Design and Analysis*, Hoboken, NJ: Wiley-IEEE Press, 2012.

[4] National Institute of Standards and Technology (NIST), ‘NIST Smart Grid Standards Roadmap’, National Coordinator for Smart Grid Interoperability, Gaithersburg, MD, 2009.

[5] U.S. Department of Energy, ‘Smart Grid System Report’, DOE, July 2009.

[6] U.S. Department of Energy, ‘The Smart Grid Stakeholder Roundtable Group Perspectives’, DOE, September 2009.

[7] SMB Smart Grid Strategic Group, ‘IEC Smart Grid Standardization Roadmap’, 2010.

[8] Electric Power Research Institute (EPRI), ‘Methodological Approach for Estimating the Benefits and Costs of Smart Grid Demonstration Projects’, Electric Power Research Institute, Palo Alto, CA, 2010.

[9] Electric Power Research Institute (EPRI), ‘Guidebook for Cost/Benefit Analysis of Smart Grid. Demonstration Projects Revision 1, Measuring Impacts and Monetizing Benefits’, Electric Power Research Institute, Palo Alto, CA, 2012.

[10] E. Triantaphyllou, *Multi-Criteria Decision-Making Methods: A Comparative Study*, Dordrecht: Springer Science+Business Media, 2000.

[11] European Commission, ‘JRC REFERENCE REPORTS/Guidelines for conducting a cost-benefit analysis of Smart Grid projects’, European Commission/Joint Research Centre Institute for Energy and Transport, Luxembourg, 2012.

[12] A. Mardani, A. Jusoh, K. Khalifah, Z. Zakwan and N. Valipour, ‘Multiple criteria decision-making techniques and their applications – a review of the literature from 2000 to 2014’, *Economic Research-Ekonomska Istraživanja*, vol. 18, no. 1, pp. 516–571, 2015.

[13] M. Uslar and J. Masurkewitz, 'A survey on application of maturity models for smart grid: review of the state-of -the-art', in *Proceedings of EnviroInfo and ICT for Sustainability 2015*, Atlantis Press, pp. 261–270, 2015.
[14] A. Flore and J. M. Gomez, 'Development and comparison of migration paths for smart grids using two case studies', *Heliyon*, vol. 6, pp. 1–18, 2020.
[15] A. Flore and J. M. Gomez, 'Requirements on dimensions for a maturity model for smart grids based on two case studies: disciplined vs. agile approach', in M. Mora, J. Gómez, R. O'Connor, and A. Buchalcevová, (eds), *Balancing Agile and Disciplined Engineering and Management Approaches for IT Services and Software Products* (pp. 271–292), Hershey, PA: IGI Global, 2021.
[16] Carnegie Mellon University, 'Electricity Subsector Cybersecurity Capability Maturity Model (ES-C2M2), version 1.0', Carnegie Mellon University, 2012.
[17] S. Widergren, A. Levinson, J. Mater and R. Drummond, 'Smart grid interoperability maturity model', IEEE PES General Meeting, Minneapolis, MN, USA, 2010, pp. 1–6, doi:10.1109/PES.2010.5589785.
[18] M. Knight, S. Widergren, J. Mater and A. Montgomery, 'Maturity model for advancing smart grid interoperability', in *IEEE PES Innovative Smart Grid Technologies Conference (ISGT)*, Washington, DC, USA, 2013, pp. 1–6, doi:10.1109/ISGT.2013.6497915.
[19] The GridWise Architecture Council, 'Smart Grid Interoperability Maturity Model Beta Version', 2011.
[20] Software Engineering Institute, 'SGMM Model Definition. A framework for smart grid transformation, Version 1.2', CERT® Program Research, Technology, and System Solutions Program Software Engineering Process Management Program Carnegie Mellon University, 2011.
[21] S. Rohjans, S. Lehnhoff, S. Schütte, F. Andrén and T. Strasser, 'Requirements for Smart Grid simulation tools', in *IEEE 23rd International Symposium on Industrial Electronics (ISIE)*, Istanbul, Turkey, 2014, pp. 1730–1736, doi:10.1109/ISIE.2014.6864876.
[22] J. Bélanger, P. Venne and J. Paquin, 'The what, where, and why of real-time simulation', in *Proc. IEEE PES Gen. Meeting*, vol. 1, pp. 37–49, 2010.
[23] J. Khazaei, L. Piyasinghe, V. R. Disfani, Z. Miao, L. Fan and G. Gurlaskie, 'Real-time simulation and hardware-in-the-loop tests of a battery system', in *2015 IEEE Power & Energy Society General Meeting*, Denver, CO, 2015, pp. 1–5, doi:10.1109/PESGM.2015.7285993.
[24] V. H. Nguyen, Y. Besanger, Q. T. Tran, *et al.*, 'Real-time simulation and hardware-in-the-loop approaches for integrating renewable energy sources into smart grids: challenges & actions', in *Proceedings of IEEE PES Innovative Smart Grid Technologies ISGT Asia*, 2017.

Chapter 3

Load management and demand response for urban grids

Load management and demand-side management are terms commonly applied to the load control on the user side, in order to improve the operating condition of distribution systems. This concept was later complemented with another one called demand response (DR) to improve the efficiency of the distribution systems. Both concepts and methods to implement them are discussed in this chapter.

3.1 Demand-side management

Demand-side management (DSM) is the process of balancing energy demand, primarily derived from electricity, with its supply. DSM encompasses various techniques and strategies aimed at modifying consumer behaviour or managing appliance operations to reduce or shift energy demand during peak hours. This, in turn, diminishes the need for additional generation capacity by optimising the use of the existing power grid infrastructure. DSM has gained significant relevance in recent years, driven by the increased integration of intermittent renewable energy sources, such as wind and solar power, which introduce supply-demand imbalances. These imbalances necessitate the deployment of energy storage systems or flexible demand management to ensure grid stability. Furthermore, DSM can assist utilities in reducing carbon emissions by optimising the use of existing generation assets.

The application of DSM can extend the capabilities of the power system and facilitate the integration of distributed generation. One of the primary advantages of DSM is its cost-effectiveness, achieved by controlling load rather than constructing additional power plants. Originally, this concept was more utility-centric; however, with the introduction of smart metering, it has become increasingly customer-driven. According to the impact of measurements and the sample timing, DSM is categorised as energy efficiency (EE), time of use (TOU), DR and spinning reserve (SR) [1]. Figure 3.1 shows the graphical representation of these categories.

Within the categories of DSM, it is defined as energy efficiency in the lower boundary, which includes modifications on equipment (such as the replacement of inefficient ventilation systems), as well as enhancements to the physical attributes of systems (such as augmenting building insulation). These measures yield immediate and enduring reductions in energy consumption and associated

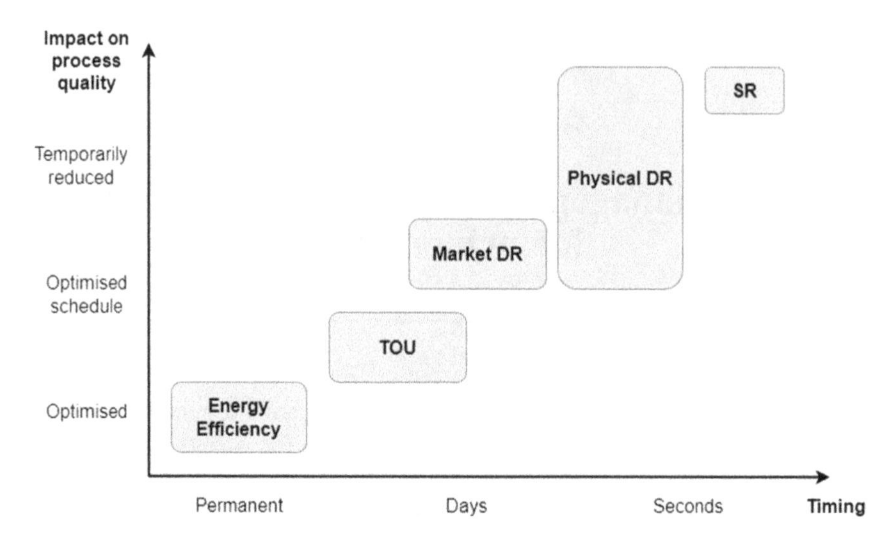

Figure 3.1 Categories of demand-side management

emissions, which highlights this option as very suitable in most cases. In some cases, the energy conservation (EC) can also be related in this category, which mainly focuses on inducing behavioural shifts of energy consumption and leads to more judicious energy utilisation.

TOU is another category of DMS, in which energy tariffs are imposed during specific time intervals, for instance, prompting customers to reconfigure their operational schedules by a higher cost of the energy between 17:00 and 19:00. Normally, a change in the TOU price-schedule is subject to a contract or predefined tariffs, which makes changes occurring infrequently. This strategy significantly enhances the cost-effectiveness, security and emission reduction potential, particularly those with a substantial proportion of wind power [1].

Dynamic DSM is a different category, in which the temporal patterns of consumption are influenced. It is required that the process to be controlled does a 'catch up' if, for some reason, it encounters interruptions, it may subsequently need to restart once it resumes normal operation. For example, a water pumping system with storage tanks, which can be temporarily for 30 minutes, replenish its tanks after following this period of inactivity, as the tanks would have been depleted during the interruption. This phenomenon is called a 'rebound effect' or 'payback', during which the creation of new consumption peaks may occur as illustrated in Figure 3.2, and it does not achieve any energy savings. Sometimes the aforementioned effect may be circumvented; however, doing so may result in a diminished quality of processes. An ideal case of 'peak shaving' is applied to any ventilation system. If it typically operates at 50% capacity and experiences a half-hour interruption, compensating for this downtime is prohibited with a subsequent half-hour of operation at 100% capacity.

Market DR relies on specific marketplaces where prices are determined and energy products are traded. Transactions in such markets typically occur a day in

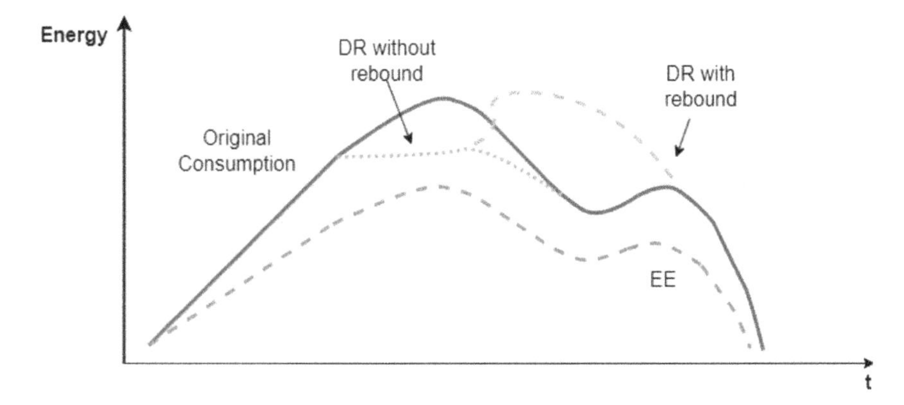

Figure 3.2 Impact of improved energy efficiency versus demand response

advance, except for real-time pricing (RTP), which immediately conveys figures from an energy spot market to end-users (e.g., EEX or European Energy Exchange in Leipzig). Although surplus wind power or adjusting prices to reflect grid congestion may influence customer behaviour and stability, it is an over-simplification to assume that monetary incentives such as RTP can solely address all energy grid challenges.

For the previous category, some situations that do not correlate directly with pricing and limited customer responsiveness require additional strategies for an effective load shedding. This is when the physical DR is applied by issuing binding requests for demand management when grid performance is compromised due to failures or maintenance in its infrastructure (such as lines, transformers and sub-stations). An optimal grid operation often requires a balanced combination of both market and physical DR approaches.

The final category called spinning reserves (SR) is mainly managed by loads and corresponds to the faster end of the DSM spectrum. However, the terminology surrounding spinning reserves is used somewhat loosely within the power community. In this context, spinning reserve encompasses primary control (active power output directly tied to control the system's frequency) and secondary control (injection of additional active power injected to restore frequency and guarantee grid stability). Typically, regulation power plants assume this role. Loads can function as 'virtual' (or negative) spinning reserves by aligning their power consumption with grid conditions using 'droop control' or similar intelligent techniques. This may occur autonomously (akin to primary control) or through coordinated efforts (similar to secondary control). The choice of technology and communication methods varies depending on the type of DSM employed.

Normally, EE is a more desirable objective over DSM when comparing both approaches. EE aims to reduce energy consumption by leading to emission savings at the same time and with positive environmental impacts in majority. On the other hand, DSM is more focused on changing or shifting the consumption behaviour through time without impacting the energy usage.

The choice of a specific dynamic DSM method should be based on the system's particular conditions and context. Factors that must be considered are the electrical environment, the structure of the energy market, policy frameworks and the capabilities of the system itself. Financial viability of automation investments and incentives plays a crucial role in determining the extent to which these methods can be effectively deployed and sustained.

3.1.1 Energy efficiency

The process to improve buildings or industrial buildings in terms of energy efficiency must start with data and information gathering as well as the processes involved. Energy losses are always present and are unavoidable in every building type such as compressed air leakages, control failures or improper settings, dirty filters and poor equipment states among others. Very often, these issues are not thoroughly looked into unless more advanced instrumentation is used like the one depicted in [2].

A typical energy information system, as depicted in Figure 3.3, consists of the following components:

- Data acquisition infrastructure: which includes sensors, data loggers, gateways and other hardware for collecting consumption data.
- An application server: which is equipped with a database, calculation and analysis algorithms, alarming systems and reporting functionalities.
- User interfaces: for visualisation and configuration.

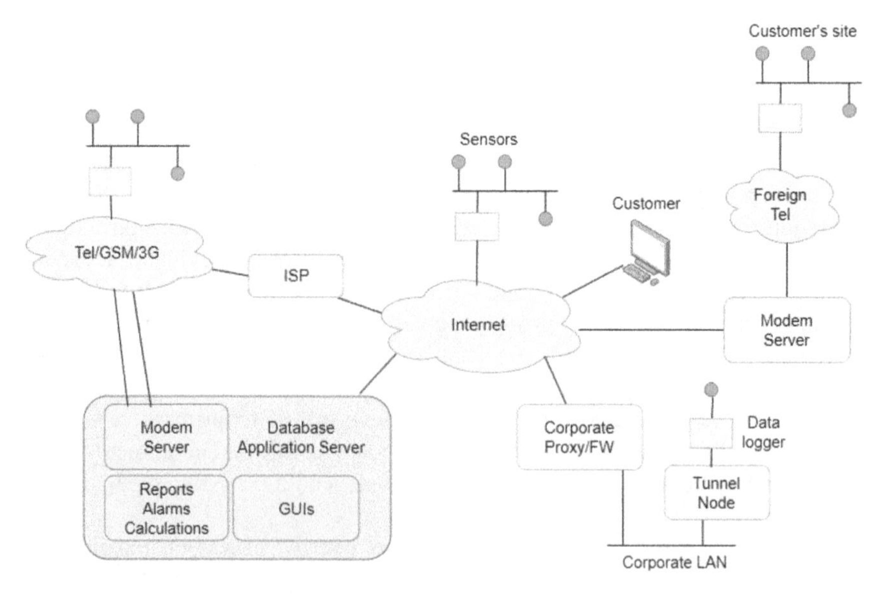

Figure 3.3 A web-based energy information system

Generally, the calculations and analyses performed by such systems may include:

- Baseline and peak load comparison: aims to identify standby power consumption and old equipment performance that may result in a high energy load.
- Weekly time series: This analysis reveals instances where lighting or ventilation systems inadvertently operate during the night or over the weekend, leading to unnecessary energy use.
- Benchmarks: allows to compare energy performance to that of others, which can be particularly useful.
- Process correlations: By examining whether energy consumption is strongly correlated with external factors like outside temperature or solar gains.

In addition to static efficiency information, the dynamics of energy consumption might be very insightful. Hence, facility or processes managers and smart algorithms can analyse consumption patterns and identify improvement opportunities to reduce peak loads or losses. The deployment of automation equipment may be useful to achieve peak load reduction and ensure compliance with energy supply agreements.

3.1.2 Energy controllers

To facilitate consumption-driven adjustments for equipment operations, an energy controller can be employed. This device continuously monitors energy consumption. If a trend indicates consumption levels that are undesirable or exceed predefined thresholds, the energy controller can autonomously switch off specific equipment or define a new setting based on priorities and predefined rules. This is depicted in Figure 3.4.

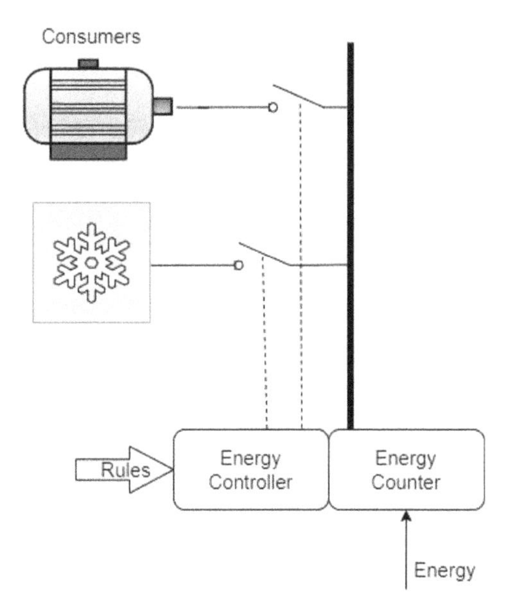

Figure 3.4 An energy controller switches (off) devices

Configuration of such controllers might be complex depending on the particular application and considerations. Changes on loads and behaviours might affect the stability and performance of the controller. One simple example of how to determine the priority level depending on the consumption trend is shown in Figure 3.5.

Figure 3.5 depicts a single measurement period used for billing purposes. During each period, the energy consumption trend starts from zero and steadily increases. As the consumption lines cross any of the upper thresholds, specific consumers are switched off or subjected to duty cycling, based on their priority levels.

Figure 3.5 also illustrates three device classes:

- C1 as the most relevant
- C2, and C3 as the least relevant.

All devices are initially allowed to operate when the line tendency is steep. It is steeper than the ideal (dashed) curve, so it eventually crosses the 'C3 off' line, resulting in the deactivation of this device class. This action proves insufficient and the trajectory later crosses the 'C2 off' level, leading to the deactivation of class 2 devices, leaving C1 devices under operation only. As the trajectory intersects 'C2

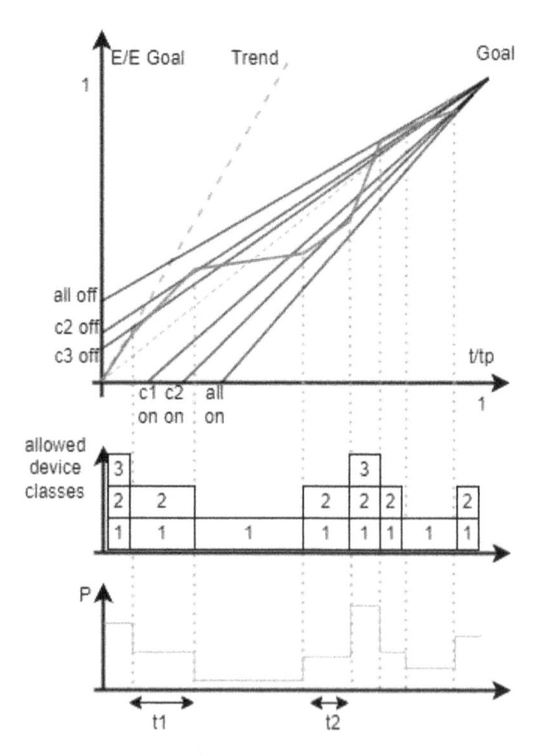

Figure 3.5 Selection of priorities in a maximum demand monitor

on', class 2 devices are allowed to operate again. If the trajectory crosses the 'all off' line, it results in a power consumption of zero and a horizontal energy trajectory.

With this system, the consumption trajectory will reach its goal by the end of the analysis period. The priority-based control ensures that devices are activated or deactivated as needed to maintain a manageable and cost-effective energy profile.

3.1.3 Demand response

A much quicker response to energy management is provided by various DR mechanisms. A signal from a distribution or transmission system operator can contain information such as pricing or commands for load shedding or load shifting. The response deadline is not instantaneous and could pertain to a future situation.

Classical direct load control (DLC) operates under the assumption that loads are fully controllable. The intelligence resides in the main controller, which utilises load models to make informed decisions. In [3], a stochastic state-space model was employed for loads simulating an urban power system. The results showed cost savings and losses reduction.

One modern system for automated DR is OpenADR, developed by the leading research group on DR, the Demand Response Research Center (DRRC) at the Lawrence Berkeley National Laboratory. OpenADR is an open specification and an open-source reference implementation of a distributed, client-server-oriented DR infrastructure using a publisher-subscriber model. Its primary components include the following, as depicted in Figure 3.6:

- Demand Response Automation Server (DRAS).
- DRAS Clients at the customers' sites.
- The Internet as communication infrastructure.

The client side of OpenADR is often a communication library that manufacturers can integrate into their products. Clients using this library can subscribe to DR programs such as critical peak pricing or demand bidding. The DRAS serves as a simple market platform and subscription manager, where it maintains a database of the participating clients and the programs to which they are subscribed.

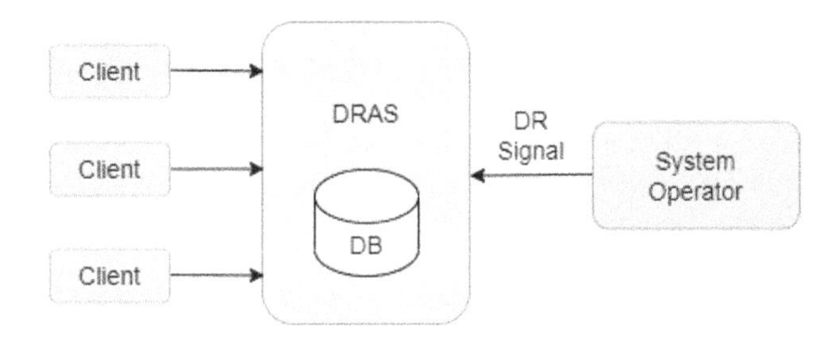

Figure 3.6 OpenADR clients and the system operator connect to the DRAS

The system acts as an open-loop control system. When a utility or system operator issues an emergency message to the DRAS, the server forwards the message to all clients participating in the emergency program. Transactions are recorded by the DRAS because financial incentives are tied to reacting to such events.

Another approach combines ripple control with a wide area phasor measurement system using global positioning system (GPS) timestamps with distributed voltage/current measurement equipment. This introduces a feedback loop into the control system, resulting in wide area control for energy systems. With approximately 10% of loads controllable via ripple control, the authors estimate around 30% savings in transmission corridor losses and approximately 40% in control power savings.

3.1.4 Distributed spinning reserve

Distributed spinning reserve aims to replicate the behaviour of traditional ancillary service providers by responding to changes in grid frequency. On the consumer side, this involves the ability to decrease or increase load when the grid frequency falls or rises.

In a distributed control framework, the response is still based on classical control principles: the system reacts when grid frequency deviates and stops once the desired frequency is restored. A more advanced variation involves model-predictive control, where devices equipped with load models can predict how much load they can shed and for how long, taking into account process constraints. These load models contribute to system stability by determining the magnitude of the response required to address anticipated issues and identifying which devices can provide it.

3.1.5 Demand shifting

Load models play a crucial role when demand needs to be shifted to different time periods. In scenarios where forecasts predict a grid emergency, informed consumers can proactively plan their tasks to align with grid requirements. This may involve adjusting the timing of processes to accommodate the grid's needs such as precooling or production scheduling.

Processes that are amenable to shifting typically are among the following categories:

- Inert thermal processes: These involve heating or cooling systems.
- Inert diffusion processes: This category includes processes like ventilation and irrigation.
- Mass transport: It covers activities like pump operation with tanks and conveyor belts.
- Logistics: This pertains to scheduling, interdependencies and considerations like meal breaks.

Shifting the load to a later time is a straightforward approach, where the load is reduced during the critical or peak periods. However, issues may arise if there are insufficient resources during the time of load reduction. Therefore, a more effective approach is to reschedule the peak before the load reduction time, ensuring preparedness. In this context, load models predict the duration for which processes can be halted, the capacity required to fill a 'virtual storage' and the associated costs.

3.1.6 Load sharing

Load sharing denotes the concurrent operation of multiple generators within a facility. Technically, it pertains to the equitable distribution of both active power load and reactive power load among multiple generators. Parallel operation and load sharing are intricately linked, as the parallel operation of a set of generator units necessitates load sharing among alternators. Various operations leverage parallel generator sets to enhance their capacity and fulfil heightened energy output demands. Failure to distribute the load among a network of generator sets poses the risk of causing damage to the generator sets or the power grid.

The electrical load on a power system experiences fluctuations throughout the day and across different seasons. Mechanisms facilitating load sharing play a pivotal role in disseminating the load among available generators based on their capacity and operational attributes. During periods of heightened demand, multiple generators may concurrently operate to meet the escalated load. Fluctuations in load on a generator, be it an increase or decrease, can alter the generator's speed and frequency. Load sharing mechanisms ensure that generators respond to these load changes, thereby preserving the system frequency.

Automatic generation control (AGC) represents a control system integral to power systems, automatically modulating the power output of generators in response to alterations in load and system conditions. AGC contributes to load sharing by perpetually monitoring the system frequency and adjusting generator output to maintain a balance between generation and load. Synchronisation is imperative before generators can share the load. This process entails aligning the frequency, phase and voltage of generators before connecting them to the power grid. Once synchronised, generators can efficiently share the load, contributing to the overall stability of the power system.

Examples of load sharing types include droop load sharing, where reactive droop compensation permits a predetermined percentage decline in engine speed or alternator voltage as the load increases. Isochronous kW and kVAr load sharing, utilising controls to calculate load information and ascertain how the generators' load percentage compares to the system load; and cross current compensation, occurring when paralleled generators operate without intentional voltage droop, which requires identical voltage regulators on each generator unit for successful load sharing.

Load sharing increases the reliability of the power system by offering redundancy. In the event of a generator fault or maintenance requirement, the load can shift to other operational generators. This redundancy increases the system's resilience against equipment failures or unforeseen events. Load sharing proves advantageous in scenarios where load demands frequently fluctuate and uninterrupted power is paramount. Applications extend to data centres, power plants, construction sites and critical facilities in healthcare and telecommunications.

3.1.7 Loads as virtual storage power plants

Virtual power plants (VPPs) represent a collective of smaller generation units, often including renewable energy sources, that collectively operate as a single power plant when interfacing with grid management systems. In modern systems, VPPs employ SCADA systems and standards such as IEC 61850, to integrate and coordinate the various constituent parts in order to efficiently manage the distributed and diverse components.

A unique situation arises when components within the VPPs are loads, which indicates that they contribute as virtual storage units through load shifting. By aggregating numerous loads, the VPP reaches a scale where it can actively participate in power markets and rival traditional electric storage solutions.

One of the most concerning issues with VPPs is that they use load-based virtual storage power plants in order to ensure availability. This implies that when a grid operator demands a specific amount of regulation power, it must be delivered. However, many loads exhibit unpredictable behaviours. At times, a customer's processes might not be interruptible or their virtual storage may be empty. To address these challenges and provide reliability, dependable load models are essential. These models enable VPP operators to make and keep their commitments for available capacity.

One notorious example of a DSM project is the 'Low Carbon London' project, funded by Ofgem (the UK's energy regulator) and implemented by UK Power Networks (UKPN) [4,5]. The main goal of the project was to analyse the potential for DSM to reduce peak electricity demand in London by means of smart meters in more than 5,000 homes and businesses and to assess the impact of different low carbon technologies on the electricity distribution network. The project conducted tests on various DSM measures, including time-of-use tariffs, dynamic pricing and DLC.

The project was initiated in 2011 and concluded in December 2014, with the findings, conclusions and recommendations documented in a collection of 27 final reports published in 2014. Additionally, the project conducted an extensive knowledge dissemination program to share these outputs with other DNOs, industry organisations and interested stakeholders through a series of roadshows and public events. In general, the main outcomes and insights include, but were not limited to:

- Voluntary contractual reductions in demand lead to shifting enough electricity to serve 18,000 homes at peak demand.
- Precise data on electricity consumption, which guided investment in electricity networks.
- Gathered data from the households suggested a potential 10 TWh/year reduction in electricity consumption by 2020 if consumers switched to more energy-efficient appliances.

- Mass charging of electric vehicles was found to have a substantial impact on electricity networks.
- Wind-twinning tariffs, which incentivise customers to use more electricity when wind power is abundant (domestic demand-side response), were considered viable in urban areas.
- The project introduced a system called active network management, which could enable up to a third more distributed energy plants to export green power to urban networks.
- Smart grids were shown to save customers money by optimising network capacity usage.

Another example is the 'Isles of Scilly Smart Energy Islands' project [6]. This initiative focuses on establishing a smart grid infrastructure on the Isles of Scilly, an archipelago located off the coast of Cornwall in the southwestern region of England. The main objective is to seamlessly incorporate renewable energy resources and energy storage solutions while implementing DSM strategies.

The Isles of Scilly face significant carbon intensity issues and rank as the 8th highest region in England for fuel poverty. This situation is largely due to their heavy reliance on electricity and the importation of fossil fuels such as heating oil. However, the Isles of Scilly have set ambitious energy targets for their future, including a 40% reduction in electricity bills, meeting 40% of energy demand through renewable sources and ensuring that 40% of vehicles are low carbon or electric by 2025.

Supported by funding from the European Regional Development Fund, the Smart Energy Islands project aims to utilise cutting-edge technology, Internet of Things (IoT) platforms and artificial intelligence (AI). These technologies will be instrumental in lowering the islands' carbon footprint while optimising the use of locally generated renewable energy.

The initiative has already seen the installation of approximately 400 kW of solar panels on a variety of locations, including the rooftops. Furthermore, residential homes are serving as test sites for various energy technologies, such as battery systems and air source heat pumps. To manage energy consumption effectively, cloud-based Internet of Things platforms will be employed in order to learn consumption patterns, optimise power generation and utilisation within households, and facilitate integration with the broader electricity grid.

A further phase involves the introduction of electric vehicles (EVs) into the electric system. These EVs will serve as mobile batteries, allowing them to charge during periods of abundant renewable supply and discharge energy during peak demand times. This approach is referred to as vehicle-to-grid (V2G).

Furthermore, the project has spurred the creation of a not-for-profit energy community. This community sells the energy produced by the solar panels and reinvests the revenue to lower electricity costs through a dedicated energy tariff.

The Smart Energy Islands initiative demonstrates a community-centred approach that employs technology to meet the specific needs of the Isles of Scilly communities. It highlights that neither size nor location is a barrier to innovation when there is a vision for the positive impact such initiatives can bring to the environment and the community, with a shared commitment to making a meaningful difference.

3.2 Operating principle

DSM programs come in various forms, each designed to optimise energy usage and reduce peak demand considering the system's characteristics and particular purposes. These programs include the ones mentioned before: time-of-use pricing, DLC, DR and energy efficiency measures.

- Time-of-use pricing: It involves charging customers different electricity rates based on the time of day. Rates are typically higher during peak hours when demand is high and lower during off-peak hours. Aims to incentivise consumers to shift energy-intensive activities to times when electricity is cheaper, thereby reducing peak demand.
- Direct load control: Utilities remotely manage specific appliances or equipment, such as water heaters or air conditioners, during periods of high electricity demand. Allows utilities to temporarily curtail or adjust the operation of these devices under previous agreement.
- Demand response: Encourages customers to voluntarily reduce their energy consumption during peak demand periods. Participants may receive incentives.
- Energy efficiency measures: Focuses on reducing overall energy consumption by improving the efficiency of appliances, lighting and building designs. Aims to provide energy savings, reduce environmental impacts and enhance the sustainability of energy resources.

3.3 Demand response

DR is a strategy employed to effectively manage electricity demand, particularly during peak periods. DR aims to incentivise consumers to curtail their consumption at peak periods with aims to eventually benefit the power grid by avoiding construction of new power plants, decreasing greenhouse gas emissions and enhancing grid reliability by simply managing demand supply in a more adequate manner.

In the context of the United States, the Energy Commission employs the term 'demand response' to denote customers' capacity to adapt their power usage in response to triggers related to grid reliability or pricing, which may originate from their utility system operator, load-serving entity, regional transmission organisation/independent system operator (RTO/ISO) or another DR provider [7,8]. Historically, the term was associated with actions to trim peak demand and was restricted to specific hours of the year. However, following its adoption in Order No. 719, the Commission defined 'demand response' as 'a reduction in customers' electric energy consumption compared to their anticipated consumption, prompted by an increase in the price of electric energy or by incentive payments intended to encourage reduced electricity usage.

In a more detailed description, DR can be identified as dispatchable and non-dispatchable. Dispatchable DR involves deliberate modifications in electricity consumption, initiated by an entity other than the customer (as the utility) and carried out under the customer's previous agreement. This category contemplates actions such as DLC, as well as controlled reductions in consumption in exchange for lower rates (commonly referred to as curtailable or interruptible rates). The common agreement can be prompted by the customer's acceptance to sell their demand reduction at a specified price in an energy market or through arrangements with a retail provider.

On the other hand, for non-dispatchable DR, the customer has the autonomy to choose whether and when to curtail their electricity usage based on a retail rate structure that changes over time. This type is often known as retail price-responsive DR and encompasses dynamic pricing programs that impose higher rates during periods of high demand and lower rates during other times.

In essence, DR is considered as a technology-facilitated economic rationing system for electricity supply. It aims to encourage voluntary curtailing through market incentives. Consequently, users who do not reduce their usage during peak hours will face elevated unit prices, whether directly or incorporated into general rate structures.

DR in a more general application benefits smart appliances or devices within customer areas that can automatically and nearly instantaneously react to price signals or changes in the electric grid conditions, like shifts in system frequency, are more notorious. Additionally, it involves the intelligent integration of adaptable consumption with variable generation, serving to facilitate the integration of new technologies (such as wind farms and rooftop solar systems) into utility systems or customer premises. Control power flows to and from energy storage systems, like the batteries of plug-in hybrid electric vehicles (PHEVs), to offer these grid services is also a valuable feature of these services, leading to shifting consumption from peak to off-peak hours.

3.3.1 Peak load reduction

Peak load reduction is a type of DR strategy that aims to decrease electricity demand during peak hours. This approach can be implemented through various strategies, including load shedding, DSM and time-of-use pricing. The primary goal of peak load reduction is to mitigate the risk of blackouts, reduce the necessity for costly peak plants and lower greenhouse gas emissions.

Any significant imbalance between generation and consumption within the power system can lead to grid instability, voltage fluctuations and grid failures. To ensure grid stability, the total generation capacity is designed to match the total peak demand, often with a margin for contingencies such as power plants being temporarily offline during peak demand periods. Grid operators aim to utilise the least expensive generating capacity available, but as demand increases, they may need to use more expensive or less efficient plants. DR strategies are typically employed to curtail peak demand to reduce the risk of disturbances, minimising the need for additional capital investments in new plants and avoiding the use of higher-cost generation sources.

DR can also serve the purpose of boosting electricity demand during periods of excess generation and low demand. Certain types of power plants, like nuclear, typically operate near full load, while others, such as wind and solar, generate with minimal additional cost. Since energy storage capacity is often limited, DR measures can be employed to increase consumption during these periods to maintain grid stability.

For instance, it could happen that at certain period electricity prices turn negative for specific users, indicating an excess of available electricity. Energy storage solutions offer a method to elevate electricity consumption during low-demand periods for later use. The use of DR to boost demand during periods of surplus supply is less common but may become necessary or efficient in systems where there is a significant amount of generating capacity that cannot be readily adjusted or shut down.

Some electricity grids employ pricing mechanisms that may not be real time but are easier to implement such as higher tariffs during the day and lower tariffs at night. These mechanisms aim to provide some of the benefits of DR with fewer technological demands [1,7–10]. For example, in the United Kingdom, schemes like Economy 7 have been in operation since the 1970s, attempting to shift demand linked to electric heating to overnight off-peak periods. In 2006, Ontario initiated a 'smart meter' program, implementing TOU pricing. This system categorises pricing based on on-peak, mid-peak and off-peak schedules. During the winter, on-peak corresponds to morning and early evening, mid-peak spans midday to late afternoon and off-peak encompasses night-time. In the summer, the on-peak and mid-peak periods are reversed, reflecting air conditioning as the primary driver of summer demand. Starting from 1 May 2015, most Ontario electrical utilities had successfully transitioned all customers to 'smart meter' time-of-use billing, with on-peak rates approximately 200% and mid-peak rates about 150% of the off-peak rate per kilowatt-hour (kWh).

Australia has established national standards for DR, documented in the AS/NZS 4755 series. These standards have been adopted and implemented by electricity utilities across the country for many years. They involve the control of various devices like storage water heaters, air conditioners and pool pumps. In 2016, these standards were expanded to incorporate guidelines for managing electrical energy storage systems.

Arizona, a state characterised by hot summers and high demand for central air conditioning, to mitigate the growth of peak loads, the two major utilities in the state, Arizona Public Service Company and the Salt River Project, have been offering TOU pricing programs to their residential customers for over 20 years. Over this time, these programs have attracted 30%–40% of the residential market. For instance, the Salt River Project has more than 222,000 TOU customers, constituting approximately 25% of its electric customers and making it the third-largest TOU initiative in the United States. Customers enrolled in this program typically reduce their electricity bills by 7%. It's noteworthy that these results were achieved without the use of advanced enabling technologies. This case study illustrates that with well-designed and effectively marketed time-based rates, substantial customer participation and peak demand reduction can be accomplished, even without sophisticated technological interventions.

Florida Power & Light (FPL) operates a substantial DR program called 'On Call', which ranks among the largest load management systems in the United States. As of the conclusion of 2008, FPL aimed to deliver load control of 973 MW during peak demand periods on its electrical grid. This program employs over 900,000 load control transponders that connect to more than 780,000 users. The load control capacity stands at 984 MW. FPL utilises a power line communications system with two-way communication capabilities, enabling a variety of control strategies. When effectively managed, this system can serve as an economically viable alternative compared to the overall cost of constructing new power plants to meet peak demand.

Michigan has a long history of conducting demand reduction initiatives, with Detroit Edison operating a substantial Direct Load Control Interruptible Air Conditioning (IAC) program. This program was established over two decades ago during the era of DSM. The IAC program has grown to encompass more than 280,000 enrolled customers out of a total customer base of 1.9 million and there is a 79.1% saturation of central air conditioning among these customers.

Participants in the IAC program receive a $0.02 reduction in the hourly rate during the cooling months of June to October in exchange for allowing the utility company to potentially interrupt their air conditioning service through remote control relays. This pricing structure differs from conventional programs in that it offers a discount on the hourly rate rather than providing a fixed monthly credit.

Currently, the average load reduction per customer is 0.85 kW, a decrease from roughly 1.2 kW a decade ago. This reduction in load reduction is attributed to the improved efficiency of central air conditioning units over the years. The IAC program serves the dual purpose of curbing peak demand during emergency or high-price periods and acting as a capacity reserve, with a response time of 10 minutes. It operates on a cycling schedule with 15-minute on/off intervals and can shed up to 230 MW of load on days when the temperature exceeds 90 °F. IAC may be cycled for a maximum of hours within a 24-hour period.

Dynamic demand aims to advocate institutional change, stimulate research and foster discussion to promote the implementation of 'dynamic demand control' technologies within the UK power grid. The adoption of demand control technologies has the potential to significantly enhance grid stability and manage peak demand effectively, thereby resulting in substantial reductions in carbon dioxide emissions. Furthermore, these technologies can ease the integration of greater quantities of intermittent renewable energy sources like solar and wind power. Dynamic demand represents both a semi-passive technology designed to support DR through demand adjustments and an independent not-for-profit organisation in the United Kingdom. The organisation receives charitable support from the Esmée Fairbairn Foundation and is committed to advancing this technology. Established in January 2005 by a group of independent academics, campaigners and engineers dedicated to addressing the challenges of climate change.

The main idea involves intermittent loads monitoring the power grid's frequency and their internal controls. They autonomously switch on and off at optimal times to balance the grid's load with generation, thus mitigating critical imbalances. These switching actions only cause minor adjustments in the operation cycles of appliances, rendering them imperceptible to end-users. In 1982 in the United States, a patent was granted to power systems engineer Fred Schweppe, based on this concept [11]. Additional patents have also been issued, drawing from this idea.

Dynamic demand shares similarities with DR mechanisms used to manage residential and industrial electricity consumption in reaction to supply conditions such as customers reducing their electricity use during peak times or in response to pricing signals. The key distinction is that dynamic demand devices passively shut off when they detect stress on the grid, while DR mechanisms act in response to explicit requests to curtail electricity usage.

3.3.2 Frequency response

Frequency response is a form of DR that focuses on regulating the supply of electricity to uphold the frequency of the power grid. Among the various DR techniques, frequency response is the quickest, making it ideal for locations capable of reacting within seconds to counteract disruptions at major power stations. This includes applications as energy storage, diesel rotary uninterruptible power supplies (DRUPS) and specific industrial loads.

When the demand surpasses the available supply, the power grid's frequency decreases. Conversely, when the supply of electricity exceeds the demand, the frequency increases. Frequency response mechanisms function by adjusting the output of generators and other power sources to ensure the grid maintains a stable and consistent frequency.

A technical assessment of some practical challenges associated with incorporating DR into power system frequency control has been presented in [12]. These challenges encompass aspects such as synchronising electrical loads and evaluating the merits and drawbacks of centralised versus decentralised structures.

The role of DR in frequency control within power systems and isolated microgrids has been explored in the literature. These studies incorporate frequency-sensitive load controllers with various frequency-time characteristics, saturable reactors and selective load control schemes. Also, research has delved into the coordination of DR and local frequency control, addressing the impact of communication delays on frequency stability, as documented in [2].

In order to bolster the stability of the power system, the implemented DR features are distributed across each area in the power system under consideration. A more detailed explanation of the relationship between frequency and power is presented in Section 3.4.2. When an unforeseen disturbance takes place, the DR system becomes activated. It works by adjusting P_{DRi}, which represents the power consumed by specific electrical appliances. This alteration in P_{DRi} impacts the overall power demand, ultimately contributing to the broader effort to maintain stable frequency control.

Each DR appliance designed for frequency control comprises two essential components: the electrical appliance itself, which can be items like electric water heaters, among others, and the controller. The controller has the capability to automatically vary the on/off state or to adjust the power consumption based on a predefined control strategy that aims to support frequency control. While a single DR appliance may make only a minimal contribution to the overall system frequency, the collective impact of numerous DR appliances can significantly influence the system frequency. The effectiveness of DR in frequency control is closely tied to the specific control strategy applied.

Over the past several years, numerous DR frequency control strategies have been proposed. Most of these methods rely on monitoring the frequency deviation [13]. When the detected frequency deviation falls below a defined threshold, a portion of the DR appliances will be activated. By effectively managing the consumption of a large number of DR appliances, the aggregate frequency response characteristics of the system can closely resemble those of conventional thermal generators. It's worth noting that individual DR appliance responses are typically discrete, as most of them can only be turned on or off. Nevertheless, with proper management of the frequency thresholds, the collective DR system can deliver a smooth frequency response characteristic, as illustrated.

In Figure 3.7, the total amount of the activated DR resource P_{DRi} can be formulated as follows:

$$P_{\mathrm{DR}i} = \begin{cases} -P_{\mathrm{DRmaxi}} & \Delta f_i > \Delta f_{\mathrm{maxi}} \\ -k_{\mathrm{DR}i}\Delta f_i & \Delta f_{\mathrm{maxi}} \leq \Delta f_i \leq \Delta f_{\mathrm{maxi}} \\ P_{\mathrm{DRmaxi}} & \Delta f_i < -\Delta f_{\mathrm{maxi}} \end{cases} \tag{3.1}$$

where k_{DRi} is a predefined coefficient; $D_{f\max i}$ is the frequency regulation range by DR and $P_{DR\mathrm{max}i}$ is the maximum available DR for the frequency control. The value of k_{DRi} can be determined by:

$$k_{DRi} = \frac{P_{DR\mathrm{max}i}}{\Delta f_i} \tag{3.2}$$

For these control methods, the only input signal for the DR controller is the frequency deviation and the tie-line swing between the areas is not considered. Therefore, these DR control strategies aim at only restoring the area frequency. These control strategies are not good for damping inter-area oscillations and stabilising the frequency in different areas.

3.3.3 Load shifting

Load shifting entails placing electricity consumption from peak times to off-peak hours [14]. This involves motivating users to use electricity during periods of lower demand by providing incentives. Load shifting aims to alleviate the demand during peak times, thereby mitigating the risk of blackouts, diminishing the necessity for costly peak plants and lowering greenhouse gas emissions.

Load shifting, as part of DSM strategy, is a widely employed approach, influenced by consumer behaviour and their electricity usage patterns. This

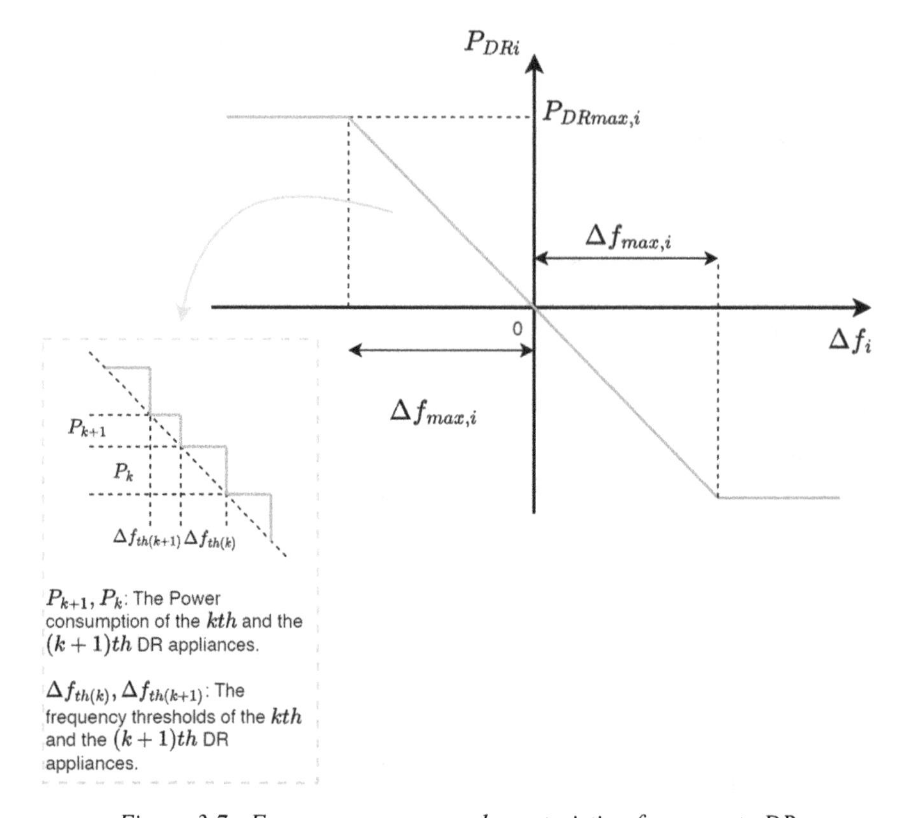

Figure 3.7 Frequency response characteristic of aggregate DR

algorithm is designed to curtail energy demands by redistributing it to off-peak hours. The methodology is clearly outlined in Figure 3.8, which provides a visual representation of how the load is transitioning between different times of the day.

When the generation capacity falls below the demand, i.e., load loss, utilities may implement load shedding measures to control the contingency. Load shedding is activated via targeted blackouts or by entering agreements with specific industrial consumers to deactivate equipment during periods of system-wide peak demand.

As mentioned, incentives are put into place for users to allow modification of usage times. These incentives can take either formal or informal approaches. A utility may establish a tariff-based incentive by transferring short-term increases in electricity prices or they may enforce mandatory cutbacks for selected high-volume users, compensating them for their participation. For instance, California implemented its Emergency Load Reduction Program, wherein enrolled customers received a credit for reducing their electricity consumption ($1 per kWh in 2021, $2 in 2022).

Commercial and industrial users may implement load shedding independently and without any request from the utility. Some businesses produce their own energy in accordance with their consumption capacity to avoid purchasing power from the

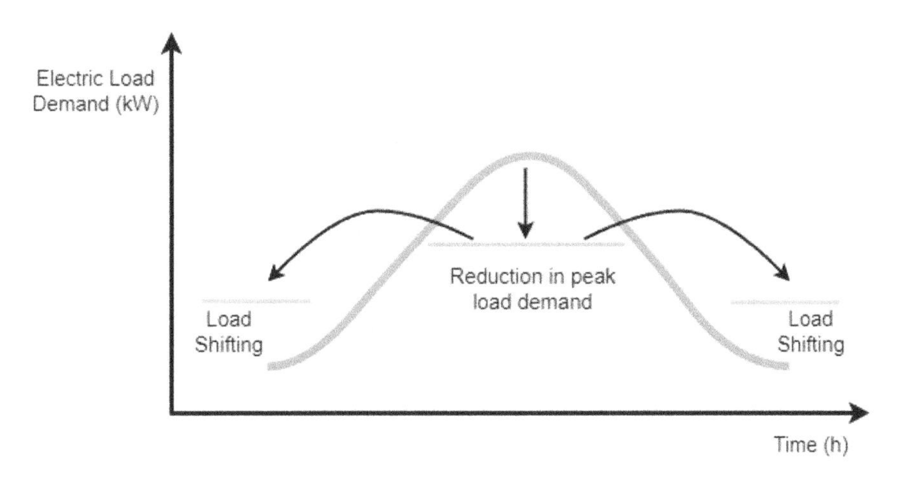

Figure 3.8 Load shifting-based demand-side management strategy

grid. Certain utilities employ commercial tariff structures that determine a customer's power costs for the month based on the moment of highest use or peak demand. This approach encourages users to flatten their energy demand, a practice known as energy demand management, which may involve temporarily reducing services.

In some regions, smart metering has been introduced to offer RTP for users of all kinds, moving away from fixed-rate pricing during the demand period. In this scenario, users receive a direct incentive to decrease their consumption during high-demand, high-price periods. However, some users may face challenges in effectively reducing their demand at different times, or the peak prices may not be sufficiently high to prompt a change in demand during short periods, thus indicating low price sensitivity or low elasticity of demand. While automated control systems are effective, they may be costly for certain applications to be implemented.

3.4 Load control

Load control involves the strategic management to maintain a balance between electricity production and demand, ensuring the stability of the power system. Various load control methods, such as load shedding, ripple control and frequency-based control, are employed to fulfil this purpose.

When deciding to reduce load, utilities prioritise system' reliability. Load shedding is executed only when the stability or reliability of the distribution system is at risk. Effective load management is designed to be non-invasive and should not impose any undue hardship on consumers.

In contrast, DR techniques put the control in the hands of consumers, as mentioned. While many residential consumers pay a flat rate for electricity year-round, the actual costs for utilities fluctuate based on demand, the distribution network and the composition of the electricity generation portfolio. In a free market, the wholesale price of energy varies throughout the day. DR programs, facilitated by smart grids, aim to incentivise consumers to limit usage based on cost considerations. As costs rise during peak hours, a free-market economy would allow prices to increase and a corresponding decrease in demand should follow. However, for unforeseen equipment failures that develop within seconds, a quick resolution is crucial to avoid power blackouts. Some utilities interested in DR also seek load control capabilities to operate the 'on-off switch' swiftly before price updates reach consumers.

The application of load control technology is expanding, with the sale of radio frequency and powerline communication-based systems. Certain smart meter systems can also function as load control systems. Charge control systems prevent the recharging of electric vehicles during peak hours and vehicle-to-grid systems enable the return of electricity from electric vehicle batteries to the utility or regulate the recharging rate.

3.4.1 Load shedding

Load shedding is a method of load control used to manage the demand for electricity during peak hours or when the power supply is insufficient. In load shedding, the power supply is deliberately interrupted to certain areas or customers to reduce the overall load on the system. This is done to prevent a complete blackout of the power grid.

Load shedding can be done manually or automatically. Manual load shedding is done by trained operators who use a load shedding plan to determine which areas or customers to disconnect. Automatic load shedding is done by a control system that monitors the power grid and disconnects areas or customers when the demand exceeds the supply.

When a total or partial loss of generation occurs within the system, the first indicators are a drop in voltage and frequency. However, given that voltage drops can also be caused by system faults, it is generally recognised that a drop in frequency is a more reliable indication of loss of generation. A sudden loss of generation in the system will result in a reduction in the frequency at a rate of change that depends on the size of the resultant overload and the inertia constant of the system.

To design an automatic load shedding system, a model that represents the different generating machines should first be defined and then the load parameters and the criteria for setting the frequency relays.

3.4.1.1 Simple machine model

Within the scope of this book, a single machine has been used in the power system model to illustrate a load shedding system. This is equivalent to assuming that the generator units are electrically connected with negligible oscillations between them, and with a uniform frequency across the whole of the system, ignoring the effect of the regulating equipment. The load is represented as a constant power, which implies that there is no reduction in load as a result of the voltage and frequency drops after a contingency situation. This model provides a pessimistic simulation of the system since the reduction of the load due to the frequency drops and the effect of the speed regulators are neglected.

When a total or partial loss of generation occurs within the system, the first indicators are a drop in voltage and frequency. However, given that voltage drops can also be caused by system faults, it is generally recognised that a drop in frequency is a more reliable indication of loss of generation. A sudden loss of generation in the system will result in a reduction in the frequency at a rate of change that depends on the size of the resultant overload and the inertia constant of the system.

The relationship which defines the variation of frequency with time, following a sudden variation in load and/or generation, can be obtained, starting from the equation for the oscillation of a simple generator:

$$\frac{GH}{\pi f_o} \cdot \frac{d^2\delta}{dt^2} = P_A \tag{3.3}$$

where

G is the nominal MVA of machine under consideration
H is the inertia constant (MWs/MVA = MJ/MVA)
δ is the generator torque angle
fo is the nominal frequency
PA is the net power accelerated or decelerated (MW)

The speed of the machine at any instant (W) can be given by the following expression:

$$W = W_0 + \frac{d\delta}{dt} = 2\pi f \tag{3.4}$$

where Wo is the synchronous speed, i.e., the nominal speed at rated frequency.

Differentiating (3.4) with respect to time:

$$\frac{dW}{dt} = \frac{d^2\delta}{dt^2} = 2\pi \frac{df}{dt} \tag{3.5}$$

Replacing (3.5) in (3.3) gives:

$$\frac{df}{dt} = \frac{P_A f_0}{2GH} \tag{3.6}$$

Equation (3.6) defines the rate of the variation of the frequency in Hz/s and can be used for an individual machine or for an equivalent which represents the total generation in a system. In such a case, the inertia constant can be calculated from:

$$H = \frac{H_1 MVA_1 + H_2 MVA_2 + \ldots + H_n MVA_n}{MVA_1 + MVA_2 + \ldots + MVA_n} \tag{3.7}$$

In using this model, the inertia constant of the system is calculated using (3.7). The rate of change of the frequency is calculated from (3.6) with the following assumptions:

- The mechanical power entering the generators does not vary and is equal in magnitude to the prefault value;
- The magnitude of the load does not vary with time, voltage or frequency. It is only reduced by disconnecting part of the load as a result of the automatic load shedding system.

This simple machine model, with loads modelled as a constant power, is used to determine the frequency relay settings and to verify the level of minimum frequency attained before a contingency situation is reached, for the following reasons:

- The ease of using an iterative process to design the load shedding system;
- Consideration of suitable security margins, since the fact that the load diminishes with loss of voltage is neglected. This implies introducing much more severe rates of frequency variation, thus achieving more rapid settings.

3.4.1.2 Considerations of implementing a load shedding system

The following aspects need to be defined to implement the load shedding system.

Maximum load to be disconnected

Generally, one of the most drastic conditions corresponds to the total loss of interconnection between the public electricity supply network and the internal electrical system within the industrial plant. In this case, the unbalance between generation and load will be equal to the maximum import and should be compensated by the disconnection of a similar amount of load from the plant system.

Starting frequency of the load shedding system

The disconnection system should be set so that it will initiate operation at a value of frequency below the normal working system frequency. Taking into account variations in frequency caused by oscillations inherent in the public system, this value is normally selected at approximately 93% of nominal system frequency. However, if there is a possibility of more severe oscillations occurring in the system, then it is recommended that a supervisory control arrangement using overcurrent relays that can detect the outages of circuits connecting the industrial plant to the public system should be installed to avoid incorrect operations.

Minimum permissible frequency

A steam turbine is designed so that, when operating at nominal mechanical speed and generating at nominal system frequency, excessive vibrations and stresses in its components, e.g., resonance of turbine blades, are avoided. However, when running below normal speed at a reduced system frequency, cumulative damage could be produced by excessive vibration. It is recommended, therefore, that the time limits given in Table 3.1 should not be exceeded. However, during transient operation and

Table 3.1 Typical times for the operation of turbines (full load)

Percentage of rated frequency at full load	Maximum permissible time (min)
99.0	Continuously
97.3	90
97.0	10
96.0	1

with load below nominal, in most of the cases reduction of frequency down to 93% of rated frequency can be permitted without causing damage either to the turbine or to the turbogenerator auxiliary lubrication and cooling systems.

3.4.1.3 Criteria for setting frequency relays

The determination of the frequency relay settings is an iterative process and is carried out in such a way that the final settings satisfy the requirements of both speed and coordination. In this process, the coordination between relays that trip successive stages of load should be checked in order to ensure that the least amount of load is shed, depending on the initial overload condition.

3.4.1.4 Operating times

When selecting the settings, it is necessary to consider the time interval between the system frequencies decaying from the relay pick-up value to the point in time when the load is effectively disconnected. The relay pick-up time is included in this time interval, plus the preset time delay of the relay, if this is required, and the breaker opening time.

The following values are typically used for industrial systems:

- relay pick-up time: 50 ms
- breaker opening time: 100 ms

The frequency variation required to calculate the settings is obtained by using a simple machine model for the system and a constant power model for the load. This assumes that the load connected to the generators is the same before, and after, the contingency, neglecting any form of damping. Given this, the calculated rate of loss of frequency in the system is pessimistic, and the settings determined on this basis thus provide an arrangement that rapidly restores the frequency to its normal value, thereby ensuring a secure system.

The procedure for calculating the settings of the frequency relays at a typical industrial plant is presented. The single-line diagram is shown in Figure 3.9, together with the principal data for the study.

The initial load conditions are summarised below:

- total load: 24.0 MW
- in-house generation: 8.0 MW
- total import: 16.0 MW

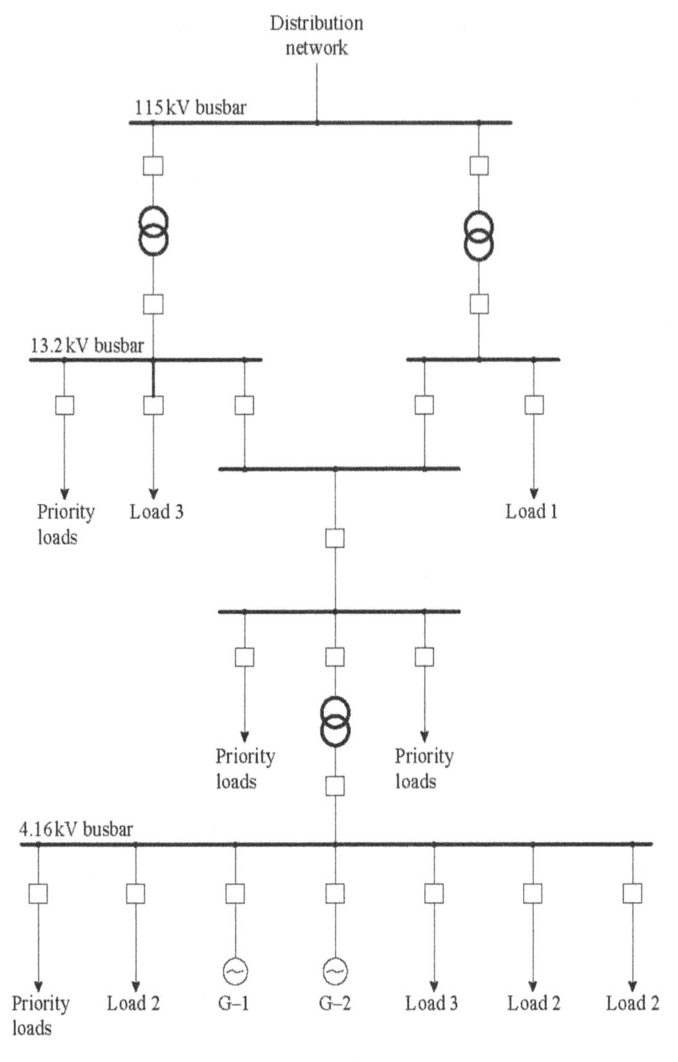

Figure 3.9 System arrangement for example of calculating frequency relay settings

- GH constant: 35.43
- Rated system frequency: 60 Hz

In the eventuality of the loss of the incoming grid supply, the in-house generators will experience the following overload:

$$\%overload = \frac{24.0 - 8.0}{8.0} \cdot 100\% = 200\%$$

With the loss of the grid supply, 16 MW of capacity is lost, which will have to be borne initially by the in-house generators. A load equal to, or greater than, this amount must be disconnected in order to relieve the overload. It should be noted that there are high priority loads totalling 8.02 MW which cannot be disconnected.

Therefore, the load to be shed is 24 − 8 = 16 MW.

Disconnection of the load is initiated each time the system frequency falls to 59 Hz, which indicates that a loss of generation has taken place. A minimum frequency of 56 Hz is acceptable.

Assume that three stages of shedding have been set up which will off load the incoming circuit by 15.98 MW, as detailed in Table 3.2.

With this arrangement, the in-house generation can then supply the 8.02 MW of high priority load.

3.4.1.5 Determination of the frequency relay settings

The settings of the frequency relays shown in Table 3.2 are determined in such a way that each stage is disconnected only when the system frequency falls to a predetermined value. This value is obtained by calculating the reduction in the system frequency due to an overload equal to the stage considered, as described as follows.

- First-stage setting

The first stage is disconnected when the frequency reaches 59 Hz.

- Second-stage setting

An overload equal to the first stage is considered and the subsequent rate of frequency drop is determined from:

$$\frac{df}{dt} = \left(\frac{-6.03}{2GH}\right) \cdot 60 = -5.106\frac{Hz}{s}$$

The frequency as a function of time is given in Figure 3.10 and $f = (60 - 5.106\,t)$. The opening time for the first stage is:

$$t_{\text{trip}} = t_{\text{pick-up}} + t_{\text{breaker}} + t_{\text{relay}}$$

$$t_{\text{pick-up}} = (60 - 59)/5.106 = 0.196\ \text{s}$$

$$t_{\text{trip}} = 0.196 + 0.100 + 0.05 = 0.346\ \text{s}$$

Table 3.2 Load shedding stages

Priority	Description	MW
1	Load 1	6.03
2	Load 2	4.73
3	Load 3	5.22
Total load to be shed		**15.98**

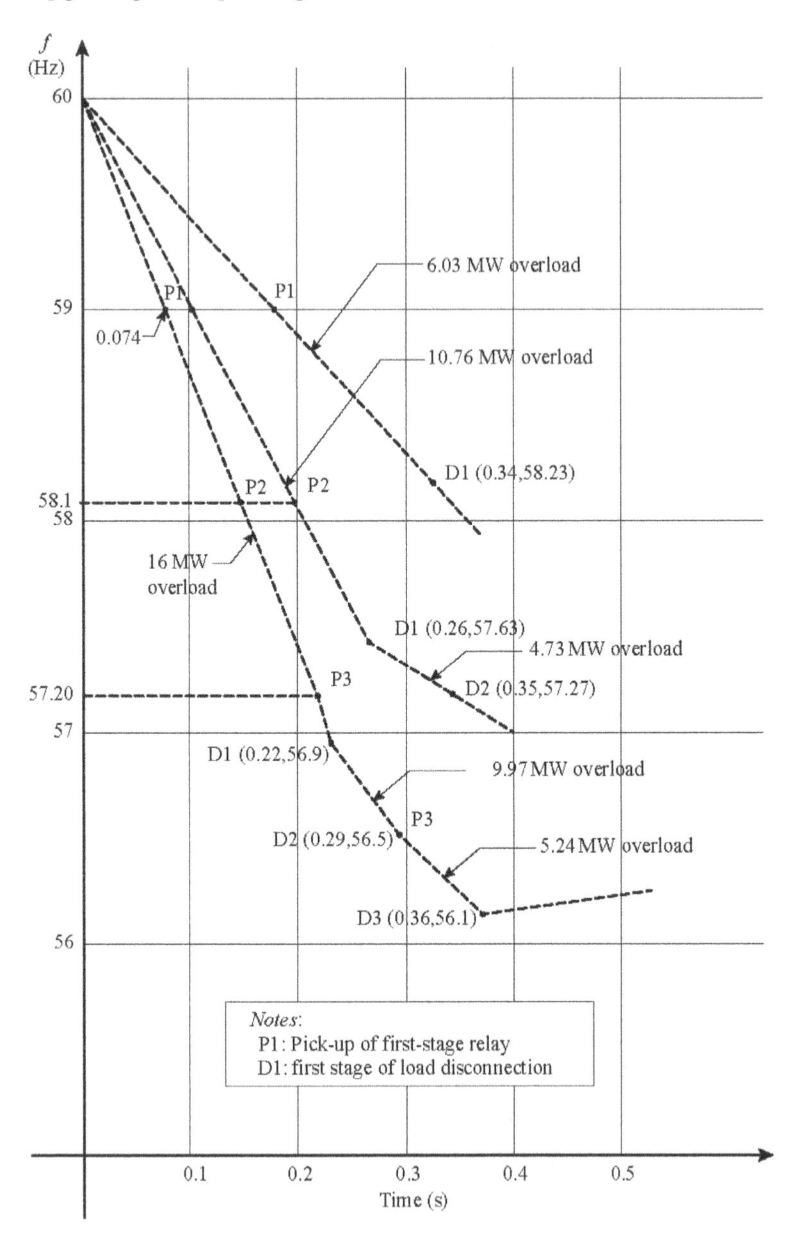

Figure 3.10 Calculation of settings of frequency relays

The frequency drop-up to the operation of the first stage is f = [60 − 5.106 (0.346)] = 58.233 Hz

The second stage is set below this value, i.e., at 58.15 Hz.

Table 3.3 Frequency relay settings

Stage	Frequency setting (Hz)	Delay time (s)
First	59.00	Instantaneous
Second	58.15	Instantaneous
Third	57.20	Instantaneous

- Third stage setting

These are set so that they will not operate for overloads below (6.03 + 4.73) = 10.76 MW, which are disconnected by stages 1 and 2.

$$\frac{df}{dt} = \left(\frac{-10.76}{2GH}\right) \cdot 60 = -9.111 \frac{Hz}{s}$$

$$f = 60 - (9.111)t$$

The pick-up time for the first stage is $t = (60 - 59)/9.111 = 0.110$ s and the tripping time for this stage is $t_{\text{trip1}} = 0.110 + 0.05 + 0.100 = 0.260$ s. Thus, the frequency drop, f, = 60 - 9.111(0.26) = 57.631 Hz.

The value which produces pick-up at the second stage is $t = (60 - 58.15)/9.111 = 0.203$ s, so that $t_{\text{trip2}} = 0.203 + 0.05 + 0.100 = 0.353$ s.

The frequency equation shows that in $t_{\text{trip1}} = 0.260$ s, there is a variation of slope as a consequence of the disconnection of the first stage (6.03 MW), see Figure 3.10. After this, the accelerating power is only 4.73 MW.

Therefore:

$$\frac{df}{dt} = \left(\frac{-4.73}{2GH}\right) \cdot 57.631 = 3.847 \frac{Hz}{s}$$

and, from the frequency equation:

$$f - 57.632 = -3.847 \cdot (t - 0.26), \quad t > 0.26s$$

$$f = 58.632 - 3.837t$$

The frequency drop-up to the stage 2 tripping is given by $f = 58.632 - 3.847t_{\text{trip2}}$, from which $f = 57.274$ Hz. Therefore, a third stage setting of 57.20 Hz is selected.

The final settings are given in Table 3.3.

The operation of the system in the presence of a total loss of connection with the distribution network is represented by the lower characteristic in Figure 3.10.

3.4.1.6 Verification of operation

To verify the operation of the proposed system, the reduction in frequency during the process of load shedding is studied by two different methods: modelling the load as a constant power and modelling the load with damping as a result of the voltage drop. For the latter case, a transient stability program is used.

- Modelling load as a constant power

For this case, it is assumed that the magnitude of the load is constant and therefore does not depend on the voltage level. This consideration is pessimistic with regard to the actual situation and therefore provides some margin of security. The frequency analysis is carried out using (3.6), the settings of the frequency relays given in Table 3.3 and assuming a maximum circuit breaker opening time of 100 ms.

Figure 3.11 shows the behaviour of the frequency from $t = 0$ s, when loss of supply occurs at a total system load of 24 MW. Under such conditions, the sequence of events is as given in Table 3.4. The drastic initial overload is

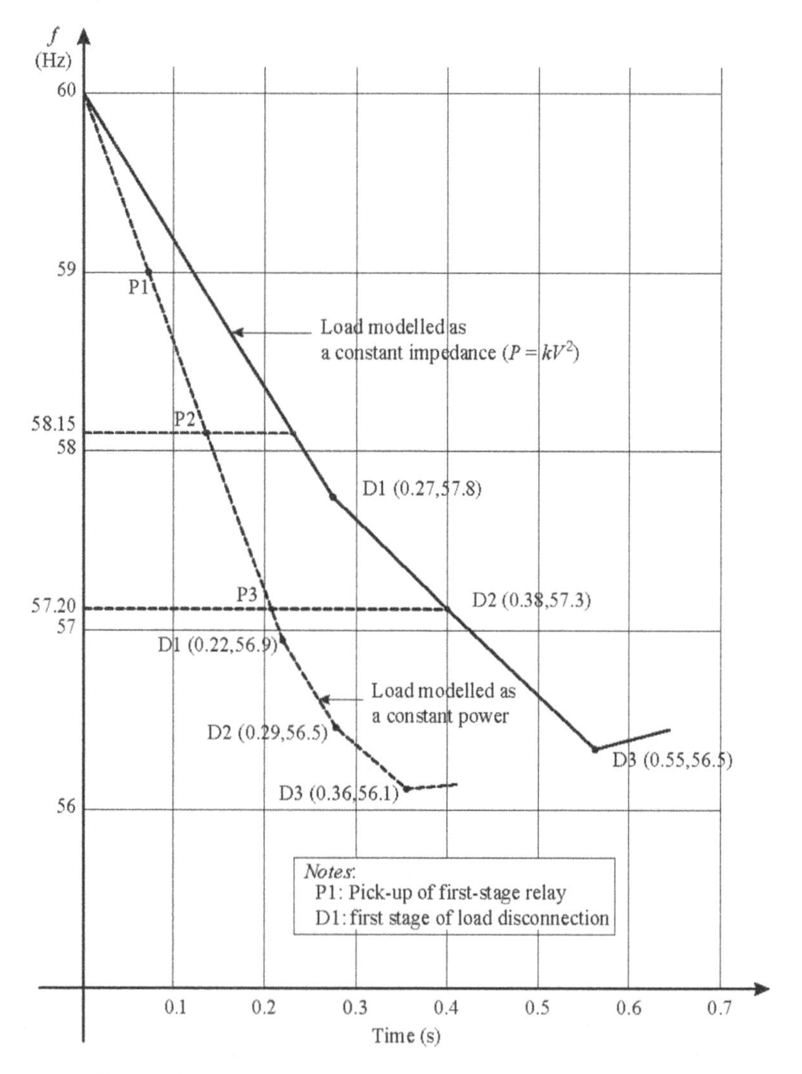

Figure 3.11 Variation of frequency during load shedding

Table 3.4 Disconnection of the grid supply; sequence of events

Time	Frequency (Hz)	Event	Rate of change of frequency (Hz/s)	Rate of change of load shed (MW)	Overload remaining on generators (MW)
0.000	60.00	Disconnection of grid supply	−13.46	–	16.00
0.074	59.00	Pick-up of first-stage relay	−13.45	–	16.00
0.137	58.15	Pick-up of second-stage relay	−13.45	–	16.00
0.207	57.20	Pick-up of third-stage relay	−13.45	–	16.00
0.224	56.90	First stage of load disconnection	−8.00	6.03	9.97
0.287	56.50	Second stage of load disconnection	−4.20	4.73	5.24
0.357	56.10	Third stage of load disconnection	0.00	5.22	0.02

eventually eliminated by the load shedding system in 0.357 s. During this time, the frequency will fall to 56.17 Hz, a value which more than meets the standards accepted for industrial systems operating separately. After $t = 0.357$ s, the system frequency then starts to recover.

To obtain the values in Table 3.4, it must be borne in mind that, since the initial overload is equal to 16 MW, then:

$$\frac{df}{dt} = \left(\frac{-16}{2GH}\right) \cdot 60 = -13.45\frac{Hz}{s}$$

Therefore, the pick-up time for the first stage is:

$$t_p = \frac{60 - 59}{13.45} = 0.074s$$

The opening time of the breakers associated with the first stage is $t_\text{d} = 0.074 + 0.100 + 0.05 = 0.224$ s and the frequency at that moment is $f = 60 - 13.45t_\text{d} = 56.9$ Hz. The values for the other stages can be calculated in a similar way and are given in Table 3.4.

- Modelling load with damping as a result of the voltage drop (P = kV2)

  A stability program should be used to set the initial condition of the load defined for this study, as this can simulate the disconnection of the infeed at $t = 0$ s and subsequently the disconnection of the loads at the time at which the frequency relays are set to operate.

  Figure 3.11 also illustrates the frequency characteristic obtained by computer, modelling the loads by constant admittances. This method is less drastic than the constant power model since, in this case, the power of the load is damped by the voltage drop. As the overload is reduced, the frequency drop is less than that obtained using the first model. As seen in Figure 3.11, the frequency relay settings, and therefore the tripping of the breakers, are much slower although the minimum value of the frequency eventually reaches 56.55 Hz, which is above the predetermined minimum limit of 56 Hz. From

Figure 3.11 and the results of the stability program, it can be seen that the frequency after the loss of the third stage gives recovery at a rate of 0.67 Hz/s. This implies that it should take approximately 2.345 s for recovery to the normal level of 60 Hz.

The modelling of the loads by these two methods makes it possible to obtain graphs of the frequency variation as a function of time which correspond to the most adverse and most favourable extremes in the system. In this way, the curves obtained for the two models delineate the operating area of the system under study, before the loss of connection with the grid supply network.

- Analysis of voltage with total loss of infeed supply and operation of the load shedding system

  The loss of generation in the system not only causes loss of frequency but also a drop in voltage. The automatic load shedding scheme should prevent system voltages falling to such a level as to cause tripping of the contactors on the motors serving the plant.

  A check should be made on the voltage levels on the system as follows:
- Determine the initial voltage at each busbar, using a load flow program.
- Determine the variations in voltage at each busbar after the loss of connection with the grid supply and while the load shedding scheme is in operation. For this, a transient stability program is used to obtain the voltages at each busbar for each stage of the analysis.
- Produce the curves of voltage versus time for the busbars feeding the priority loads.

3.4.1.7 Load shedding using logic schemes

There is an alternative strategy for load shedding in which the behaviour of the system is known beforehand and the trip is not conditioned only to the monitoring of change of frequency, as was the main requirement for the protection scheme presented before. This includes an analysis of steady state (load flow analysis to check the balance consumption-generation) and a dynamic analysis of the system. The simplest criterion that can be used under this scheme is to disconnect the demand that cannot be supplied by the available capacity of the system or when there is an overload condition. That amount of load should be shed by using logic schemes additional to the frequency schemes, which require a comprehensive analysis of the operational conditions of the system. An example of an industrial plant is presented in Figure 3.12, which shows a possible approach using both frequency based and logic schemes.

When logic schemes are used, a proper understanding of the system is required to take preventive actions before a system failure occurs. This is done through the measurement of key variables, simulations and analyses of critical conditions by the operator.

Programmable logic controllers (PLCs) are used in this application, which initiate a trigger signal to the appropriate power switches. Variables such as total amount of load, number of generators and conditions associated with low frequency

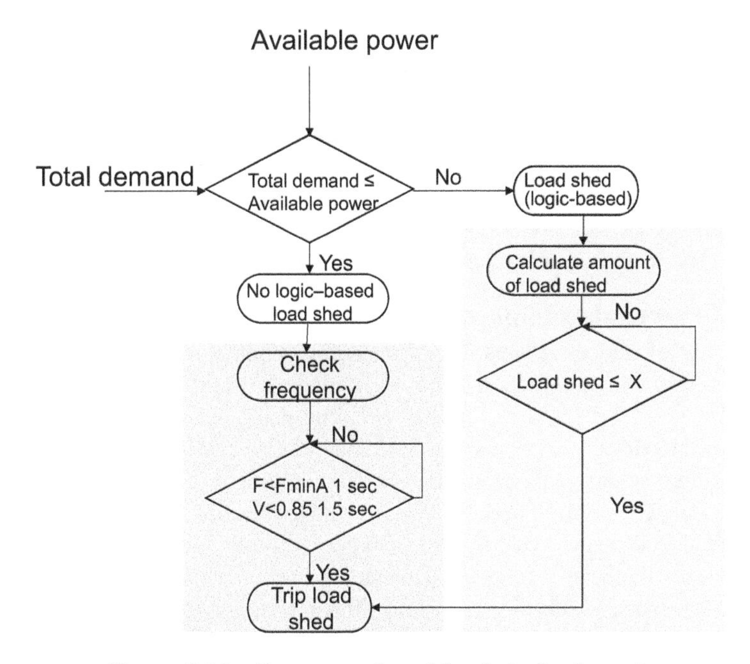

Figure 3.12 Frequency-based load shed schematic

are reviewed. This requires continuous monitoring of the status of the generator, the positions of the circuit breakers and the general power flows in the system. The load shedding required is assembled according to the contingency assessment performed for that system.

These systems can be faster and more accurate than systems based on low frequency relays. However, PLCs also use low frequency relays as a backup because their processing time leads to a slow response of load shedding.

For the load shedding based on the PLC scheme to operate properly, it is required to continuously monitor at least the available generation and plant load. This information might be enough for a simple system with few modes of operation. The logic of this PLC scheme is similar to that of a hard-wired system. For other complex systems with many modes of operation, the scheme could be a lot more complicated and costly.

The PLCs must continuously monitor the status of the entire installation, including the power flow in many motor starters and switches to ensure the proper operation of the system. The logic of the PLC must be programmed to continuously know the operating conditions for any power load, generation level and plant generation. The priorities of the plant must also be defined to program in the corresponding logic, the loads that should be triggered first and the ones that should not be triggered considering that they are essential loads.

The load shedding system is designed for rapidly changing conditions in a situation that requires quick responses. When a major overload occurs suddenly in an isolated generator, the generator must have a reserve that can be used; otherwise some loads must be removed within a period of time from fractions of a second up to a couple of seconds. Consequently, the definition of 'fast enough' requires the same analysis as for frequency-based load shedding. If the PLC can instantly react to utility disconnections, this system may have some speed advantage over the frequency-based relay system. The PLC begins to react the moment the network connection is lost, just when the overload has started, but the frequency detecting devices respond shortly after the overload has started.

3.4.1.8 Practical example

An example of under frequency load shedding schemes characteristics and performance criteria is presented in Table 3.5, which illustrates the type of scheme that is currently used by TSOs in Europe [15]. The percentage of load shed is taken from national load. It is composed of four under-frequency load Shedding steps and this traditional scheme typically allows a relatively large inaccuracy in frequency measurement. The relative maximum load that can be shed is 60%, which allows a larger imbalance in the system. This scheme is highly sensitive to the inertia of machines and the load frequency dependence, because of its inaccuracy in frequency measurement. However, the maximum post-shedding frequency reached with this scheme is relatively low.

3.4.2 Ripple control

This method is used to control load demand remotely by setting power consumptions [13,16]. In ripple control, a high-frequency signal is used to send the control action that trigger devices. This signal (usually of 100–1600 Hz) is superimposed on the standard power network signal (50–60 Hz). The devices are designed to receive this signal and connect/disconnect the power consumption. These devices to be controlled can be non-essential residential or industrial loads such as water heaters, air conditioning units, crop-irrigation pumps or pool pumps and this method is widely used in several countries around the world.

In a distribution network equipped with ripple control, the associated devices are equipped with communicative controllers capable to restrict the duty cycle of the controlled equipment. Typically, consumers receive incentives for participating in the load control program, often in the form of a reduced rate for energy consumption [13]. Effective load management by the utility enables the

Table 3.5 Typical example of under-frequency load shedding schemes characteristics and performance criteria

Frequency threshold (Hz)	Percentage of load shed	Total of load shed
49	15	15%
48.5	15	30%
48	15	45%
47.5	15	60%

implementation of load shedding strategies to avert rolling blackouts and concurrently mitigate operational costs.

Some of the distribution system operators have conducted comprehensive measurements of their ripple control systems (RCS) groups and have acquired intricate data pertaining to the power consumption of each group. These findings align with a theoretical conjecture, which posits that the individually configured shutdown temperatures and the specific heat stored in each electric boiler contribute to an observable, nearly exponential decay in the power consumption of a controlled group over time. The time function of the power consumption of an RCS controlled group can be approximated as follows:

$$P(t) = P_0 e^{-\frac{t-t_0}{\tau_1}} - P_0 e^{-\frac{t-t_0}{\tau_2}} \tag{3.8}$$

where t_0 is the switch-on time, τ_1 determines the fall time constant of the double exponential curve and τ_2 determines the rise-time constant. When there are no RCS measurements available, an undistorted load curve is assumed to be linear during a certain time-period that is, the total power consumption can be expressed as follows:

$$P(t) = P_0 e^{-\frac{t-t_0}{\tau_1}} - P_0 e^{-\frac{t-t_0}{\tau_2}} + mt \tag{3.9}$$

where t_0 is the switch-on instant or the moment when there is an abrupt increase of the total consumption and the other four parameters are determined product of the curve-fitting method by minimising the following expression:

$$f(P_0, \tau_1, \tau_2, m) = \sqrt{\sum_{t_0}^{t_{\max}} \left[P_{measured}(t) - P(t)\right]^2} \tag{3.10}$$

where $t_{\max}$ can represent either the switch-off time of the same group or the switch-on time of the next group. An undistorted load curve can be approximated as the difference of the measured load curve and $P(t)$ as in (3.8). However, this method is inaccurate since $t_{\max}$ could be close to t_0 and the linear assumption in (3.9) cannot take place anymore. An example of a possible result is shown in Figure 3.13.

The reception of ripple control signals can encounter challenges, contributing to the occasional unpopularity of such systems. The reliability of modern electronic receivers surpasses that of antiquated electromechanical systems. Additionally, contemporary systems often incorporate mechanisms to retransmit telegrams, ensuring the initiation of comfort devices. Responding to consumer demand, many ripple control receivers are equipped with a manual override switch to force the activation of comfort devices.

Modern RCS employ digital telegrams ranging from 30 to 180 seconds in duration. In the past, electromechanical relays received these telegrams. Nowadays, microprocessors are commonly employed for this purpose. To guarantee the activation of comfort devices, some systems employ telegram repetition. The broadcast frequencies, falling within the range of human hearing, may induce audible vibrations in wires, filament light bulbs or transformers. Various regions adhere to

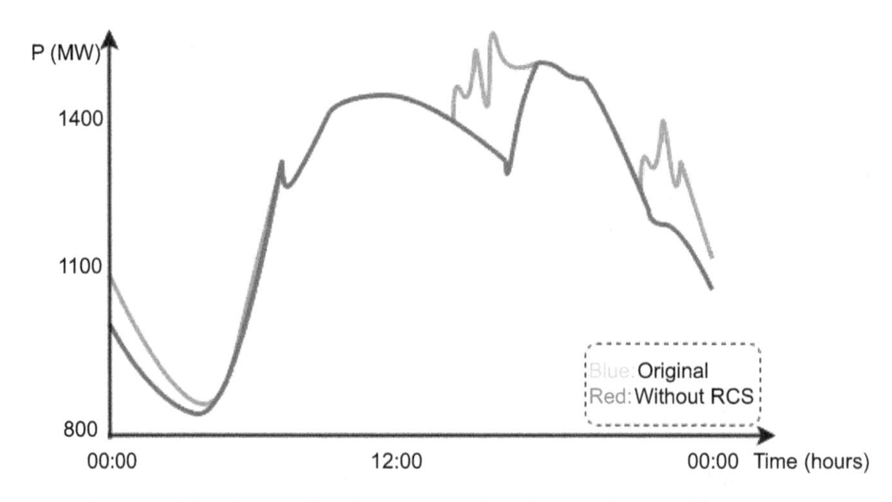

Figure 3.13 Example of a measured and an undistorted load curve

different standards for telegrams such as 'ZPA II 32S', 'ZPA II 64S' and Versacom. ZPA II 32S, for example, transmits a sequence comprising a 2.33-second on signal, followed by a 2.99-second off signal and then 32 one-second pulses, with a one-second 'off time' between each pulse. ZPA II 64S, with a shorter off time, allows for the transmission or omission of 64 pulses. Each data pulse within a telegram has the potential to double the number of commands, theoretically allowing for 2^{32} distinct commands with 32 pulses. However, practical implementations link specific pulses to types of devices or services. Some telegrams serve unique purposes such as setting clocks in connected devices to midnight in most RCS.

3.4.3 Control of power system based on frequency

A different alternative in which the electricity demand is managed by adjusting the frequency of the system is called frequency-based control [17,18]. In this method, the frequency of the system is inversely proportional to the power demanded. Therefore, the frequency increases when the power consumption is reduced, while the frequency decreases when the power consumption is increased.

The operation for larger power systems with several interconnected areas becomes more complex, because it is required to provide the power throughout the system and maintain the voltage and frequency within acceptable operational limits [19–21]. Therefore, it implies the need to maintain equilibrium between the generated power and the demand at the load side. Several operational facets of the entire power system are impacted due to the inherent uncertainty nature associated with load demand over time. These effects manifest as fluctuations in both the scheduled power exchange within the system and frequency, and these variations produce adverse consequences. In severe cases, potentially culminating in system instability can bring a complete blackout of the entire system.

The most common method to deal with frequency fluctuation is the hierarchical control, which is divided into primary, secondary and tertiary control levels. Under standard operating conditions, minor frequency fluctuations are mitigated by the primary control, characterised by a time response of only several seconds. To address more substantial frequency deviations during off-normal operation, the secondary control loop is implemented, also known as load frequency control (LFC). The corrective action taken by the LFC loop may extend up to a duration of 10 minutes. However, in the event of a pronounced imbalance between power generation and load demand resulting from a major fault, the restoration of frequency to its nominal level through the LFC loop may prove unattainable. In such instances, tertiary control mechanisms come into play. An emergency control loop also is required to restore the frequency of the power system, based on the deviation of the frequency from its nominal. These services are additionally deployed to diminish the likelihood of cascading failures, as shown in Figure 3.14. Given these scenarios, LFC or AGC emerges as a pivotal service with a fundamental role in power systems, ensuring their reliable and successful operation [22,23].

Frequency control is employed to maintain the frequency of the power system within a predefined range by adjusting the supply of electricity. This involves regulating the output of generators and other power sources. Maintaining a stable frequency is crucial for the power company to guarantee a consistent and reliable supply service.

LFC has been a longstanding challenge in centralised power systems, with its primary objectives being the maintenance of uniform frequency, equitable load sharing among generators and effective control of tie-line interchange schedules. AGC and LFC have traditionally operated within a centralised framework. AGC is

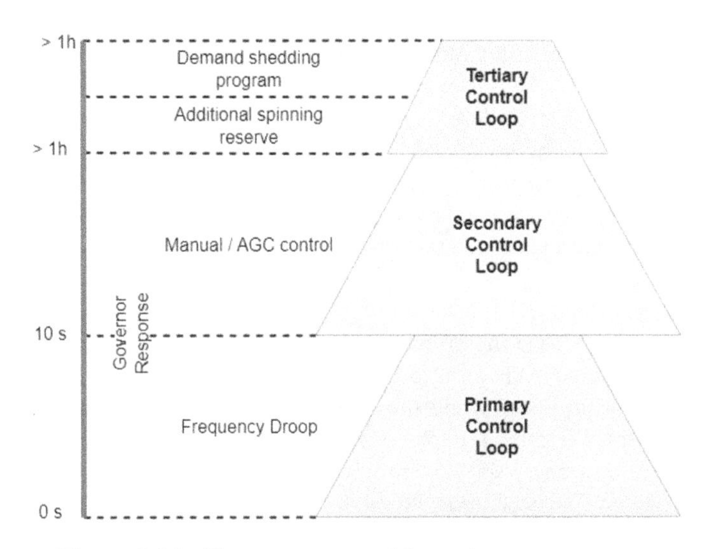

Figure 3.14 Frequency control loops in a power system

responsible for achieving load frequency and utilises frequency deviation to detect changes in load demand. A dedicated communication link is implemented for signal transmission. As the number of ancillary services increases, the necessity for duplex and distributed communication links becomes more evident.

On the other hand, area control error (ACE) and generator control error (GCE) signals are crucial for ensuring frequency within permissible limits, as they are distributed across different areas through dedicated communication networks. ACE and GCE play a role in adjusting generated power levels. The advent of smart metering holds the potential to enhance power system efficiency, reliability and reduce CO_2 emissions.

Smart meters, equipped to measure energy and related information, provide real-time insights into the power system, including load characteristics. These devices offer customers dynamic pricing information, facilitating load peak reduction and load shaving. Moreover, smart meters can serve to disconnect loads during emergencies, a factor underscored by various authors highlighting their significance in LFC.

The frequency of a power system is related to the balance of active power. Alterations in active power demand reverberate across the entire system, leading to a shift in frequency. Consequently, system frequency serves as a valuable indicator, signalling imbalances between generation and load within the system. In cases of short-term imbalances, the system frequency undergoes instantaneous changes, initially mitigated by the kinetic energy of the rotating machines. A substantial loss in generation, without a commensurate system response, can result in extreme frequency excursions that extend beyond the operational limits of the generators.

The regulation of active power by a generator is contingent upon the mechanical power output of its prime mover, which can be a steam turbine, gas turbine, hydro turbine or diesel engine. In instances involving steam or hydro turbines, the control of mechanical power is facilitated through the adjustment of valves that regulate the influx of steam or water into the turbine. It is imperative to continuously regulate the steam or water input to generators to align with the demands of real power. Failure to maintain this synchronisation results in variations in machine rotating speed, leading to corresponding variations in frequency.

As a complement to primary frequency control, most of the large synchronous generators incorporate a secondary frequency control loop. The schematic block diagram of a synchronous generator featuring frequency control loops is depicted in Figure 3.15.

As shown in Figure 3.15, the speed governor senses the change in frequency reflected in the speed, via the primary and secondary control loops. The hydraulic amplifier supplies the mechanical forces to position the main valve against the high steam (or hydro) pressure, while the speed changer ensures a consistent power output setting for the turbine. Each generating unit is equipped with a speed governor that fulfils a primary speed control role. All generating units contribute to the overall alteration in generation through their speed governing mechanisms, regardless of the location of a load change. However, primary control action alone is often insufficient to restore the system frequency, particularly in the context of an

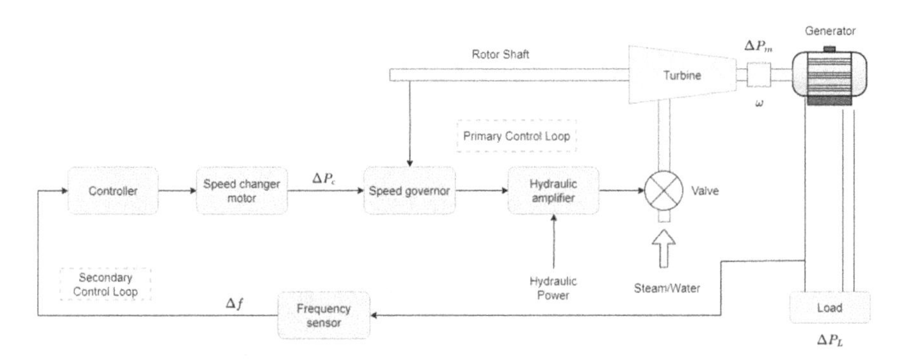

Figure 3.15 Schematic block diagram of a synchronous generator with basic frequency control loops

interconnected power system. The secondary control loop becomes essential, requiring adjustments to the load reference set point through the speed changer motor.

The secondary loop uses the frequency deviation to perform feedback and adds it to the primary control loop through a dynamic controller. The system frequency is, therefore, regulated by the resulting signal (ΔP_C) that is used to regulate the system frequency. The frequency experiences a transient change (Δ), when there is a change in load (ΔP_L). Then, the feedback mechanism generates an appropriate signal for the turbine to make generation of mechanical power (ΔP_m) by tracking the load power and re-establishing the system frequency.

A simplified frequency response model for the described schematic block diagram with one generator unit is shown in Figure 3.15. The overall generator-load dynamic relationship between the mismatch power ($\Delta P_m - \Delta P_L$) and the frequency deviation (Δf) is obtained after the swing differential equation:

$$\Delta P_m(t) - \Delta P_L(t) = 2H_{eq} \cdot d\Delta f(t)dt + D\Delta f(t) \quad (3.11)$$

where *Heq* the inertia constant and *D* is the load damping coefficient. The damping coefficient is usually expressed as a percent change in load for a 1% change in frequency. For example, a typical value of 1.8 for *D* means that a 1 % change in frequency can cause a 1.8 % change in load. Equation (3.11) can be rewritten after applying Laplace transform as follows:

$$\Delta P_m(s) - \Delta P_L(s) = 2H_{eq}s\Delta f(s) + D\Delta f(s) \quad (3.12)$$

Equation (3.12) is represented in a block diagram, as shown in Figure 3.16. The generator-load model reduces the schematic block diagram of a closed-loop synchronous generator as shown in Figure 3.15, to obtain a new representation, as shown in Figure 3.17.

A simplified governor-generator model of a generic power system is shown in Figure 3.18, which is used for power system frequency analysis and control design. First-order transfer functions are used to model the governor-turbine.

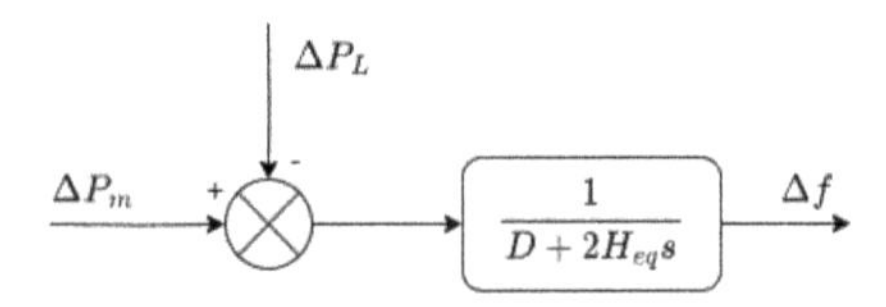

Figure 3.16 Generator-load model block diagram representation

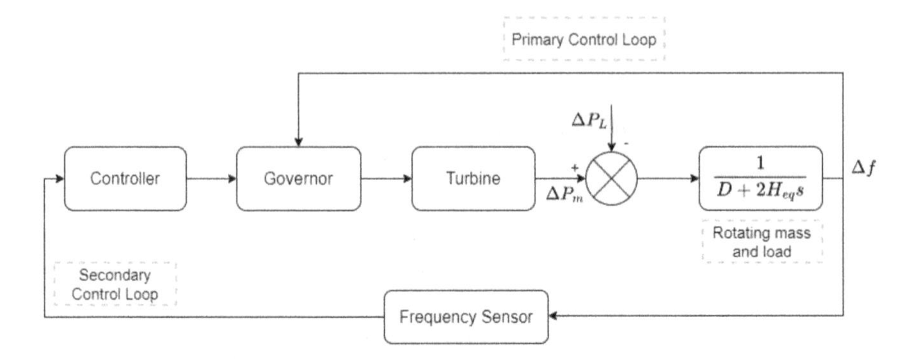

Figure 3.17 Reduced block diagram of Figure 3.15

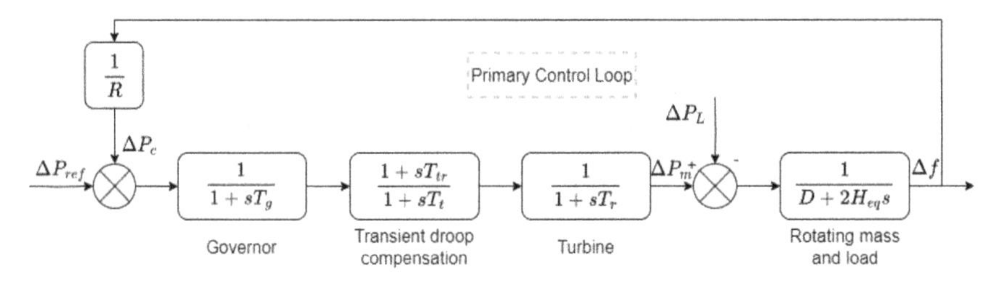

Figure 3.18 Power system primary frequency control model

The droop gain (R) corresponds to the ratio of frequency change Δf and the change of generator power output ΔP. This relationship is illustrated in Figure 3.19. The purpose of the turbine-governor control is to maintain the desired system frequency by setting the mechanical output power of the turbine ΔPm. The frequency–power relationship of turbine-governor control is defined by the following expression:

$$\Delta P_c = \Delta P_{ref} - \frac{1}{R}\Delta f \tag{3.13}$$

The term $\Delta P_c - \Delta P_{ref}$ is denoted by ΔP_1 and the droop gain is defined as follows:

$$-R = Slope = \Delta f \Delta P \tag{3.14}$$

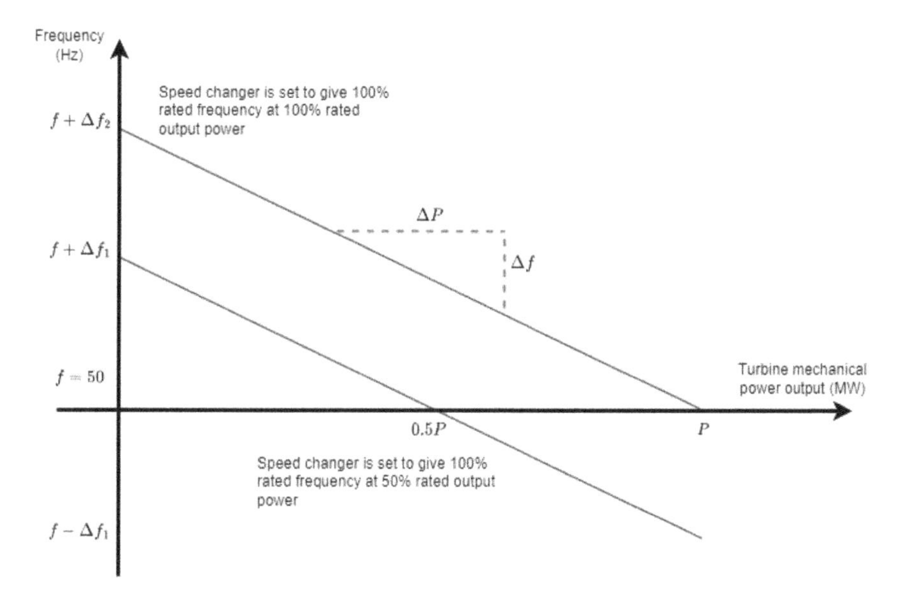

Figure 3.19 Steady-state frequency-power relationship of a turbine-governor control

The governors employ droop control to manage the power output of generators in response to frequency variations. This mechanism of primary frequency control is automatically supplied by the governors. In practical terms, if there is an increase in demand power or a decrease in generation power leading to a decline in frequency, the governors automatically provide a low-frequency response. Conversely, in the event of a decrease in demand (resulting in a frequency rise), the governors offer a high-frequency response service.

Normally, a large power system is represented as a multiarea power system that is interconnected by high-voltage transmission lines. The frequency trend measured within each control area serves as an indicator of the mismatch power trend across the interconnection, rather than within the confines of an individual control area. In an interconnected multiarea power system, the secondary frequency control system of each area is requested to regulate the local frequency and the interchange power with other control areas. Figure 3.20 illustrates this control system by depicting a power system with N control areas. The power flow through the tie-line from area 1 to area 2 is expressed in (3.15).

The frequency trend monitored in each control area serves as an indicator for the trend of mismatched power across the interconnection point. The secondary frequency control system within each control area should manage both the interchange power with other control areas and its local frequency. To illustrate this, Figure 3.21 depicts a power system with N control areas. The power flow through the tie-line from area 1 to area 2 can be expressed by (3.15):

$$P_{tie,12} = V_1 V_2 X_{12} \cdot Sin(\delta_1 - \delta_2) \quad (3.15)$$

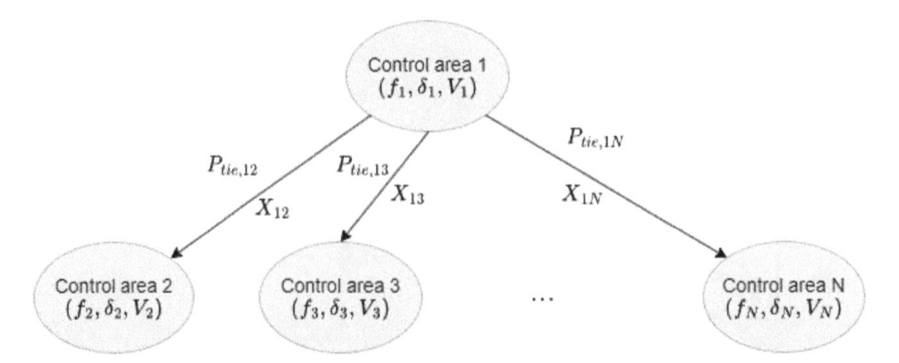

Figure 3.20 A schematic diagram of an N-area interconnected power system

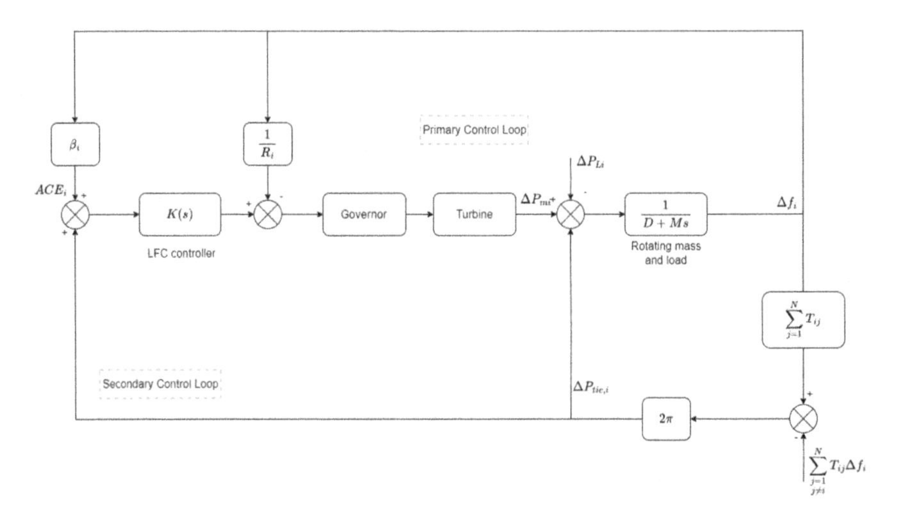

Figure 3.21 A simplified interconnected power system with LFC controller

Here, V_1 and V_2 are the p.u. voltages at the machine's terminals for area 1 and 2, respectively, as well as δ_1 and δ_2 are the power angles of equivalent machines. X_{12} represents the line reactance between both areas.

In a linearised model, the equilibrium point $(\delta_{10}, \delta_{20})$ can be expressed as:

$$\Delta P_{tie,12} = T_{12}(\Delta\delta_1 - \Delta\delta_2) \tag{3.16}$$

where T_{12} represents the synchronising torque coefficient presented by

$$T_{12} = |V_1||V_2|X_{12} \cdot Cos(\delta_{10} - \delta_{20}) \tag{3.17}$$

Equation 3.17 can be re-written considering the angle and frequency relationships between areas, where Δf_1 and Δf_2 are frequency changes in area one and two, respectively. The Laplace transform of (3.18) results in (3.19):

$$\Delta P_{tie,12} = 2\pi T_{12} \cdot \left(\int \Delta f_1 - \int \Delta f_2 \right) \tag{3.18}$$

$$\Delta P_{tie,12}(s) = 2\pi T_{12} \cdot (\Delta f_1(s) - \Delta f_2(s)) \tag{3.19}$$

Similarly, the net power interchange between A1 and A3 is given in (3.20):

$$\Delta P_{tie,13}(s) = 2\pi T_{13} \cdot (\Delta f_1(s) - \Delta f_3(s)) \tag{3.20}$$

The total line power change between area 1 and the other areas can be calculated as:

$$\Delta P_{tie,1} = \Delta P_{tie,12} + \Delta P_{tie,13} = 2\pi s \cdot \left[\sum_{j=2,3} T_{1,j} \Delta f_1 - \sum_{j=2,3} T_{1,j} \Delta f_j \right] \tag{3.21}$$

Similarly, for N-control areas, the total tie-line power change between area 1 and other areas is:

$$\Delta P_{tie,i} = \sum_{i \neq j} \Delta P_{tie,ij} N_{j=1} = 2\pi s \cdot \left[\sum_{i \neq j} T_{ij} \Delta f_i - \sum_{i \neq j} T_{ij} \Delta f_j N_{j=1} \right] \tag{3.22}$$

Equation (3.22) can be illustrated with a block diagram, considering the mechanical power mismatch ($\Delta P_m - \Delta P_L$) described in Figure 3.18. Hence, Figure 3.21 depicts a simplified block diagram of the interconnected power system.

In the secondary control loop, the tie-line power flow change ($\Delta ptie,i$) is added to the frequency change (Δfi) through a secondary feedback loop. The area control error (ACE_i) signal is then computed as follows and applied to the controller $K(s)$:

$$ACE_i = \Delta P_{tie,i} + \beta_i \Delta f_i \tag{3.23}$$

where β_i is a bias factor, which can be obtained according to (3.24):

$$\beta_i = \frac{1}{R_i} + D_i \tag{3.24}$$

If a drop in area frequency occurs, the LFC controller $K(s)$ will adjust the value of ACE_i, bringing it back to zero, and transmitting the control signal to the governor. This action is done to regulate the frequency of the area and sustain the power interchanged at the scheduled level.

Proposed exercises

1. Consider the system presented in Figure 3.22

   The initial load conditions are summarised below:

   - Total load: 35.0 MW
   - In-house generation: 9.5 MW
   - Total import: 25.5 MW
   - GH constant: 42.34
   - Rated system frequency: 60 Hz

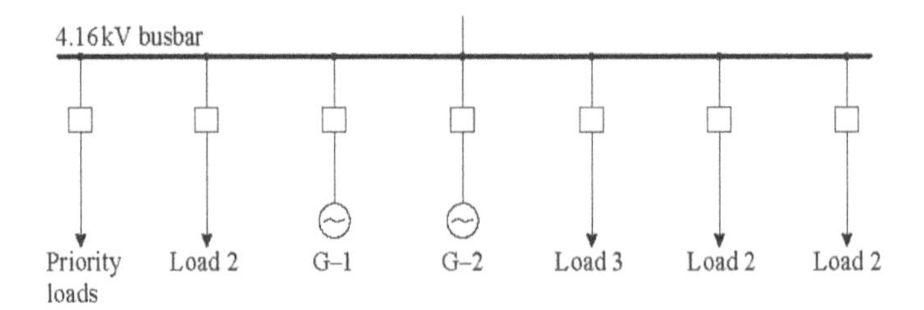

Figure 3.22 Substation for proposed exercise 3.1

a) Evaluate the total perceptual overload in case of loss of half and then full of the in/house generators.
b) With the loss of the total grid supply, determine the amount of load that must be disconnected in order to relieve the overload. It should be noted that there is a high priority load totalling 11.3 MW which cannot be disconnected.

2. A 150 MVA 60 Hz turboalternator operates at no load at 3600 rpm. A load of 42.5 MW is suddenly applied to the machine and the steam valves to the turbine commence to open after 0.6 s due to the time-lag in the governor system. Assuming inertia constant H of 5.7 kW-s per kVA of generator capacity, calculate the frequency to which the generated voltage drops before the steam flow commences to increase to meet the new load.
3. Two turbo alternators rated for 136 MW and 308 MW have governor drop characteristics of 4.8% from no load to full load. They are connected in parallel to share a load of 320 MW. Determine the load shared by each machine assuming free governor action.

References

[1] P. Palensky and D. Dietrich, 'Demand side management: demand response, intelligent energy systems, and smart loads', *IEEE Trans. Ind. Informat.*, vol. 7, no. 3, p. 381, 2011.
[2] Y. Izumi, T. Senjyu, and A. Yona, 'Load frequency control by using demand response with $\mathcal{H}\infty$ in isolated power systems', in *IEEE 15th International Conference on Harmonics and Quality of Power*, Hong Kong, China, pp. 656–661, 2012. doi:10.1109/ICHQP.2012.6381227.
[3] C. Alvarez, A. Gabaldon, and A. Molina., 'Assessment and simulation of the responsive demand potential in end-user facilities: application to a university customer', *IEEE Trans. Power Syst.*, vol. 19, no. 2, pp. 1223–1231, 2004.
[4] UK Power Networks (UKPN), 'Low Carbon London Project Closedown Report V1.0', London, 2015.

[5] UK Power Networks (UKPN), 'LCNF Successful Delivery Reward Application Low Carbon', 2019.
[6] Hitachi, 'Smart Energy Islands – Isles of Scilly: sharing locally-produced energy using the IoT: Social Innovation: Hitachi', [Online]. Available: https://social-innovation.hitachi/en-eu/case_studies/smart-energy-islands-isles-of-scilly/ [accessed 23 March 2024].
[7] The Federal Energy Regulatory Commission Staff, 'National Action Plan on Demand Response. Docket No. AD09-10', 2010.
[8] Department of Energy, 'Demand Response', [Online]. Available: https://www.energy.gov/oe/demand-response [accessed 20 March 2024].
[9] D. P. Chassin, M. K. Donnelly, and J. E. Dagle, 'Electrical power distribution control methods, electrical energy demand monitoring methods, and power management devices', US, 2006. [Online]. Available: https://www.osti.gov/biblio/957481 [accessed 20 March 2024].
[10] M. K. Donnelly, D. P. Chassin, J. E. Dagle, *et al.*, 'Electrical appliance energy consumption control methods and electrical energy consumption systems', U.S. Patent 7420293, 2008. [Online]. Available: https://www.freepatentsonline.com/7420293.html [accessed 24 March 2024].
[11] F. C. Schweppe, 'Frequency adaptive, power-energy re-scheduler'. U.S. Patent 4317049A, 1979.
[12] T. Kennedy, S. M. Hoyt, and C. F. Abell, 'Variable, nonlinear tieline frequency bias for interconnected systems control', *IEEE Trans. Power Syst.*, vol. 3, no. 3, pp. 1244–1253, 1988.
[13] D. Raisz and A. M. Dan, 'Ripple control as a possible tool for daily load balancing in an open electricity market environment', *IEEE Russia Power Tech*, St. Petersburg, Russia, 2005, pp. 1–6, doi:10.1109/PTC.2005.4524451.
[14] M. Praveen and G. V. S. Rao, 'Ensuring the reduction in peak load demands based on load shifting DSM strategy for smart grid applications', *Procedia Comput. Sci.*, vol. 167, pp. 2599–2605, 2020. doi:10.1016/j.procs.2020.03.319.
[15] B. Potel, V. Debusschere, F. Cadoux, and L. de Alvaro Garcia, 'Under-frequency load shedding schemes characteristics and performance criteria', in *IEEE Manchester PowerTech*, Manchester, UK, pp. 1–6, 2017, doi:10.1109/PTC.2017.7981046.
[16] J. M. Polard, 'The remote control frequencies', vol. 21, 2011.
[17] A. Khalil and Z. Rajab, 'Load frequency control system with smart meter and controllable loads', in 2017 *8th International Renewable Energy Congress (IREC)*, Piscataway, NJ: IEEE, pp. 1–6, 2017. doi:10.1109/IREC.2017.7926033.
[18] Changhong Zhao, U. Topcu, and S. H. Low, 'Frequency-based load control in power systems', in *2012 American Control Conference (ACC)*, Piscataway, NJ: IEEE, pp. 4423–4430, 2012. doi:10.1109/ACC.2012.6315283.
[19] M. Shouran, 'Load frequency control for multi-area interconnected power system using artificial intelligent controllers'. Ph.D. Thesis, Cardiff University, Cardiff, WA, USA, 2022.

[20] P. Babahajiani, Q. Shafiee, and H. Bevrani, 'Intelligent demand response contribution in frequency control of multi-area power systems', *IEEE Trans. Smart Grid*, vol. 9, no. 2, pp. 1282–1291, 2018, doi:10.1109/TSG.2016.2582804.
[21] Y.-Q. Bao, L. I. Yang, B. Wang, H. U. Minqiang, and P. Chen., 'Demand response for frequency control of multi-area power system', *J. Mod. Power Syst. Clean Energy*, vol. 5, no. 1, pp. 20–29, 2017, doi:10.1007/s40565-016-0260-1.
[22] P. Palensky, 'The JEVIS service platform – distributed energy data acquisition and management', In *The Industrial Information Technology Handbook* (pp. 111–121). CRC Press LLC, 2005.
[23] J. Gers, *Distribution System Analysis and Automation*, 2nd ed. London: The Institution of Engineering and Technology, 2020.

Chapter 4

Reliability and FLISR schemes for urban grids

Knowing that the most important benefit of a smart grid is the increase in the reliability indices, a proper examination of this concept is important. The term 'reliability' refers to the notion that the system performs its specified task correctly for a certain duration of time. The term 'availability' refers to the readiness of a system or component to be immediately ready to perform its task. Both terms have precise definitions within the reliability engineering field and have specified equations and methods to provide quantitative metrics for them.

Availability is the probability of something being energised. It is the most basic aspect of reliability and is typically measured in percent or per-unit. The complement of availability is unavailability.

Unavailability can be computed directly from interruption duration information. If a customer experiences 24 hours of interrupted power in a year, unavailability is equal to 24/8760 = 0.00274 = 0.274%. Availability is equal to 100% − 0.2739% = 99.7260%.

Each distribution system component can be described by a set of reliability parameters. Simple reliability models are based on component failure rates and component repair times, but sophisticated models make use of many other reliability parameters. Some of the most common reliability parameters are the following:

- Failure rate, indicated by the symbol λ, describes the number of times per year that a component can experience a failure.
- Mean time to repair (MTTR) indicated by the letter r represents the expected time it will take for a failure to be repaired (measured from the time that the failure occurs). A single value of MTTR is typically used for each component, but separate values can be used for different failure modes.
- Probability of operational failure (POF) is the conditional probability that a device will not operate if it is supposed to operate. For example, if an automated switch fails to function properly five times out of every 100 attempted operations, it has a POF of 5%. This reliability parameter is typically associated with switching and protection devices.

4.1 Network modelling

The reliability analysis in physical networks is done by using the concepts of series and parallel circuits in electrical systems, which are clear and flexible tools to handle. The solutions involve mathematical methods that accurately represent the models under analysis.

Network modelling is a component-based technique rather than a state-based technique. Each component is described by a probability of being available, P, and a probability of not being available, Q. Since components are assumed to be either available or not available, Q and P are arithmetic complements: $Q = 1 - P$. It is important to highlight that the symbol P also represents the non-supplied power, especially for some software packages, which is of course different from the probability of being available.

If a component is described by an annual failure rate λ and a MTTR r in hours, the probability of being available can be computed as follows:

$$P = \frac{8760 - \lambda r}{8760} \tag{4.1}$$

Therefore:

$$Q = \frac{\lambda r}{8760} \tag{4.2}$$

As mentioned before, networks are composed of series and parallel components. Two components are in series if both must be available for the connection to be available. Two components are in parallel if only one of the components needs to be available for the connection to be available.

For a series network, the probability of an available path is equal to the product of the individual component availabilities. For a parallel network, the probability of an unavailable path is equal to the product of the individual component unavailability. Components in series reduce availability and components in parallel improve availability.

$$P_{series} = \prod P_{component} \tag{4.3}$$

$$Q_{parallel} = \prod Q_{component} \tag{4.4}$$

The failure rate λ is a measure of unreliability. The product λr (failure rate $\times$ average downtime per failure) is equal to the forced downtime hours per year and it can be considered as a measure of forced unavailability since a scale factor of 8760 converts one value to the other. The average downtime per failure r could be called restorability. The procedure and formulas for calculating the reliability indices are given in IEEE Std. 493-2007 [1] and are reproduced in (4.5)–(4.10). A schematic demonstrating these formulas is shown in Figure 4.1 for two components numbered '1' and '2' connected in series and in Figure 4.2 for two components '3' and '4' connected in parallel.

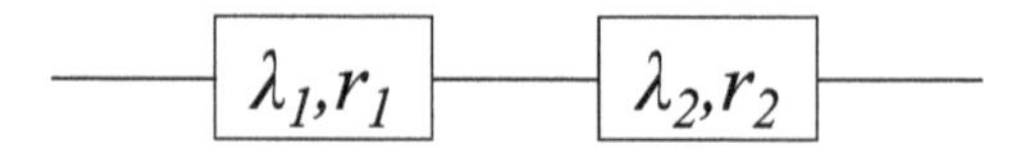

Figure 4.1 Repairable components in series – both must work for success [1]

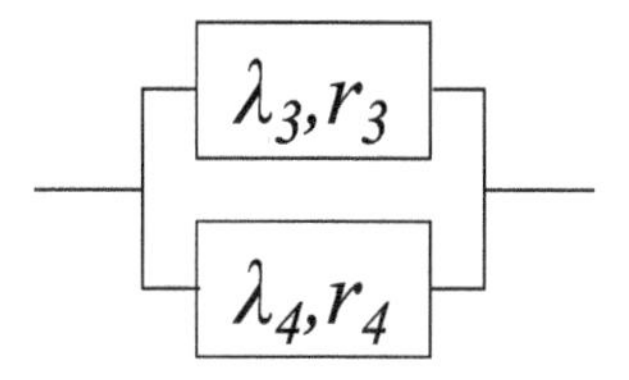

Figure 4.2 Repairable components in parallel – one or both must work for success [1]

In these schematics, scheduled outages are assumed to be zero and the units for f, λ and r are, respectively, frequency of failures for the whole system, failure rate per year of each component and hours of downtime per failure. The equations associated to Figures 4.1 and 4.2 assume the following:

a) The component failure rate is constant with age.
b) The outage time after a failure has an exponential distribution.
c) Each failure event is independent of any other failure event.

$$f_s = \lambda_1 + \lambda_2 \tag{4.5}$$

$$f_s r_s = \lambda_1 r_1 + \lambda_2 r_2 \tag{4.6}$$

$$r_s \cong \frac{\lambda_1 r_1 + \lambda_2 r_2}{\lambda_1 + \lambda_2} \tag{4.7}$$

$$f_p = \frac{\lambda_3 \lambda_4 (r_3 + r_4)}{8760} \tag{4.8}$$

$$f_p r_p = \frac{\lambda_3 r_3 \lambda_4 r_4}{8760} \tag{4.9}$$

$$r_p = \frac{r_3 r_4}{r_3 + r_4} \tag{4.10}$$

4.2 Network reduction

Complex systems have a high number of series and parallel connections. To process them, a network reduction is required. This is accomplished by simplifying parallel and series components into equivalent network components until a single component is obtained. The availability of the last component is equal to the availability of the original system. An example of network reduction is shown in Figure 4.3.

In this system, a line section (component 1) provides power to a distribution transformer (component 2). The distribution transformer can also receive power from an alternate path consisting of three-line sections in series (components 3, 4 and 5). Three applications of network reduction are required to compute the availability of the system.

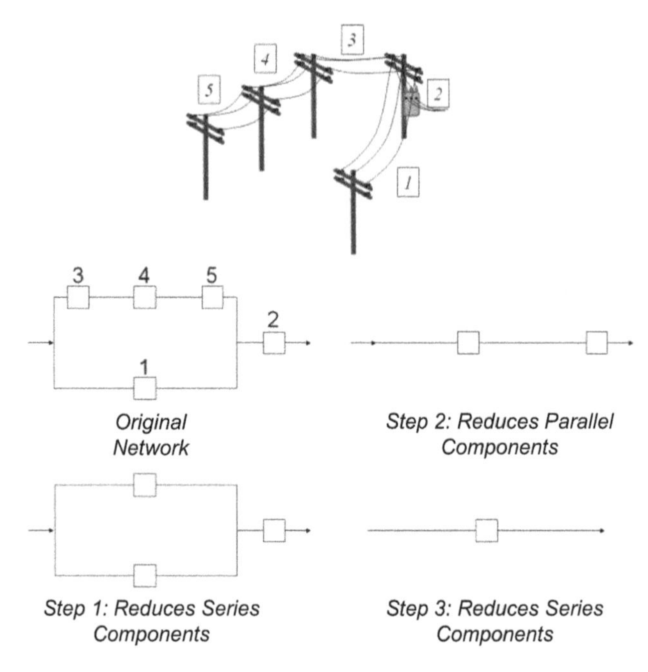

Figure 4.3 An example of network reductions

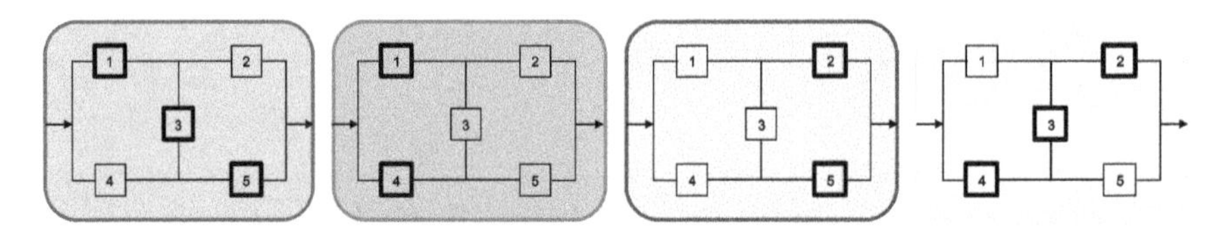

Figure 4.4 Minimal cut sets of a simple system

A good approach for network reduction is the so-called minimal cut set method. In this method, it is required to determine the minimal number of elements needed to keep the system running.

Figure 4.4 shows an example of a portion of a network that has five switches. There are five scenarios, each corresponding to a minimal cut set. Each one requires only three switches at a time for the system to keep running. The unavailability of the system is equal to the sum of the individual component of these minimal cut sets.

4.3 Quality indices

Quality indices, also called reliability indices, measure the performance of a power system. One of the most important benefits of smart grids is that they help to improve reliability performance of power systems. Given the importance of these indices to enhance a grid, a proper consideration to their meaning and application is explained in this section.

Several key definitions relating to distribution reliability include:

- Fault: an abnormal operating condition of an electrical system which normally develops a short circuit. It can be caused by natural events, rough weather conditions, animal presence, equipment failure and even vandalism. Faults can be categorised as self-clearing, temporary and permanent. A self-clearing fault will extinguish itself without any external intervention. A temporary fault will clear if de-energised and then re-energised. A permanent fault lasts until repaired by human intervention.
- Contingency: an unexpected event such as a fault or an open circuit. Another term for a contingency is an unscheduled event.
- Outage: an outage occurs when a piece of equipment is de-energised. Outages can be either scheduled or unscheduled. Scheduled outages are known in advance (e.g., outages for periodic maintenance). Unscheduled outages result from contingencies. Different utilities have different criteria to define an outage. Some utilities consider outages, those interruptions that exceed 1 minute. Others use 2 minutes and others up to 5 minutes.
- Open circuit: a point in a circuit that interrupts load current without causing fault current to flow. An example of an open circuit is the false tripping of a circuit breaker.
- Momentary interruption: a momentary interruption occurs when a customer is de-energised for just a few minutes. Most momentary interruptions result from reclosing or automated switching. Multiple reclosing operations result in multiple momentary interruptions.
- Sustained interruption: a sustained interruption occurs when a customer is de-energised for more than a few minutes. Most sustained interruptions result from open circuits and faults.

Poor reliability on the part of the electrical utility is penalised or rewarded based on quantification by reliability indices. Some utilities also pay bonuses to utility personnel, based in part on indices. Commercial and industrial customers enquire about reliability indices when locating a new facility. Most regulatory bodies have established targets for quality indices. If utilities do not fulfil them (have figures higher than those defined), they can be penalised. The three main indices are the SAIDI, SAIFI and CAIDI which are explained in the following paragraphs.

SAIDI (system average interruption duration index)

SAIDI is defined as the average interruption duration for customers served during a specified time. It is calculated by summing the customer-minutes off for each interruption during a specified period of time and dividing the sum by the average number of customers served during the period. The unit is minutes.

The index enables the utility to report the time (normally in minutes) customers would have been out of service if all customers were out at one time. The benchmarking survey SAIDI average for the United States is around 90 minutes.

$$SAIDI = \frac{\sum Customer\ Interruption\ Durations}{Total\ Number\ Customer\ Served} \min/yr \tag{4.11}$$

SAIFI (system average interruption frequency index)

SAIFI is defined as the average number of times that a customer is interrupted during a specified period of time. It is calculated by dividing the total number of customers interrupted in a period of time by the average number of customers served.

The index is calculated for both interruptions of more than one minute (SAIFI-long) and less than 1 minute (SAIFI-short). The resulting unit is 'interruptions per customer'. The benchmarking survey SAIFI average for the United States is around 1.2 interruptions per customer per year.

$$SAIFI = \frac{Total\ Number\ of\ Customer\ Interruptions}{Total\ Number\ of\ Customer\ Served}/yr \tag{4.12}$$

CAIDI (customer average interruption duration index)

CAIDI is defined as the average outage duration that a given customer would experience. CAIDI is the ratio of the SAIDI over the SAIFI and is calculated with the following expression:

$$CAIDI = \frac{\sum Customer\ Interruption\ Durations}{Total\ Number\ of\ Customer\ Interruptions} hr \tag{4.13}$$

Example 4.1

Figure 4.5 shows a power system with one transformer, two lines and the load. Each line experiences 3 failures per year and 0.75 hours downtime per failure. It has a total load of 20 MW and each line is capable of supporting that value. Calculate the frequency of failures, the minutes per year of unreliability and the non-supplied power. First assume that each line can support the whole load; then calculate again, assuming that each line cannot support the total load.

a) For the first situation, where each line can support the whole load, it is valid to assume a parallel connection.

$$f_p = \frac{3 * 3 * (0.75 + 0.75)}{8760} = 0.0015411/yr$$

$$r_p = \frac{0.75 * 0.75}{0.75 + 0.75} = 0.375h$$

$$Q_p = f_p * r_p = 0.0015411 * 0.375 = 0.000578\ h/yr = 0.03467\ \text{min}/yr$$

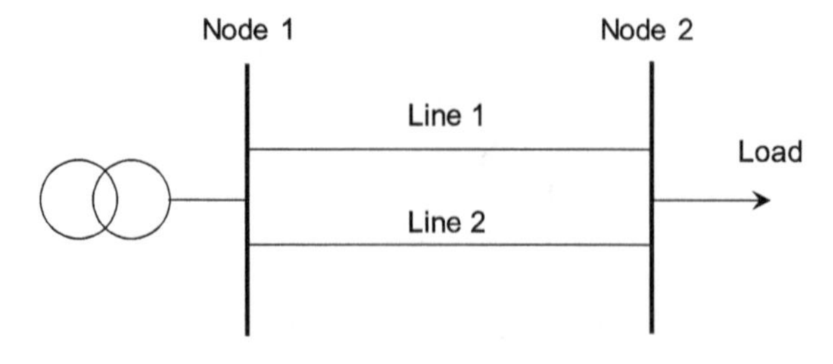

Figure 4.5 Diagram system for Example 4.1

or

$$Q_p = Q_1 * Q_2 = (Q_1)^2 = (135\ \min/yr)^2 = (0.0002568)^2$$
$$= 0.03467\ \min/yr$$

The non-supplied power per year is:

$$p_p = p_{load} * f_p = 0.031 MW/yr$$

These results were validated with power system calculation software. The result is shown in Figure 4.6.

b) For the second case where it is assumed that each individual line can't support the total load, the reliability analysis is developed by treating the system as a series connection, although both lines are physically connected in parallel.

$$f_s = 3 + 3 = 6/yr$$

$$r_s \cong \frac{3 * 0.75 + 3 * 0.75}{3 + 3} = 0.75h$$

$$Q_s = f_s * r_s = 6 * 0.75 = 4.5h/yr = 270\ \min/yr$$

or

$$1 - Q_s = P_s = P_1 * P_2 = P_1{}^2 = (1 - Q_1)^2$$

$$Q_s = 1 - (1 - Q_1)^2 = 1 - (1 - 3 * 0.75/8760)^2 = 1 - (1 - 0.0002568)^2$$
$$= 270\ \min/yr$$

The non-supplied power is:

$$p_s = p_{load} * f_s = 20 * 6 = 120 MW/yr$$

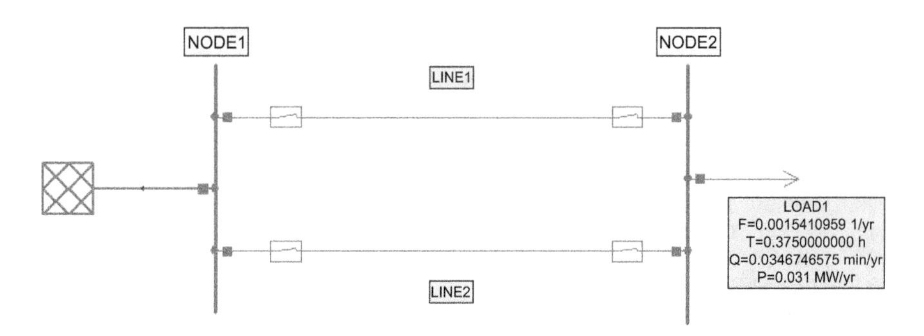

Figure 4.6 Diagram system results for the first part of Example 4.1

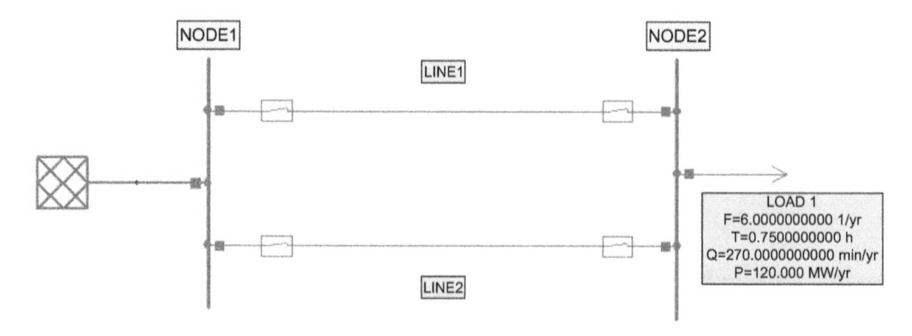

Figure 4.7 Diagram system results for the second part of Example 4.1

These results were validated with the software NEPLAN. The result is shown in Figure 4.7.

Example 4.2

Table 4.1 shows an excerpt from one utility's customer information system (CIS) database for feeder 25, which serves 1500 customers with a total load of 3 MW. In this example, Feeder 25 constitutes the 'system' for which the indices are calculated. More typically the 'system' combines all circuits together in a region or for a whole company.

S indicates a sustained interruption and M a momentary interruption. Applying the previous equation, it is possible to find these quality indexes.

$$SAIFI = \frac{250 + 368 + 23 + 87 + 1000}{1500} = 1.152/yr$$

$$SAIDI = \frac{(15.37 \times 250) + (18.03 \times 368) + (41.22x23) + (15.7 \times 87) + (45.2 \times 1000)}{1500}$$

$$= 38.66\ \text{min}/yr$$

$$CAIDI = \frac{SAIDI}{SAIFI} = \frac{38.66}{1.152} = 33.56\ \text{min}$$

Table 4.1 Example of outage data

Date	**Time**	**Time on**	**Total time [min]**	**Circuit**	**Event code**	**Number of customers**	**Load [kVA]**	**Interruption type**
2/15	14:10:17	14:25:39	15.37	25	51	250	431	S
4/20	17:20:39	17:38:41	18.03	25	306	368	800	S
5/1	06:33:36	07:14:49	41.22	25	468	23	150	S
6/2	23:18:10	23:18:57	0.78	25	522	590	1200	M
6/8	02:39:52	03:55:34	15.7	25	634	87	200	S
9/29	09:29:05	09:30:02	0.95	25	811	1500	3000	M
11/14	17:15:49	17:16:18	0.48	25	963	700	1500	M
12/4	12:16:32	13:01:44	45.2	25	1021	1000	1800	S

Now, it's worth noting that interruption costs can be evaluated. In this case, a tool developed by Lawrence Berkeley National Laboratory (LBNL) and Resource Innovations, named The Interruption Cost Estimate (ICE) Calculator, which is an electric reliability planning tool [2], is used.

By running the model with the total number of customers of 1500 (assuming 65% as residential and 35% as non-residential) and the quality indices already computed, we obtain Figure 4.8.

Momentary average interruption frequency index (MAIFI) and momentary average interruption event frequency index (MAIFI$_E$)

The momentary average interruption frequency index (MAIFI) as defined by IEEE Std 1366-2022 [3], indicates the average frequency of momentary interruptions. Mathematically, this is given by the following equation:

$$MAIFI = \frac{\sum Total\ Number\ of\ Customer\ Momentary\ Interruptions}{Total\ Number\ of\ Customers\ Served} \tag{4.14}$$

The momentary average interruption event frequency index (MAIFI$_E$) as defined by the same IEEE Standard, indicates the average frequency of momentary interruption events, not including the events immediately preceding a lockout. Mathematically, this is given by the following equation:

$$MAIFI_E = \frac{\sum Total\ Number\ of\ Customer\ Momentary\ Interruption\ Events}{Total\ Number\ of\ Customers\ Served} \tag{4.15}$$

To calculate these indices, the following equations can be used:

$$MAIFI = \frac{\sum IM_i N_{mi}}{N_T} \tag{4.16}$$

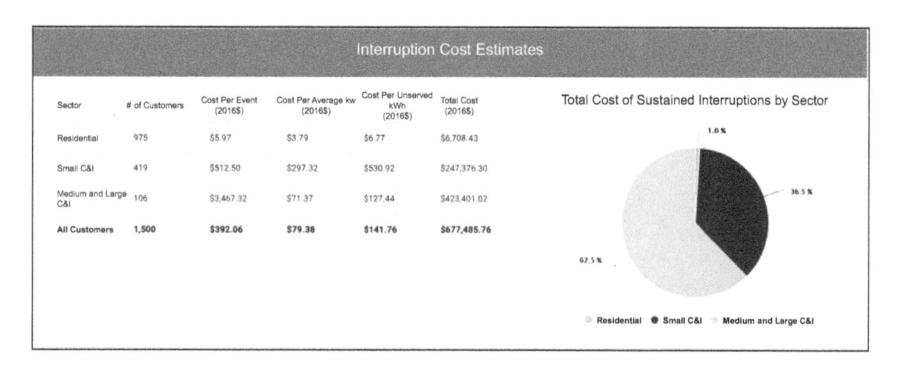

Sector	# of Customers	Cost Per Event (2016$)	Cost Per Average kw (2016$)	Cost Per Unserved kWh (2016$)	Total Cost (2016$)
Residential	975	$5.97	$3.79	$6.77	$6,708.43
Small C&I	419	$512.50	$297.32	$530.92	$247,376.30
Medium and Large C&I	106	$3,467.32	$71.37	$127.44	$423,401.02
All Customers	**1,500**	**$392.06**	**$79.38**	**$141.76**	**$677,485.76**

Figure 4.8 Interruption costs estimate from ICE calculator [2]

$$MAIFI_E = \frac{\sum IM_E N_{mi}}{N_T} \tag{4.17}$$

where:

IM_i is the number of momentary interruptions
IM_E is the number of momentary interruption events
N_{mi} is the number of interrupted customers for each momentary interruption event during the reporting period
N_T is the total number of customers served for the areas

Example 4.3

To better illustrate the concepts of momentary interruptions, sustained interruptions and the associated indices, consider Figure 4.9. The figure illustrates a circuit composed of a circuit breaker (B), a recloser (R) and a sectionaliser (S).

For this scenario, 1000 customers would experience a momentary interruption and 500 customers would experience a sustained interruption. Calculations for SAIFI, MAIFI and MAIFI_E on a feeder basis are shown in the following equations. Notice that the numerator of MAIFI is multiplied by 2 because the recloser took two shots. However, MAIFI_E is multiplied by 1 because it only recognises that a series of momentary events occurred.

$$SAIFI = \frac{500}{3000} = 0.167$$

$$MAIFI = \frac{2 \cdot 1500}{3000} = 1$$

$$MAIFI_E = \frac{1 \cdot 1500}{3000} = 0.5$$

A good comparison of reliability indices has been worked out by the Council of European Energy Regulators (CEER) in the report 'CEER Benchmarking Report 6.1 on the Continuity of Electricity and Gas Supply, data update 2015/2016' [4]. Tables 4.2 and 4.3 present the SAIDI and SAIFI data of different European from 2002 to 2016, respectively.

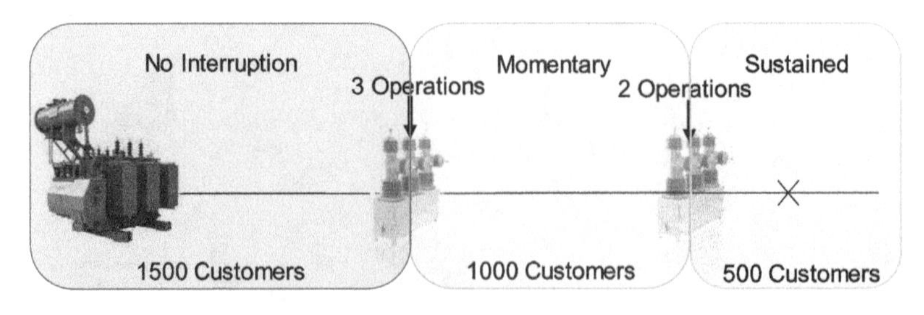

Figure 4.9 Representation of events used in calculating indices

Table 4.2 Planned and unplanned SAIDI, without exceptional events [4]

COUNTRY	2002	2003	2004	2005	2006	2007	2008	2009	2010	2011	2012	2013	2014	2015	2016
Austria											48.67	47.96	49.79	42.31	37.91
Croatia													416.49	404.42	325.25
Czech Republic			250.83	286.69	247.20	274.46	252.52	243.30	265.64	261.86	257.36	257.69	246.71	253.32	233.0
Denmark									20.37	21.04	19.44	15.90	16.65	19.55	19.34
France	46.00	56.30	57.30	60.20	79.40	68.50	82.00	90.40	86.90	71.50	75.70	84.00	66.00	67.90	66.60
Germany					36.63	33.10	30.06	26.16	24.56	25.43	27.74	22.55	19.84	19.73	23.09
Great Britain					69.61	82.99	79.92	79.91	76.74	74.63	62.14	60.39	58.94	49.89	42.30
Greece											250.00	252.00	228.00	199.00	208.00
Hungary														250.00	232.00
Ireland														158.70	144.90
Italy	186.85	177.50	139.14	124.51	107.63	98.63	102.45	93.03	103.48	105.44	111.42	97.55	100.92	112.02	115.96
Latvia											520.00	472.00	409.00	332.00	260.00
Lithuania				208.35	199.63	167.18	145.33	124.29	211.09	217.43	234.90	264.16	266.89	238.76	220.98
Luxembourg												33.64	25.09	28.81	21.72
Poland							503.56	456.77	445.96	462.15	401.32	393.97	311.17	340.04	260.36
Portugal	386.75	366.14	197.97	181.98	170.78	111.64	135.15	187.62	174.55	99.30	80.16	90.16	77.48	69.20	65.99
Romania							1025.0	959.00	963.00	880.00	876.00	697.00	591.00	519.00	474.00
Slovenia										128.44	119.68	121.44	122.96	133.43	121.77
Spain											63.24	61.26	64.68	69.12	65.82
Sweden											100.70	101.06	92.66	100.25	87.90
Switzerland									28.00	29.00	34.00	25.00	22.00	21.00	19.00

Table 4.3 Planned and unplanned SAIFI, without exceptional events [4]

COUNTRY	2002	2003	2004	2005	2006	2007	2008	2009	2010	2011	2012	2013	2014	2015	2016
Austria											0.69	0.78	0.77	0.70	0.72
Croatia													3.46	3.60	2.95
Czech Republic			2.68	2.49	2.42	2.91	2.32	2.17	2.23	2.19	2.31	2.21	2.11	2.09	1.95
Denmark									0.44	0.44	0.44	0.36	0.35	0.41	0.42
France	0.11	0.18	0.16	0.13	0.15	0.19	0.32	0.31	0.27	0.19	0.19	0.23	0.20	0.22	0.22
Germany						0.45	0.43	0.39	0.35	0.41	0.39	0.55	0.42	0.88	0.59
Great Britain					0.73	0.80	0.76	0.74	0.72	0.70	0.63	0.61	0.62	0.56	0.50
Greece											2.43	2.50	2.40	2.10	2.07
Hungary														1.64	1.52
Ireland														1.31	1.30
Italy	3.23	3.17	2.82	2.71	2.52	2.40	2.27	2.24	2.19	2.04	2.15	2.00	2.00	2.12	1.91
Latvia											4.34	3.86	3.37	2.97	2.88
Lithuania				1.18	1.17	1.38	1.20	0.96	1.22	1.29	1.32	1.26	1.23	1.25	1.20
Luxembourg												0.40	0.38	0.43	0.28
Poland							4.82	4.46	4.42	4.96	4.12	3.64	3.51	4.10	3.45
Portugal	6.22	5.11	2.92	2.90	2.82	2.10	2.38	2.78	3.15	1.95	1.63	1.76	1.57	1.46	1.46
Romania							8.32	7.87	7.38	6.90	6.48	5.74	5.15	4.96	4.48
Slovenia										1.04	1.07	1.00	1.11	1.05	0.92
Spain											3.34	1.43	1.29	1.30	1.18
Sweden											1.47	1.44	1.46	1.34	1.33
Switzerland									0.40	0.40	0.45	0.37	0.30	0.32	0.30

4.4 Optimal topology for distribution systems

For decades, distribution systems were designed with rigid topologies. In order to determine topologies that allow better operating conditions, a number of algorithms have been developed. The best topology could be used as the normal condition to run the system and determine the location of equipment flexibility to have temporary changes in topology, best referred as reconfiguration.

The optimal topology of a distribution system is also the one where operating conditions are satisfied with the lowest system losses. The conditions include all loads operating within acceptable ranges, no overloaded elements and all loads served.

To determine the optimal topology, all the poles in overhead feeders and the double dead-end poles, or switching boxes in underground feeders, are potential open points. This allows for the identification of the best boundaries among feeders to reduce the overall losses. Once the optimal topology is achieved, switches, breakers or even re-closers can be installed in the open points selected, normally called double-dead-end terminals, to allow for system reconfiguration. This will remarkably improve SAIFI, SAIDI and CAIDI indices, as depicted above.

Several methods have been developed to determine the best topology for a distribution system. These include heuristic methods, linear programming, neural networks, expert systems, fuzzy logic, simulated annealing and genetic algorithms, among others.

The heuristic method is the most used since it can produce fast results with good accuracy. Linear programming is mainly used in planning applications to minimise capital investment.

Among the heuristic methods, the feeders reconfiguration according to Civanlar *et al.* [5] is very popular. The method considers the closing of a normally open link and the opening of another normally closed in order to transfer load from a feeder with a higher voltage drop to another with a higher voltage level so as to reduce the active losses produced in the system. The reduction in losses could be easily calculated from the load flows run results for the system configurations before and after the reconfiguration takes place. The method can be illustrated with the three-feeder distribution system shown in Figure 4.10.

The A, B and C branches represent links among the feeders each with a switch normally open. It is assumed that each branch has a section switch.

The objective functions according to the Civanlar *et al.*'s method to determine the losses change as a result of transferring loads from feeder II to feeder I of a system. This is given by the following expression:

$$\Delta P = \mathrm{Re}\left\{2\left(\sum_{i\in D} I_i\right)(E_m - E_n)^*\right\} + \left|\sum_{i\in D} I_i\right|^2 \tag{4.18}$$

where:

- D is the group of nodes disconnected from feeder II and connected to feeder I.

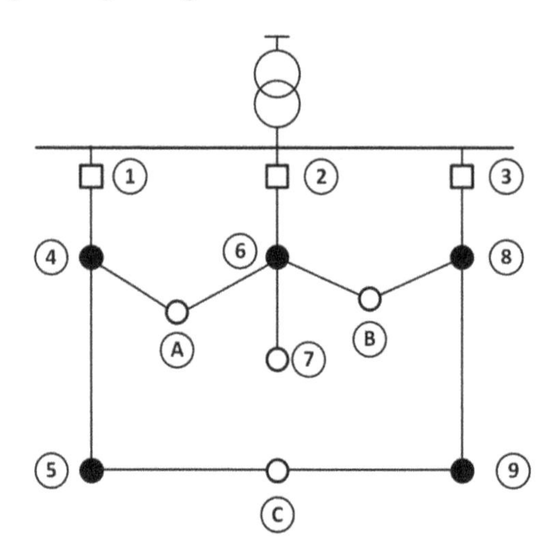

Figure 4.10 Three-feeder distribution system

- m is the node of feeder I to which the loads coming from feeder II will be connected.
- n is the node of feeder II that will be connected to node m through a link switch.
- I_i is the complex current of node i.
- R_{path} is the series resistance of the path that connects the nodes of feeders I and II through a link switch.
- E_m is the component of $E = R_{BUS}\, I_{BUS}$ corresponding to node m of feeder I.
- E_n is similar to E_m but defined for node n of feeder II.

E_m and E_n are calculated with the node currents I_i of the base case before the load transfer. Under these circumstances, the first term becomes negative if $|E_m| < |E_n|$. The effect of the capacitors in the node currents should be incorporated to have a convenient modelling. The second term of the right part is always positive. Therefore, it is possible to achieve a reduction in losses unless the first term is significantly negative.

Shirmohammadi and Hong proposed an algorithm called 'optimum flow pattern' (OFP) [6] to determine the minimum real power losses for radial networks. The OFP is computed on a purely resistive distribution network, neglecting the reactive components of the branch impedances, after closing all the connection points. Initially, the switches are closed resulting in a meshed system. Then, the switches are opened one at a time according to the OFP. The connection link to be opened in a certain loop is the one located in the branch with the lowest current flow. The method provides good results for small systems, but it is not the ideal alternative in general, since the meshed system does not represent the real operation state of the network.

The entire process, starting with the AC power flow solution of the resulting network up to the opening of the connection link with the lowest current for the optimal flow pattern, is repeated until the network becomes radial. Figure 4.11 shows the flow chart for this algorithm.

The exchange process starts by assuming a radial configuration and therefore with the distribution network operating in a radial configuration. One of the tie switches is closed, and then another switch is opened in the loop created, which restores a radial configuration. The switch pairs are chosen through heuristics and approximate formulas to create the desired change in loss reduction. The branch exchange process is stopped when no more loss reductions are possible.

Figure 4.12 taken from NEPLAN Resources [7] shows a system with four feeders. Here, the reconfiguration brings about savings on significant potential losses.

The best approach for determining the basic topology is by considering that all the poles in overhead lines, or in the connection boxes of underground feeders, can feasibly be opening points or feeder ends. This is illustrated in Figure 4.13 which has a simple system of two feeders.

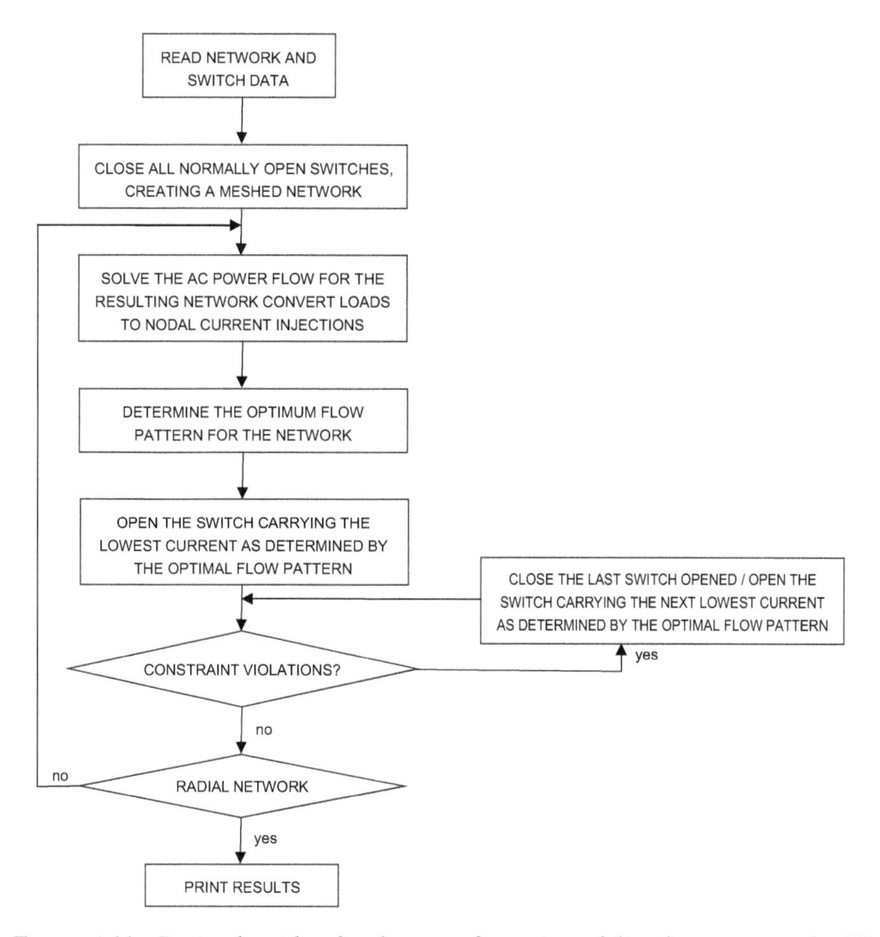

Figure 4.11 Basic algorithm for the reconfiguration of distribution networks [6]

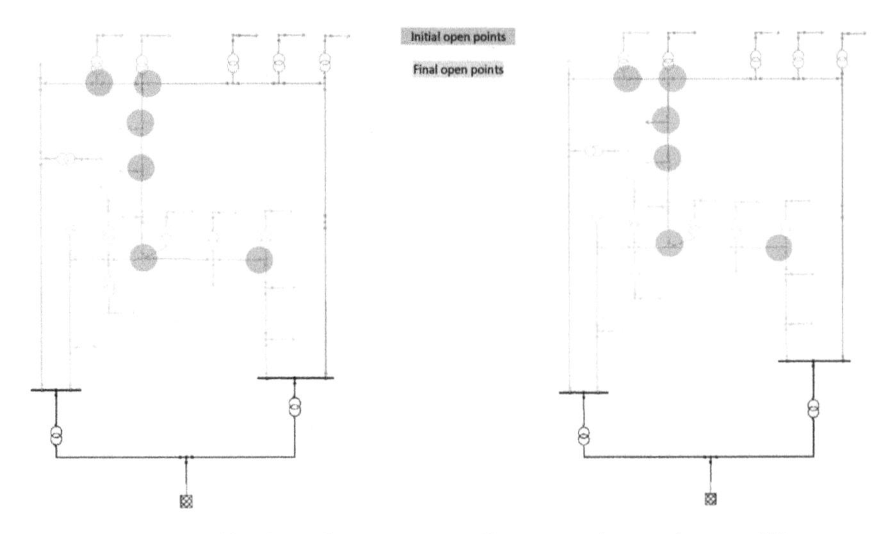

Figure 4.12 Distribution system illustrating loss reduction [7]

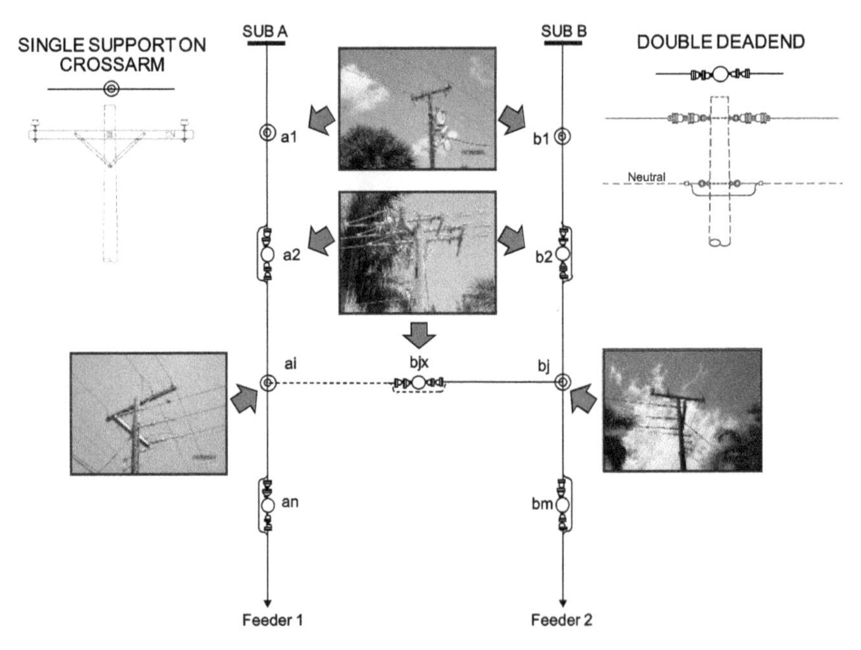

Figure 4.13 Two-feeder system showing location of potential opening points

Several specialised software packages have been developed to determine the optimal topology. The packages can have different optimisation criteria such as minimum losses, elimination of overloads and voltage control. The most common criterion is minimum losses.

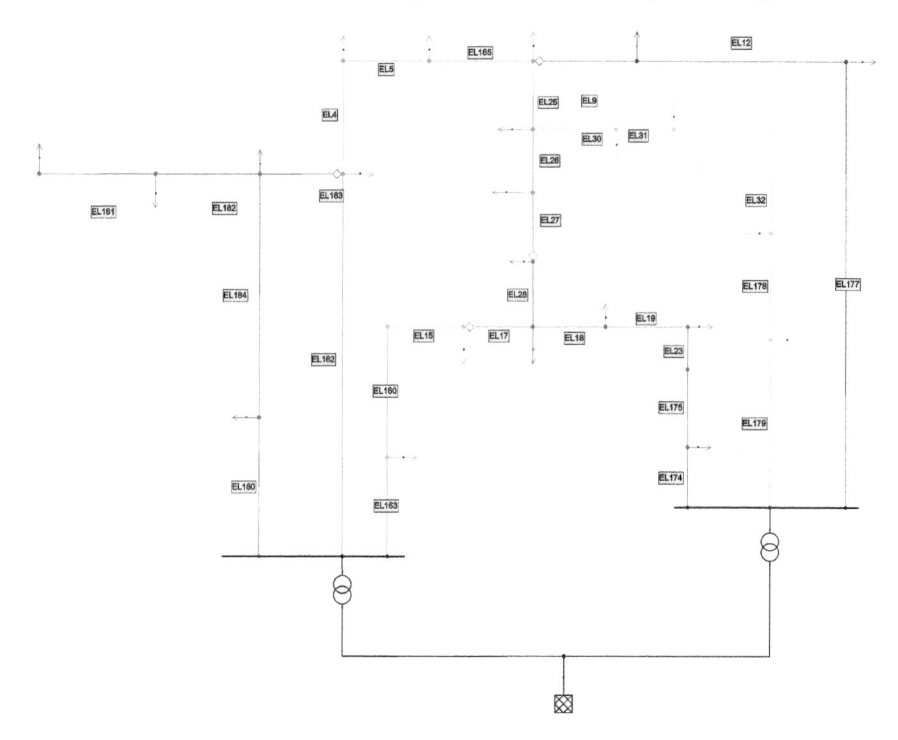

Figure 4.14 Single line diagram with the initial topology

Example 4.4

Consider the system shown in Figure 4.14. Find the new topology for this system considering loss reduction for a radial configuration. List the lines whose ends should change the status.

With a proper software program that has a module on optimal topology, it can be found that the losses are reduced significantly when the system topology is changed. Table 4.4 shows the losses for the initial topology and Table 4.5 the losses after the changes are implemented.

Using an optimal separation points (switching Optimisation) module, it can be found the changes on the topology, as indicated in Table 4.6. Figure 4.15 illustrates the configuration change based on the feeder colours.

4.5 Location of switches controlled remotely

The impact of defining an optimal topology has been illustrated, for example, to reduce losses. However, this is just merely for defining a fixed topology that remains working in a steady-state operation. Additional elements like switches in the electric grid are required to allow operators to define new topologies when a fault occurs that responds to this new operational condition in a fast and reliable

Table 4.4 Load flow results for the initial topology

Feeder	P Loss MW	Q Loss MVAr	P Imp MW	Q Imp MVAr	P Load MW	Q Load MVAr
	2.325	2.794	24.893	20.025	22.568	17.231
Feeder 1	0.698	1.667	8.36	6.544	7.662	4.877
Feeder 2	0.080	0.013	1.88	1.533	1.8	1.52
Feeder 3	0.081	0.128	2.654	3.767	2.573	3.639
Feeder 4	1.429	0.95	8.039	5.759	6.61	4.809
Feeder 5	0.003	0.02	0.823	0.379	0.82	0.359
Feeder 6	0.034	0.016	3.137	2.043	3.103	2.027

Table 4.5 Load flow results for the final topology

Feeder	P Loss MW	Q Loss MVAr	P Imp MW	Q Imp MVAr	P Load MW	Q Load MVAr
	0.759	1.889	23.327	19.12	22.568	17.231
Feeder 1	0.039	0.017	2.251	1.79	2.212	1.773
Feeder 2	0.299	−0.153	7.659	5.948	7.36	6.101
Feeder 3	0.081	0.065	2.654	3.704	2.573	3.639
Feeder 4	0.012	1.802	1.062	2.03	1.05	0.228
Feeder 5	0.071	0.025	3.191	1.962	3.12	1.937
Feeder 6	0.257	0.133	6.51	3.686	6.253	3.553

Table 4.6 Suggested changes on the topology

Name	Switch 1 initial	Switch 2 initial	Switch 1 final	Switch 2 final
EL5	Connected	Connected	Connected	Disconnected
EL9	Disconnected	Connected	Connected	Connected
EL17	Disconnected	Connected	Connected	Connected
EL19	Connected	Connected	Connected	Disconnected
EL25	Connected	Connected	Disconnected	Connected
EL27	Connected	Disconnected	Disconnected	Connected
EL30	Connected	Disconnected	Connected	Connected

way. Switches can be breakers, reclosers or sectionalisers and should have the means for remote operation to guarantee a fast reconfiguration when required. Their location on the grid should be carefully studied.

Improvements achieved by manufacturers in recent years have prompted the use of feeder reconfiguration to modify the operating condition of distribution

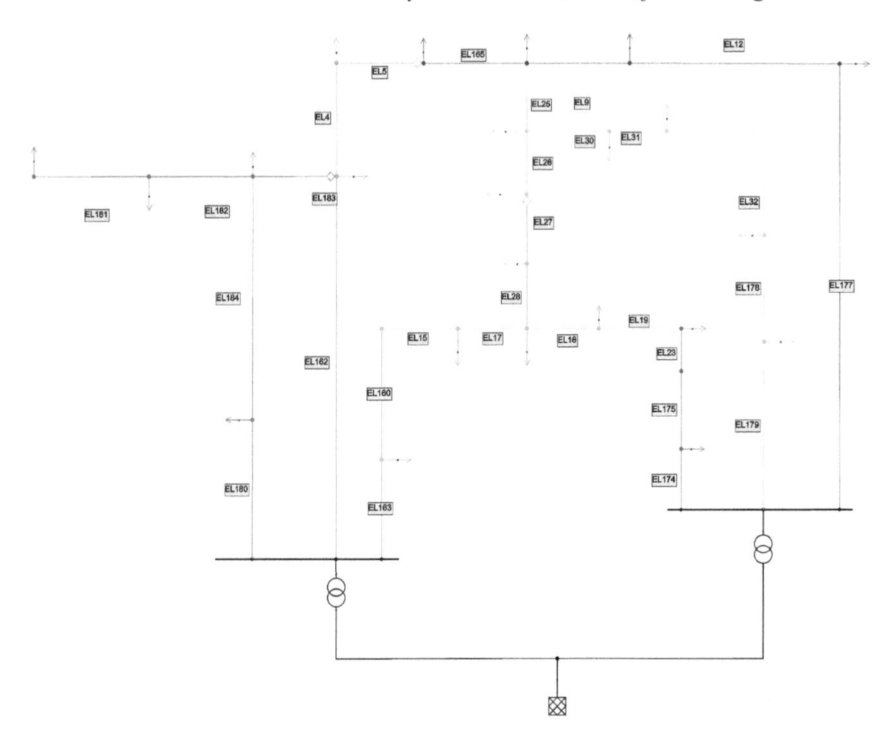

Figure 4.15 Single line diagram with the final topology

networks. These improvements include remotely controlled switches or breakers for pole installation, numerical protection and appropriate communication systems.

Under normal operating conditions, feeder reconfiguration aims for a more efficient operating condition of the network. Under faulty conditions, feeder reconfiguration aims to restore the service to the maximum number of users in the shortest time. Prior to determining the location of switches to allow changes in configuration, it is highly recommended to find the best topology for a distribution system.

The placement of switches is carried out in such a way that the reliability and flexibility criteria are considered. The reliability of distribution networks can be greatly improved by adding switches along the feeders. The benefit of adding the switches can be measured by examining the improvement in operation performance. This is explained in the following sections.

4.5.1 Considerations to increase reliability

The basic concepts of reliability were covered in the first part of this chapter. These concepts will now be used to show how the reliability of distribution networks can be greatly improved by adding switches along the feeders. Faults downstream a load causes the tripping of the corresponding protection.

If breakers with the corresponding protective equipment are installed on feeder sections downstream important loads, faults on those sections will not affect the loads upstream, thus improving the reliability indices. The benefit of adding the switches can be measured by examining the improvement in reliability performance.

4.5.2 Considerations to increase flexibility

Distribution networks should not be considered to have rigid configurations anymore since remotely controlled switches are available and allow better configuration. It is convenient that the feeders are sectionalised into equally loaded portions as far as it is possible.

Figure 4.16 evaluates the flexibility of each of the four double dead-end links that exist between the two feeders to locate the NO tie switch (or breaker). The purpose is to pick the one that offers the highest flexibility to help restore the electrical service, for faults in Sections 1, 2 and 3 of the first feeder.

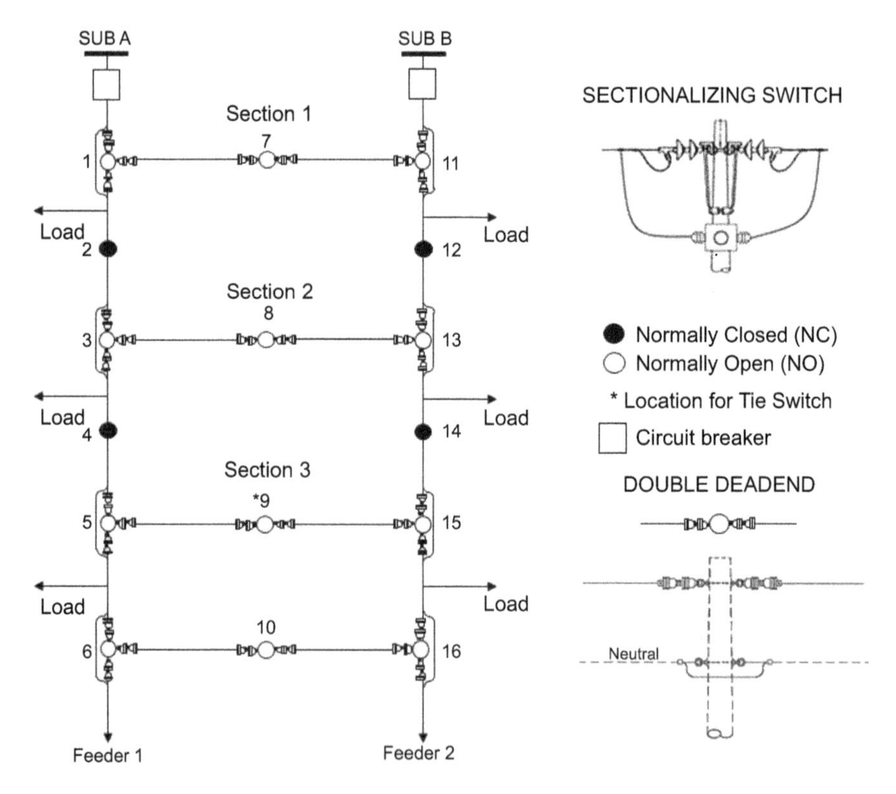

Figure 4.16 Two feeder system showing location of switches controlled remotely

A switch located in the first section of the feeders, such as number 7, would not offer the possibility of any load restoration for faults in any of the three sections of the feeder and of course, that location is discarded. The link located in the second section, number 8, would offer the possibility to restore only the first section since for faults in sections two or three, the NO switch would not be effective at all.

The links located in the third section, identified by numbers 9 and 10, allow the restoration of two-thirds of the feeder load. Therefore, the switches located in this section have the highest flexibility to carry out remotely controlled transfers. The link closest to the section switches offers a lower current path, and therefore, lower losses and better voltage regulation which could be easily validated with load flow runs. That is why the location of link 9 is the chosen one to place the NO switch.

Figure 4.17 presents some pictures for switches in distribution systems.

The system shown in Figure 4.18 has three substations with one feeder each. The section switches placed after the breaker in the three feeders are all remotely controlled and numbered from 1 to 14. For the reference system (Figure 4.18a), the best doubled-dead end (DD) location to be replaced by a switch is selected (Figure 4.18b). In this case, the selection is done considering the points in which more sections of each feeder can be transferred in case that an event occurs in the system and is required.

Table 4.7 summarises the operation of the switches for the different faults indicated in the diagram above. For some fault locations, especially at the end of the feeders, it is possible that the reconfiguration of tie switches may not be an option. In this case, the operation of the section switches is the only option to implement the upstream restoration. After the system reconfiguration takes place, the time–current characteristic settings have to be modified by recalculating the time dial and the

Figure 4.17 Location of switches in distribution systems

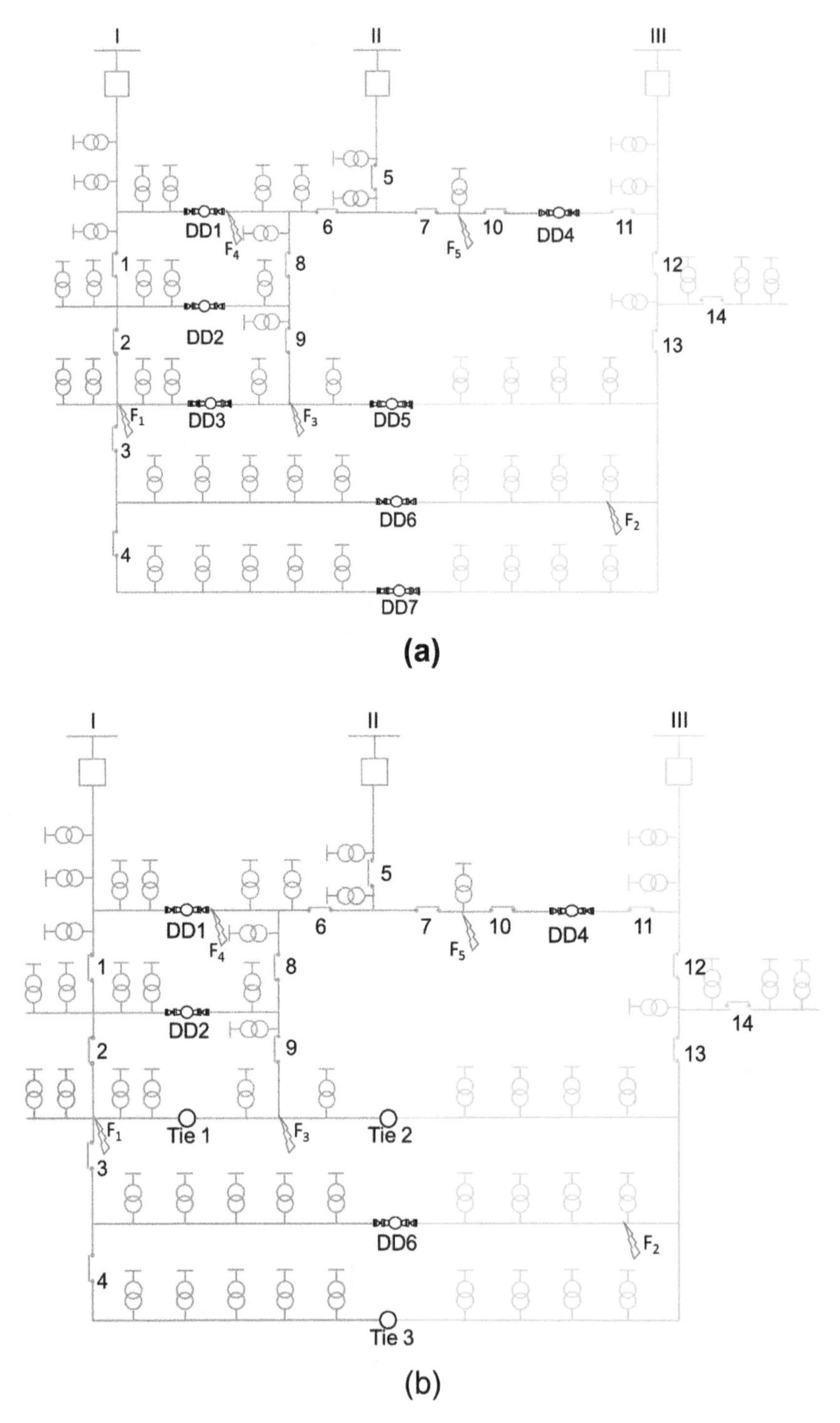

Figure 4.18 Three feeder system illustrating switch location for service restoration with no tie switches (a) and tie switches (b)

Table 4.7 Reconfiguration options for previous system

Fault	Open	Close
F1	2 and 3	T3
F2	13	—
F3	9	—
F4	6 and 8	T1 or T2
F5	7	—

pickup values according to the new topology. In order to implement these changes automatically, it is essential to have the availability of multiple setting groups.

4.5.3 System configuration

Another requirement to achieve the best system configuration is to maximise the functionality of all the equipment involved. It is crucial to bear in mind that the proper operation of switches might be affected by the following aspects:

- Customer density: customers are not evenly distributed along feeders and its density might vary among rural and urban areas, being the latest the one with higher customer density. Hence, more switches should be installed to provide more flexibility and operability.
- Fault distribution: the feeder configuration influences the fault distribution. Feeders along treed areas are more subjected to outages than underground systems; also environmental conditions must be considered.
- Source availability: for the system's reconfiguration, downstream sources must be present for load transferring and also for maintenance conditions.

When selecting installation points, positions at the far ends of the feeder are not to be included and only locations with three-phase systems are to be taken into account. With these principles in mind, optimal switch placement, therefore, optimal topology for the system's reconfiguration and restoration, considers the following steps as properly described by Robert Uluski in [8]:

1. For each possible switch location, determine the number of customers connected 'upstream' (C_{upst}).
2. Estimate the average of upstream faults (F_{upst}) and determine the upstream factor: $UF = C_{upst} \cdot F_{upst}$
3. Apply the same procedure for the downstream section and find the downstream factor: $DF = C_{dwnst} \cdot F_{dwnst}$
4. The switch placement score (SPS) for that specific node is given by:

$$\text{SPS} = UF + DF$$

5. Repeat the process for all candidate nodes.
6. The one with lower SPS is then selected for switch placements as it provides a better reliability improvement.

Example 4.5

Considering the following distribution network, the optimal switch location is addressed by the approach described. The system consists of nine candidate nodes and each line is labelled with the number of customers (#Cust.) and number of faults (#Faults). Implement the procedure for nodes N3 and N7 considering the system and information given in Figure 4.19.

For candidate node N3:

1. Number of customers connected 'upstream':

$$C_{upst} = 42$$

2. Average of upstream faults and the determine the upstream factor:

$$F_{upst} = 6$$

$$UF = 42 \cdot 6 = 252$$

3. Apply the same procedure for the downstream section and find the downstream factor:

$$C_{dwnst} = 48$$

$$F_{dwnst} = 8$$

$$DF = 48 \cdot 8 = 384$$

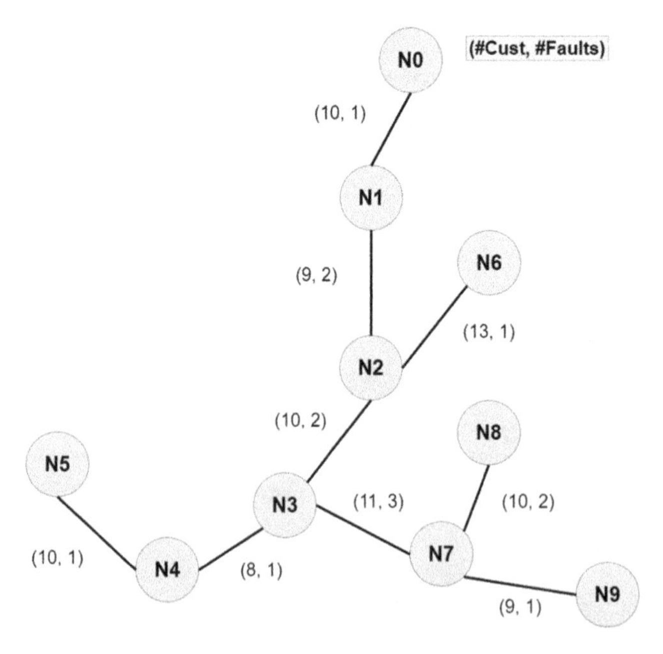

Figure 4.19 Distribution system for Example 4.5

4. The SPS for node N3 is given by:

$$SPS_{N3} = 636$$

For candidate node N7:

1. Number of customers connected 'upstream':

$$C_{upst} = 71$$

2. Average of upstream faults and the determine the upstream factor:

$$F_{upst} = 11$$

$$UF = 71 \cdot 11 = 781$$

3. Apply the same procedure for the downstream section and find the downstream factor:

$$C_{dwnst} = 19$$

$$F_{dwnst} = 3$$

$$DF = 19 \cdot 3 = 57$$

4. The SPS for node N7 is given by:

$$SPS_{N7} = 781 + 57 = 838$$

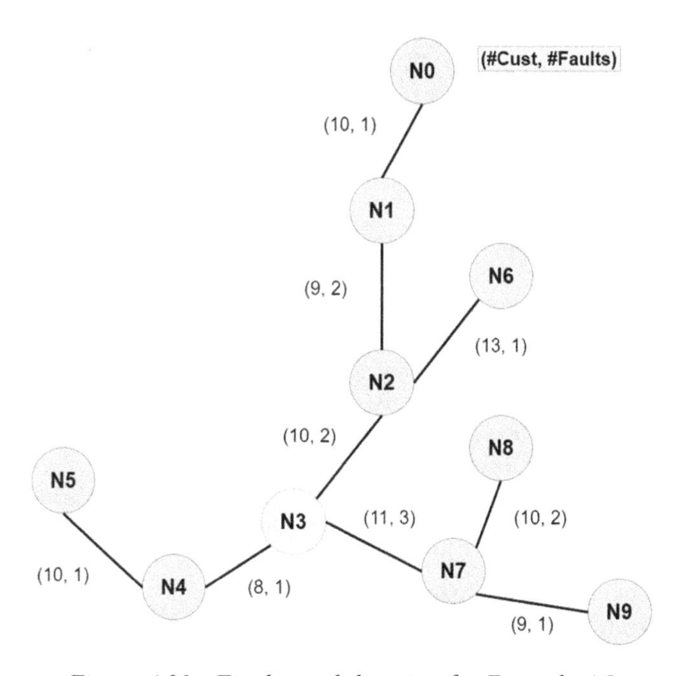

Figure 4.20 Final switch location for Example 4.5

Candidate node N3 has the lower SPS, then it is selected for switch placement as it provides a better reliability improvement than candidate node N7 (Figure 4.20).

This is an example of optimal switch placement considering trial and error strategies, which in practice has been enhanced by the use of more robust computational tools.

4.6 Feeder reconfiguration for improving operating conditions

Feeder reconfiguration consists of the modification of the topology of a network through the closing of a switch that links two feeders and the opening of another switch so as to maintain the radial condition of the feeders. The reconfiguration is carried out for better operation of the network and specifically to reduce the losses due to the Joule effect. Distribution systems should be operated at minimum cost subject to a number of constraints: all loads are served, overcurrent protective devices are coordinated, voltage drops are within limits, radial configuration is maintained and lines, transformers and other equipment operate within current capacity.

4.7 Feeder reconfiguration for service restoration

Another important function in distribution automation (DA) is service restoration in case of a fault in the primary feeder. The main goal of automatic service restoration is the execution of a series of operations of the tie (NO) and section (NC) switches, aimed at restoring the power supply to the maximum number of areas that have been impaired after a fault in a primary feeder or substation. Usually, it would reconfigure the network by transferring loads from the healthy portions of faulty feeders to neighbouring feeders that are operating normally.

Example 4.6

Consider the system that is shown in Figure 4.21 with the switches located. For each one of the faults F1, F2, F3, F4 and F5, Table 4.8 indicates the operating sequence of the tie and section switches to restore service to the highest possible number of customers.

There are multiple possibilities for reconfiguring the network for service restoration, but the group solution is finite; it is defined by the combination of binary states of the section switches and tie switches. This group grows exponentially with an increase in equipment. It can become a problem of scale. The solution consists in choosing a combination of states for the equipment in the areas not affected which satisfies some functional objective. A system reconfiguration changes the topology, line flows and short circuit values. Thus, different solutions may be feasible for the same fault.

In these circumstances, equipment, voltage profiles and equipment loading must be considered. Relay pick-up values and time dials must be rechecked to

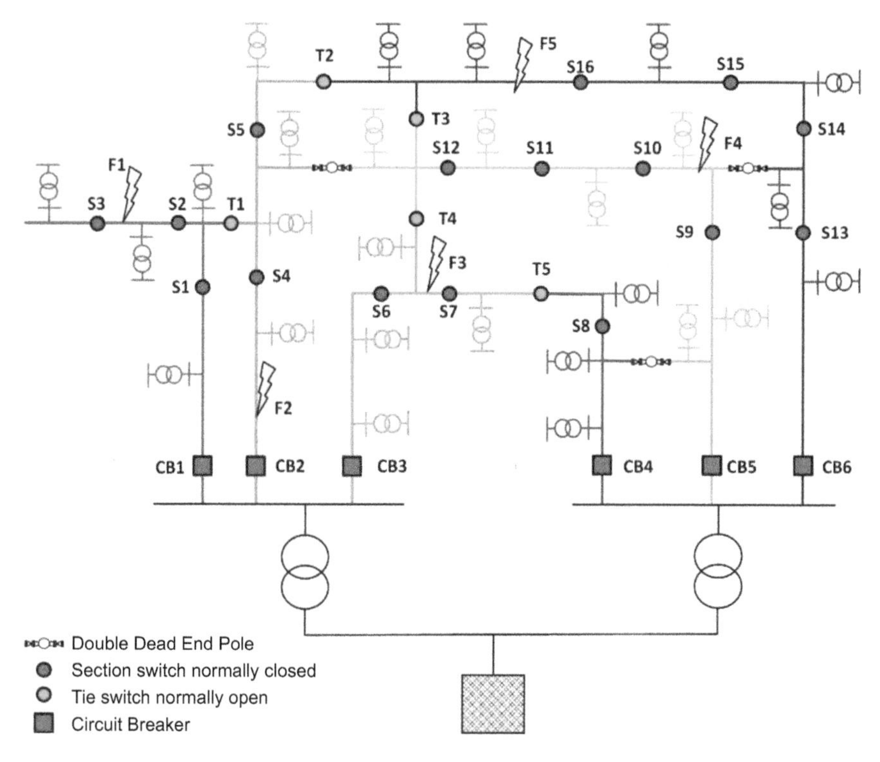

Figure 4.21 Distribution system with different fault locations

Table 4.8 Operating sequence of the tie and section switches

Fault	Open	Close	Total affected customers
F_1	S2 and S3	–	2
F_2	CB2 and S4	T1 or T2	1
F_3	S6 and S7	T5	1
F_4	S9 and S10	T3 or T4	1
F_5	S16	–	2

avoid nuisance tripping or unwanted pick-ups. This would require the use of different group settings.

4.7.1 Fault location, isolation and service restoration (FLISR)

The steps to restore service when a fault happens can be summarised as follows:

1. The corresponding relay operates and trips the breaker. If reclosing units operate and the fault remains, the feeder is open.

2. The fault is located and the associated section switches open to isolate it.
3. The feeder is re-energised up to the location of the first section switch that was open – upstream restoration.
4. The healthy sections are transferred to one or more neighbouring feeders by operating NO switches – downstream restoration.
5. The faulty section is repaired by the crew and the system is taken back to normal configuration.

The first four steps should be completed within a minute to avoid affecting SAIFI and SAIDI indices.

These steps are illustrated by using the system shown in Figure 4.22 which corresponds to the normal configuration of a distribution system with four substations, each with one feeder [9]. The system has 14 loading sections labelled Z1 to Z14 and 13 switches in total labelled S1 to S13. Switches 4, 9 and 12 are NO and the rest are NC. Figures 4.22–4.26 illustrate the first four steps already mentioned, following a fault at section Z2 in order to restore the system as far as possible. The fifth step of the process would be to take the system back to normal.

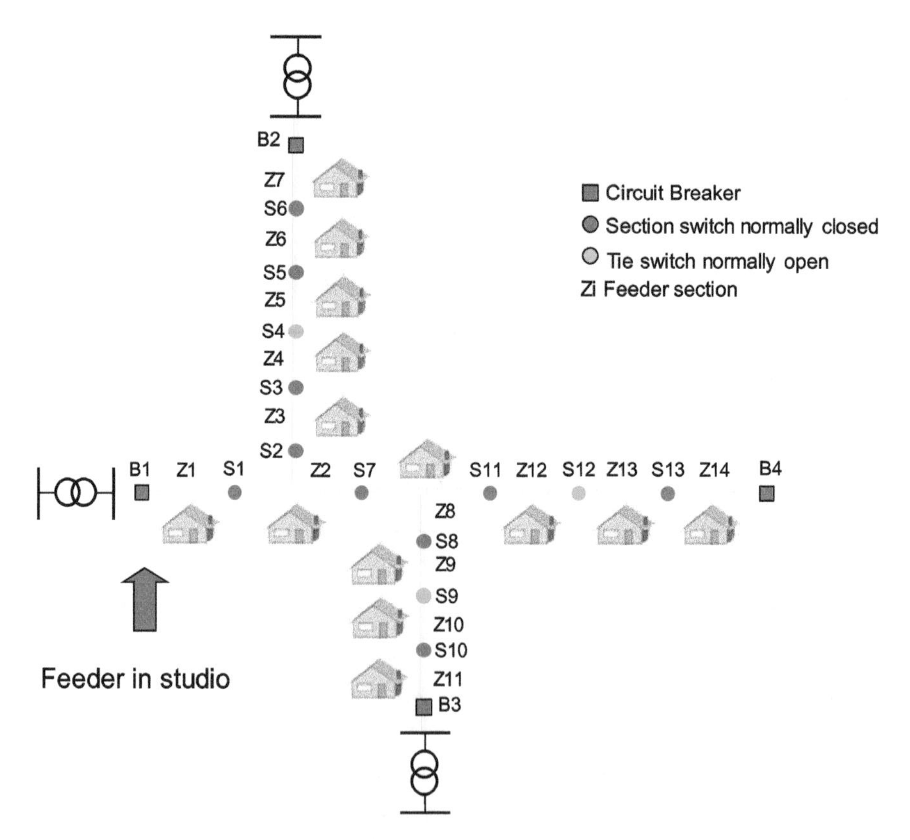

Figure 4.22 Normal configuration

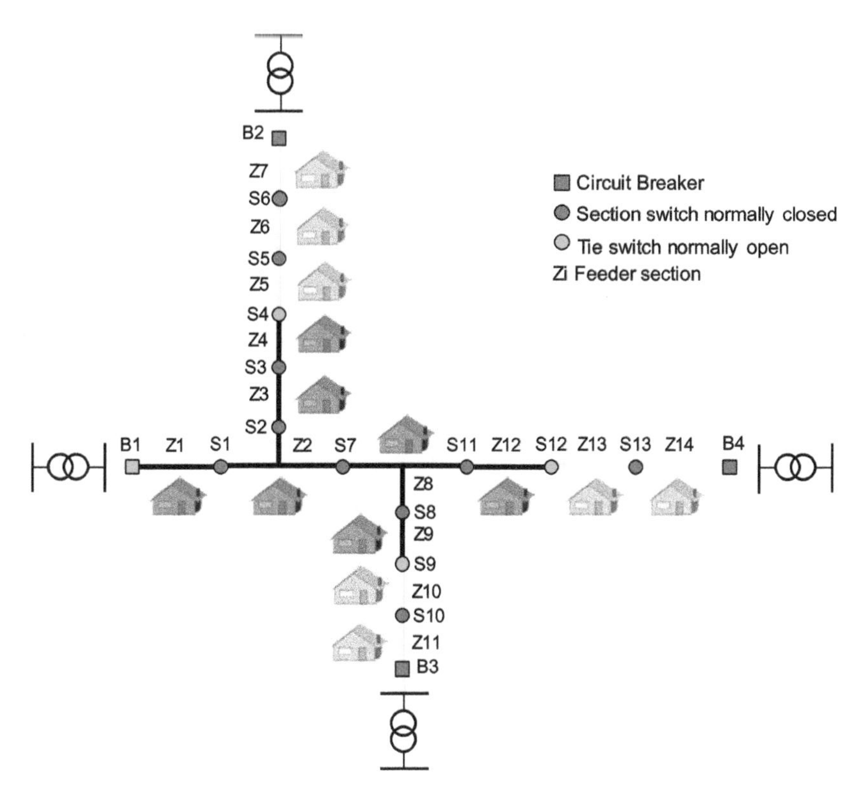

Figure 4.23 Fault clearing by relay at B1 – step 1

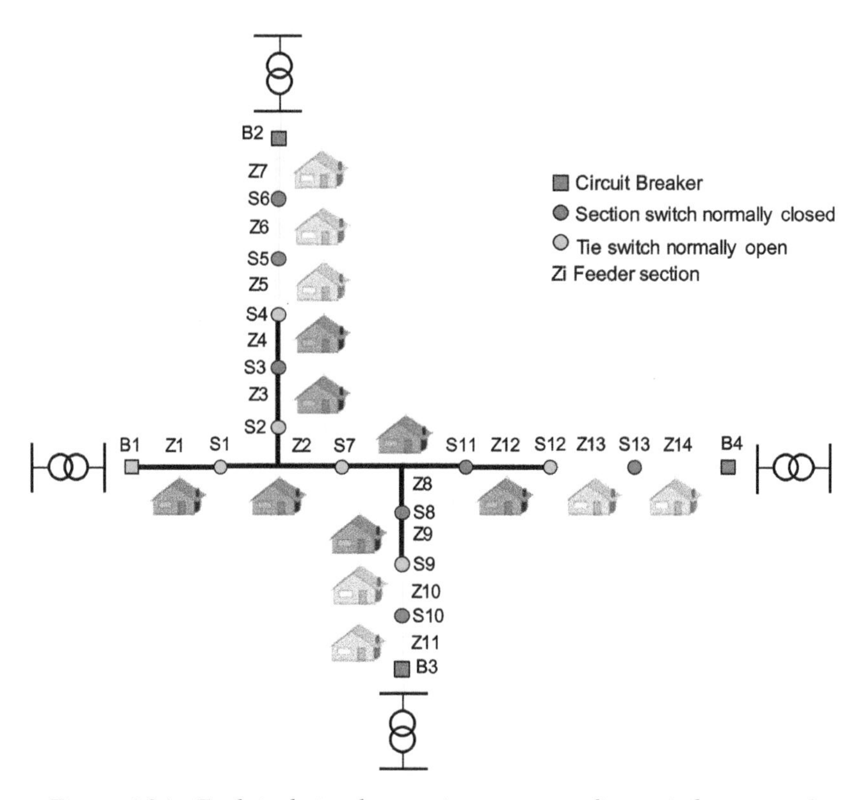

Figure 4.24 Fault isolation by opening corresponding switches – step 2

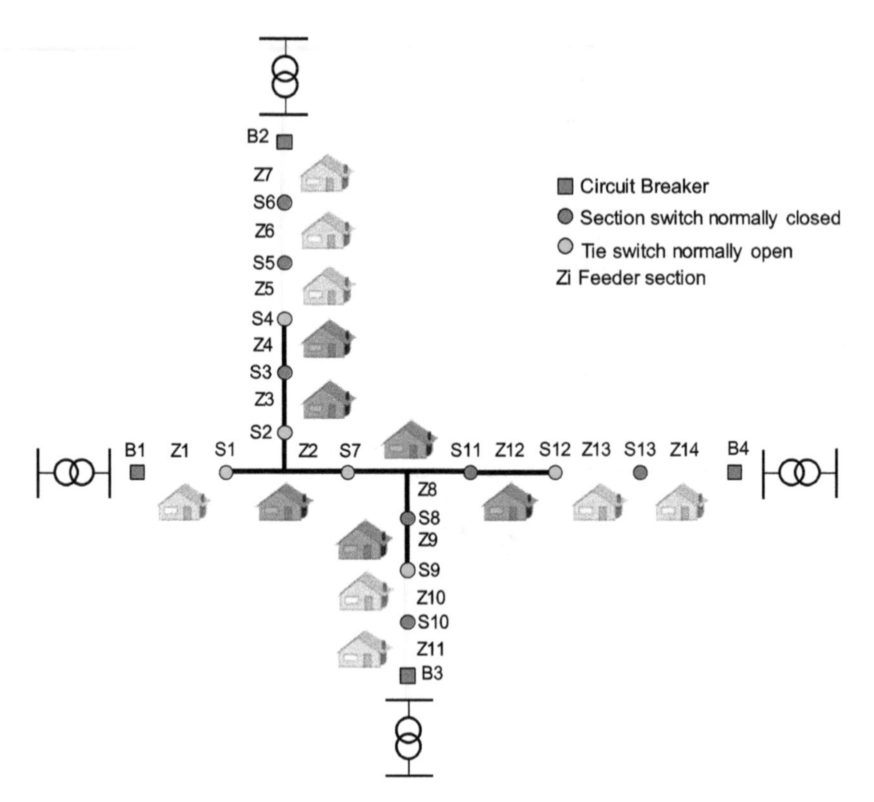

Figure 4.25 Reclosing B1 – step 3 (upstream restoration)

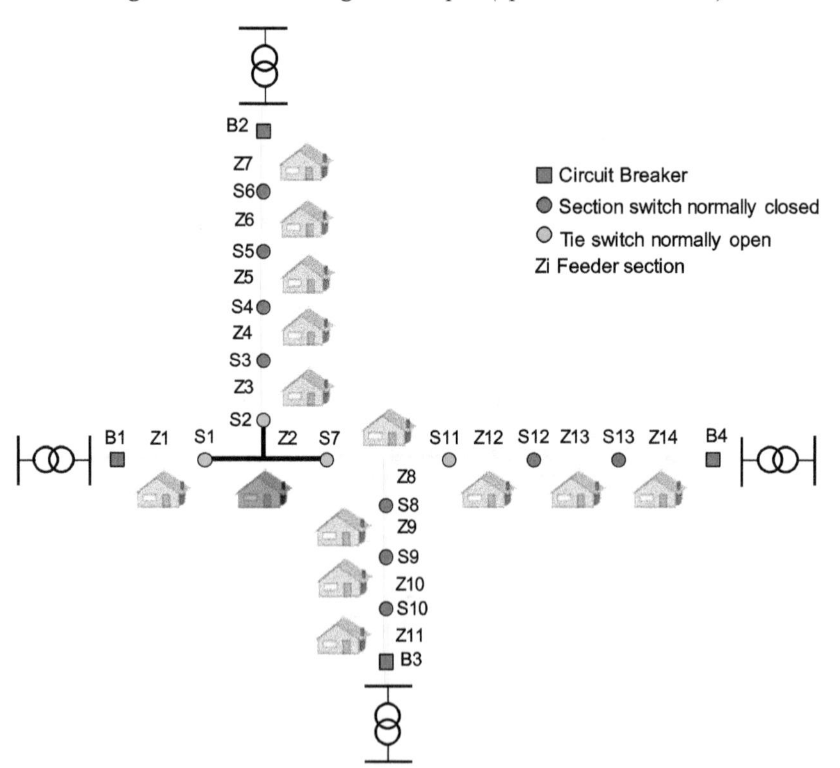

Figure 4.26 Reconfiguration by operating switches NO and NC – step 4 (downstream restoration)

4.7.2 Manual restoration versus FLISR

When a permanent fault occurs, customers on healthy sections of the feeder may experience a lengthy outage if FLISR is not available. Figure 4.27 is an illustration of the time taken at a typical distribution system to fix a fault and restore service to the users associated with these healthy sections when manual switches have been deployed without any automation. In this case, the maintenance crew must get the faulty section and perform the required switching to allow both the upstream and downstream restoration as practicable.

Figure 4.28 illustrates the improvements in quality indices as SAIDI, SAIFI and MAIFI with and without FLISR implementation, which has also monetary implications as significant number of customers are restored quickly compared to no FLISR applications [8,10–14].

4.7.3 Restrictions on restoration

When carrying out the restoration, the operations that are executed should allow the system to satisfy these restrictions:

- The current-carrying capacity of the transformers and lines should be within specified limits.

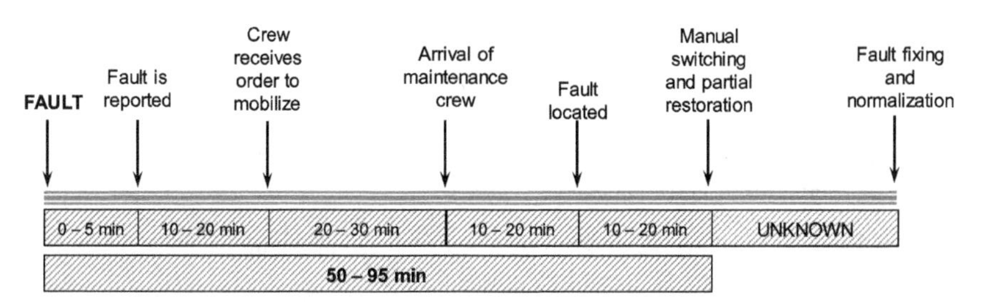

Figure 4.27 Time to fix faults for typical systems

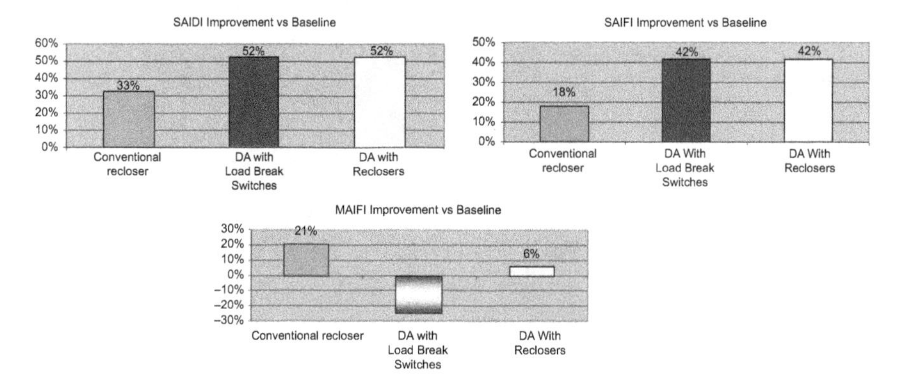

Figure 4.28 FLISR benefits in quality indices [8]

- The voltage drop should stay inside an established margin.
- The system should continue being radial.
- The number of operations of the equipment stays within limits.
- Important customers have priority.
- The system must be balanced as much as possible.
- The coordination of the protection must be maintained.

4.7.4 FLISR central intelligence

Central intelligence is still the most common method of automation to control a power system that includes substations [15–32]. Central intelligence, of course, relies on powerful and fast SCADA systems and therefore requires a very good communication infrastructure. In the case of distribution systems, central intelligence should have proper control not only of the substation elements but also of all the devices deployed on the feeders. These include breakers, switches, voltage regulators and capacitors. All of them should be monitored and controlled directly from the control system. This method is very efficient, but it requires a lot of communication.

The central intelligence in FLISR allows for the:

- Analysis of faults based upon input from real-time SCADA (fault detectors and fault currents) and crew reports.
- Isolation of the fault based upon current state of distribution network.
- Generation of recommended switching for isolation of a nominated fault and restoration of unaffected customers
- Proposal of multiple restoration plans based upon network status, topology, equipment limits and desired objectives

An illustration of feeder reconfiguration using central intelligence taken from an ALSTOM case study is given in Figures 4.29–4.34.

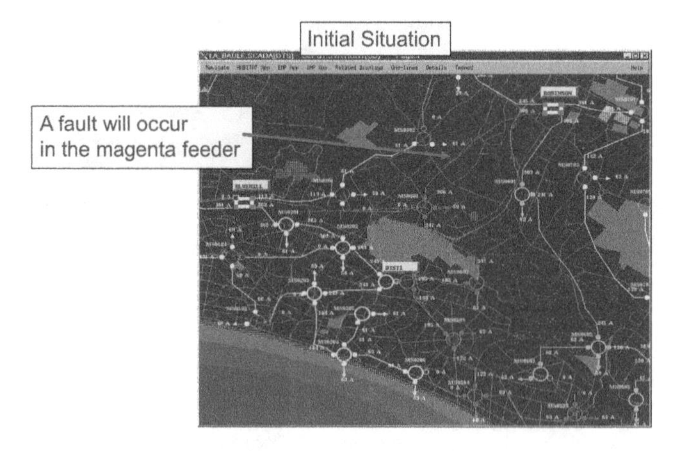

Figure 4.29 Detection of a fault in a FLISR with central intelligence (reproduced by permission of ALSTOM)

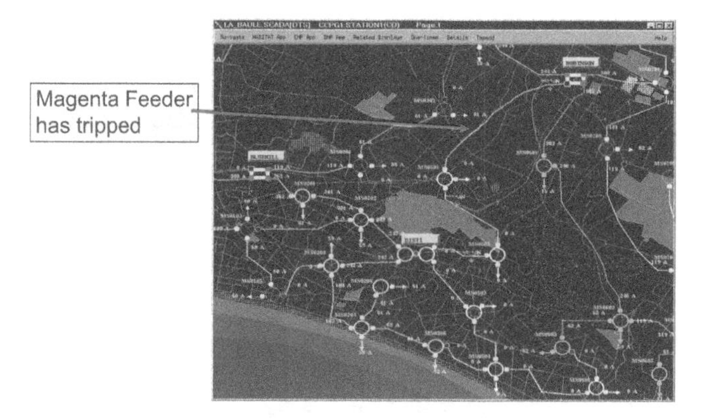

Figure 4.30 Fault isolation by opening corresponding feeder (reproduced by permission of ALSTOM)

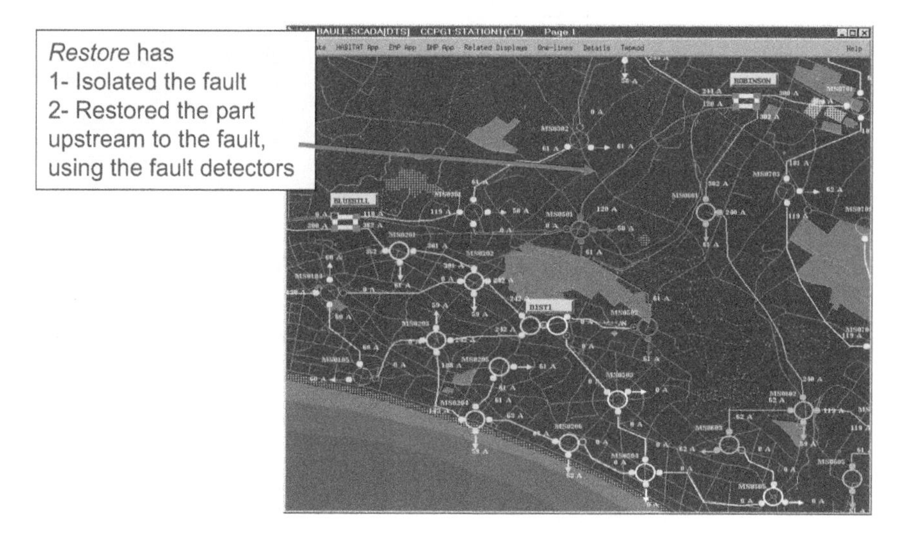

Figure 4.31 Upstream restoration (reproduced by permission of ALSTOM)

4.7.5 FLISR distributed intelligence

Distributed intelligence automatically isolates faulted distribution segments using a 'team' of peer devices [15–32]. It requires proper communication and protective relay coordination. The objective is to reconfigure adjacent feeders to restore power to customers beyond the fault (to minimise the number of outages, their size and duration).

Some software applications have been developed to find solutions using FLISR distributed intelligence. Figure 4.35 shows an application to a faulted segment. Figure 4.35(a) shows the normal configuration of the groups in the

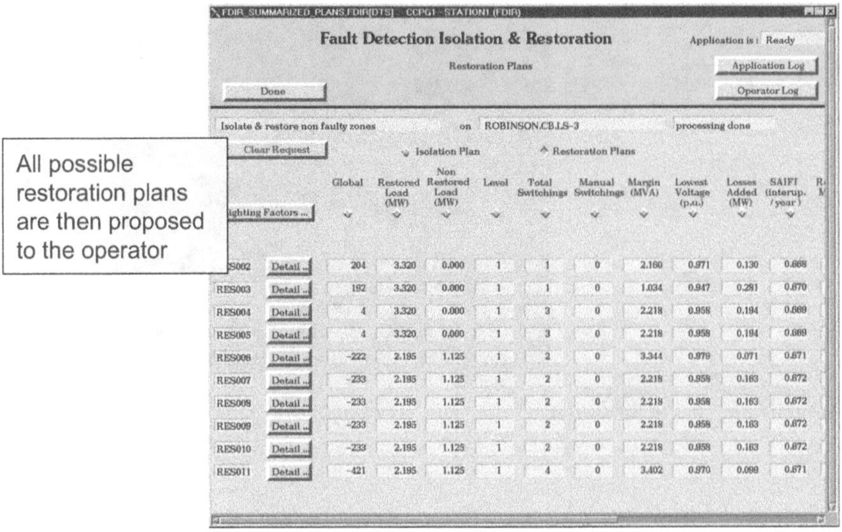

All possible restoration plans are then proposed to the operator

FDIR_SUMMARIZED_PLANS.FDIR[DTS] CCPG1 - STATION1 (FDIR)

Fault Detection Isolation & Restoration

Application is : Ready

Restoration Plans

Application Log

Done

Operator Log

Isolate & restore non faulty zones on ROBINSON.CB.LS-3 processing done

Clear Request

Isolation Plan

Restoration Plans

ighting Factors ...

		Global	Restored Load (MW)	Non Restored Load (MW)	Level	Total Switchings	Manual Switchings	Margin (MVA)	Lowest Voltage (p.u.)	Losses Added (MW)	SAIFI (interup. / year)
S002	Detail ...	204	3.320	0.000	1	1	0	2.160	0.971	0.130	0.668
RES003	Detail ...	192	3.320	0.000	1	1	0	1.034	0.947	0.281	0.670
RES004	Detail ...	4	3.320	0.000	1	3	0	2.218	0.958	0.194	0.669
RES005	Detail ...	4	3.320	0.000	1	3	0	2.218	0.958	0.194	0.669
RES006	Detail ...	-222	2.195	1.125	1	2	0	3.344	0.979	0.071	0.671
RES007	Detail ...	-233	2.195	1.125	1	2	0	2.218	0.958	0.163	0.672
RES008	Detail ...	-233	2.195	1.125	1	2	0	2.218	0.958	0.163	0.672
RES009	Detail ...	-233	2.195	1.125	1	2	0	2.218	0.958	0.163	0.672
RES010	Detail ...	-233	2.195	1.125	1	2	0	2.218	0.958	0.163	0.672
RES011	Detail ...	-421	2.195	1.125	1	4	0	3.402	0.970	0.099	0.671

Figure 4.32 Possible downstream restoration plan with FLISR (reproduced by permission of ALSTOM)

One restoration plan is selected

FDIR_DETAILED_PLAN.FDIR[DTS] CCPG1 - STATION1 (VIEW_133)

Fault Detection Isolation & Restoration

Application is : Ready

All Plans

Detailed Restoration Plan

Application Log

Done

Operator Log

Isolate & restore non faulty zones on ROBINSON.CB.LS-3 processing done

Plan RES004 Select

Automatic

Step by Step

WO Copy Plan

Check List	Global	Restored Load (MW)	Non-Restored Load (MW)	Level	Total Switchings	Manual Switchings	Margin (MVA)	Losses (MW)	Lowest Voltage (pu)	Lowest V at l
Unknown	4	3.320	0.000	1	3	0	2.218	0.184	0.958	LD MS0503_L

External Area	Substn	Device type	Device	Point	Current position	Action state	Action		Restored Load (MW)	Ma (MVA
	MS0504	CB	CB-13	STTS	is closed	Ready	OPEN	Display		
	MS0603	CB	CB-13	STTS	is open	Ready	CLOSE	Display	1.12	4.4
	MS0206	CB	CB-13	STTS	is open	Ready	CLOSE	Display	2.20	2.2

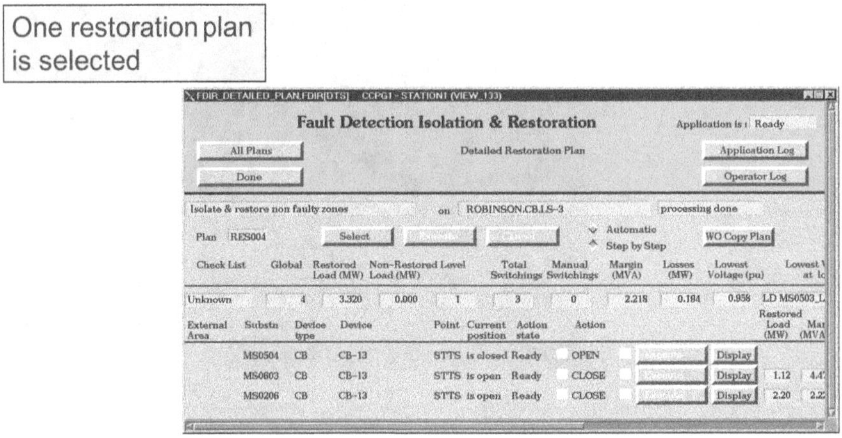

Figure 4.33 Selection of the best option for restoration (reproduced by permission of ALSTOM)

distribution system. Figure 4.35(b) presents a fault in group 6 between the switches F, G and J. All switches in the affected group experience an overcurrent followed by a loss of voltage. Therefore, switch J opens to isolate the fault. Figure 4.35(c) shows how group 6 detects the fault and initiates the opening of switches F, G and I based on voltage loss. Finally, Figure 4.35(d) shows how the service is restored to all the non-faulted segments in a short period of time: group 7 closes switch K; groups 4 and 5 close switch E; and group 2 and 3 close switch C.

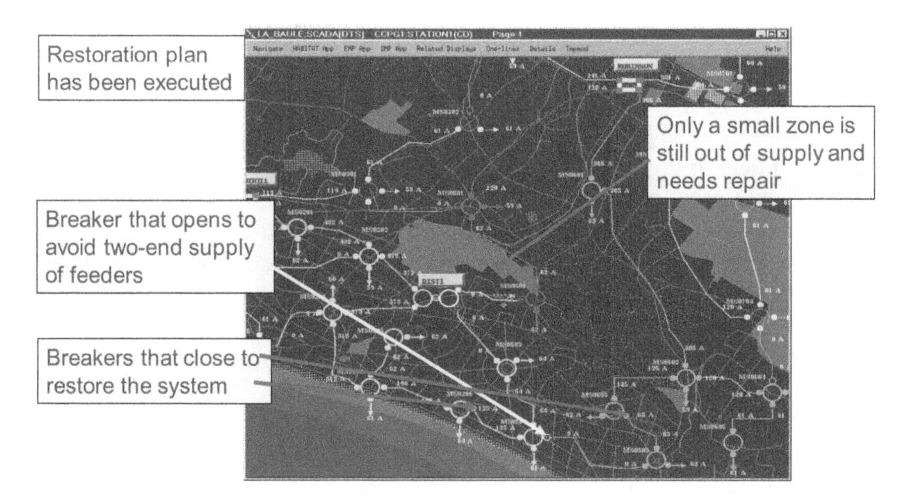

Figure 4.34 Downstream restoration of healthy distributed system section (reproduced by permission of ALSTOM)

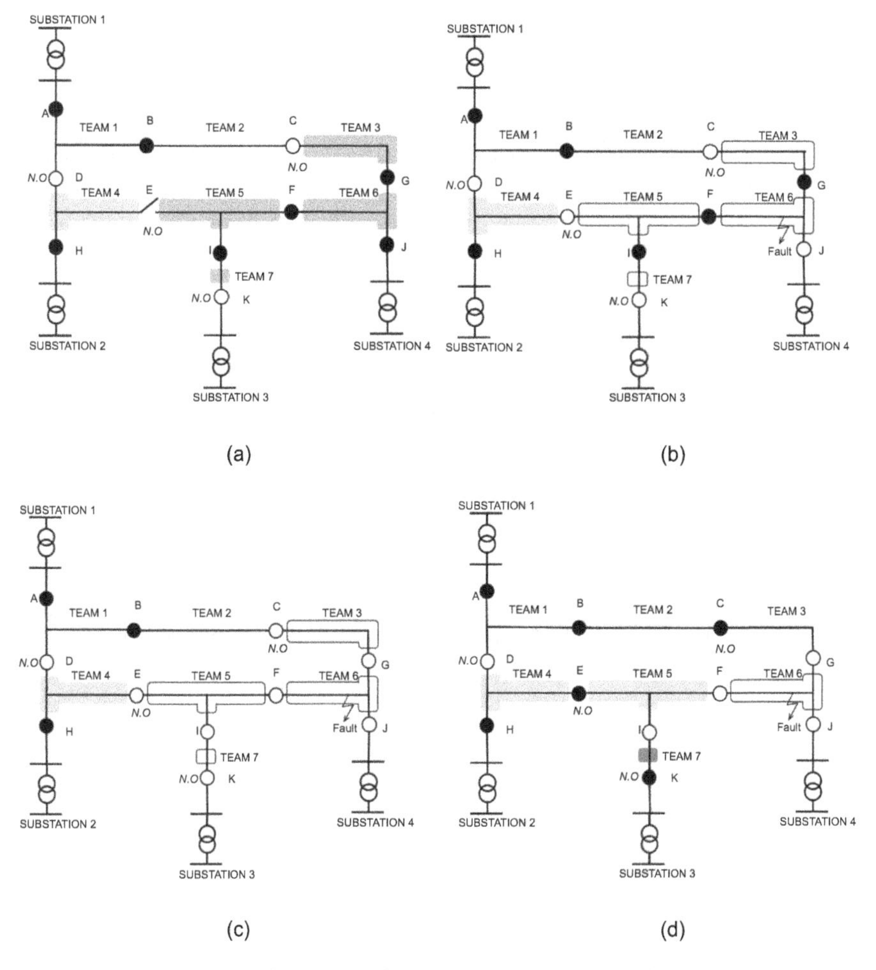

Figure 4.35 Example of a FLISR distributed intelligence solution: (a) normal configuration, (b) fault in group 6 between the switches F, G and J, (c) fault detection and switches opening, and (d) restored service

It is worth noting that other more advanced algorithms can locate the fault more accurately. If this were the case, switch I would not need to have been opened at step (c) since the opening of switches F, G and J isolate the fault.

Distributed intelligence is growing and its application is very popular for systems with long feeders such as those in the United States. Distributed intelligence avoids congesting the control centres since the communication is among elements associated only within groups. However, this method has the disadvantage of not considering the whole system. This means that the optimal topology for the whole system might not be found.

4.7.6 FLISR local intelligence

Local intelligence has been used for many years even before SCADA systems were developed [15–32]. Local intelligence is extremely simple to use, very reliable and does not require a special communication system, although in some cases, a good communication system helps to obtain better results. Local intelligence performs actions like opening or closing devices in order to satisfy previously established conditions.

To illustrate the concept of local intelligence, a system with two feeders is presented in Figure 4.36. Each feeder has two switches. The first feeder shown to the left is fed from substation A and the second feeder to the right is fed from substation B. Figure 4.37 shows that a fault between switches R03 and R04 has occurred. Switch R04 operates first. Since the fault remains, additional operations are attempted until the switch locks out. Then, the other switch operates in

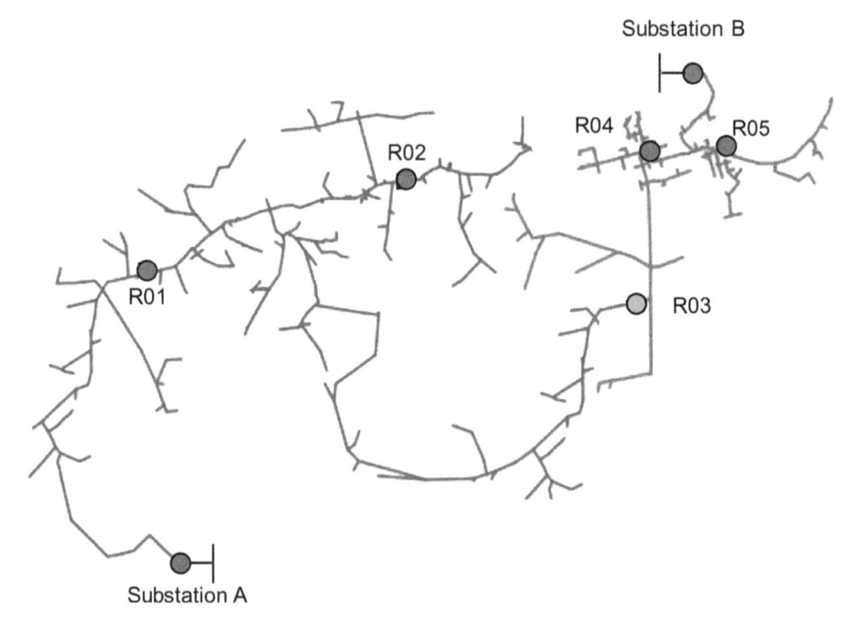

Figure 4.36 Example of a distribution system with FLISR local intelligence

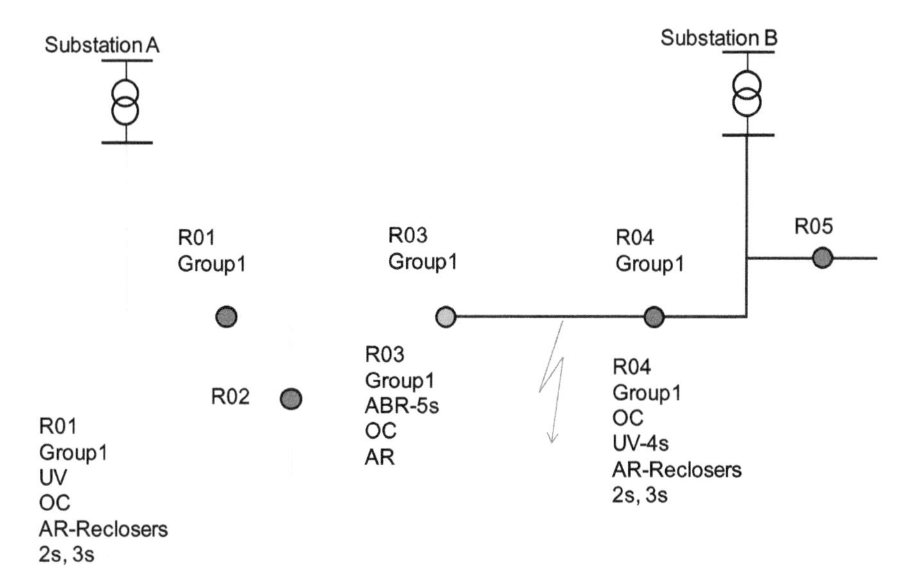

Figure 4.37 Fault between R03 and R04

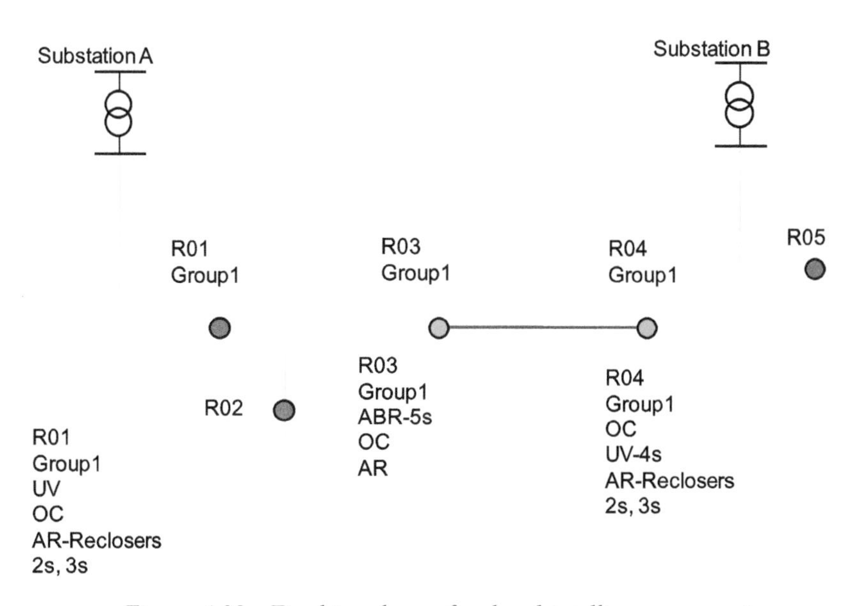

Figure 4.38 Final topology after local intelligence operation

connection with the automatic backfeed restoration (ABR) function, attempting to close. The ABR function generates a close action if supply is lost from the load side of a switch that is normally open. In this case, the fault is still present and the switch eventually locks out. In the end, the fault is cleared, as shown in Figure 4.38.

Example 4.7

A pilot project was developed based on the Civanlar approach, which considered a distribution system that has approximately 110 feeders at 13.2 kV with a total loss figure above 15%. An optimised configuration was required to reduce losses.

A sample of around 5% of the distribution network was studied. This was a good representation of the entire system.

The prototype was selected in such a way that the characteristics of the feeders were representative from the network, including several types of commercial and non-commercial users, hospitals, sport arenas and other key locations of the city. The analysis was carried out by running a load flow for each case. Table 4.9 presents the results obtained after applying the Civanlar algorithm for feeder reconfiguration.

The first columns of Table 4.10 show the identification of the poles and the feeders they correspond to. The next columns present the initial and final loss levels for the circuits receiving loads (CRL) and circuits transferring loads (CTL). The loss reduction is shown in two columns. The first corresponds to the results calculated with load flow runs and the second with the optimal topology methods. It is interesting to see the excellent accuracy of the latter. The last column shows the voltage profile to make sure that all nodes are within the established range. Table 4.11 is similar and presents the best three options for reconfiguring the system at the poles that offer the highest loss reduction.

Figure 4.39 shows the initial topology of the system and Figure 4.40 shows the topology of the system after the topology change was implemented to attain losses reduction. The numbers indicate the existing boundaries among the feeders, which are defined by double dead ends.

A simple cost saving evaluation data considering a load factor equal to 1 gives the following results:

kWh Cost: $0.10 $/kWh

Loss savings: 160 kW

Annual Saving: 160 kW * 8760 h * 0.10 $/kWh = $140,160

If the prototype represents 5% of the overall system, the total savings amount is $2,803,200 per year.

Example 4.8

Figure 4.41 shows a distribution system with three feeders modelled with NEPLAN. Table 4.12 contains the data for each load (active and reactive power) and Table 4.13 shows the reliability characteristics for each line. Calculate the frequency of failures F, the hours per year of unavailability T, the probability to being unavailable Q, the non-supplied power NSP and the non-supplied energy W at Node 11, for the following scenarios:

a) Normal operation
b) Removing the breaker at substation N11 associated with line N11-N12
c) Removing the breaker at substation N10 associated with line N2-N10
d) Removing the circuit breaker at substation N2 associated with line N2-N8
e) Removing the circuit breaker at substation N4 associated with line N4-N13

Table 4.9 Data of feeders encompassed within the prototype

Substation	Circuit name	Circuit code	Nodes	Transf.	Installed capacity	Load factor	Links	Total losses	
					(kVA)			(kW)	(kVAr)
St. Anthony	Crystals	0106	283	152	16122	0.312	10	61.08	234.53
St. Anthony	10th street	0109	100	47	9927	0.253	2	10.99	32.60
St. Anthony	St. Ferdinand	0110	152	81	7750	0.324	7	9.11	37.86
South	Britain	0513	469	251	27582	0.390	6	226.72	802.16
South	Lido	0517	405	222	27147	0.333	9	292.36	809.14
South	Cedar	0518	417	230	30760	0.283	8	99.98	323.07

Table 4.10 Result of reconfiguration analysis

Link	Location	Initial losses (kW)			Final losses (kW)			Reduction (kW)		Volt Min
#	Circuits	CLR	CTL	Total	CRL	CTL	Total	Opt. topology	Load flow	(p.u)
1	0110	9.11	292.36	301.47	46.58	163.26	209.84	90.45	91.63	0.970302
2	0110	9.11	61.08	70.19	35.87	23.21	59.08	11.05	11.11	0.973206
3	0110	9.11	61.08	70.19	43.51	16.66	60.20	9.88	9.99	0.970410
4	0110	9.11	61.08	70.19	50.48	14.45	64.93	5.18	5.26	0.967839
5	0110	9.11	61.08	70.19	9.37	60.48	69.85	0.33	0.34	0.989122
6	0110	9.11	61.08	70.19	9.11	61.08	70.19	0.00	0.00	0.989240
7	0518	99.98	292.36	392.34	101.83	289.43	391.26	1.10	1.08	0.955976
8	0518	99.98	292.36	392.34	100.64	291.12	391.76	0.61	0.58	0.956378
9	0518	99.98	292.36	392.34	130.15	252.65	382.80	9.11	9.54	0.947586
10	0518	99.98	292.36	392.34	99.98	292.36	392.34	0.04	0.00	0.956510
11	0106	61.08	99.98	161.06	74.64	84.55	159.19	1.90	1.87	0.958511
12	0106	61.08	99.98	161.06	82.34	73.40	155.74	5.37	5.32	0.958584
13	0106	61.08	99.98	161.06	89.36	69.60	158.96	2.14	2.10	0.957062
14	0513	226.72	292.36	519.08	229.28	288.57	517.85	1.20	1.23	0.932889
15	0109	10.99	226.72	237.71	56.66	118.02	174.68	62.63	63.03	0.968333
16	0109	10.99	226.72	237.71	64.87	111.53	176.40	60.63	61.31	0.963412
17	0106	61.08	292.36	353.44	127.54	189.47	317.01	35.62	36.43	0.945557
18	0110	9.11	226.72	235.83	45.35	138.54	183.89	51.67	51.94	0.968140
19	0518	99.98	61.08	161.06						
20	0513	226.72	292.36	519.08	255.70	255.93	511.63	7.32	7.45	0.928727
21	0513	226.72	292.36	519.08	250.19	266.32	516.51	2.41	2.57	0.928589

Table 4.11 Options where the highest losses reductions are obtained

Link	Location	Initial losses (kW)			Final losses (kW)			Reduction (kW)		Volt min
#	Circuits	CLR	CTL	Total	CRL	CTL	Total	Opt. topology	Load flow	(p.u)
1	0110	9.11	292.36	301.47	46.58	163.26	209.84	90.45	91.63	0.970302
12	0106	61.08	99.98	161.06	82.34	73.40	155.74	5.37	5.32	0.958584
15	0109	10.99	226.72	237.71	56.66	118.02	174.68	62.63	63.03	0.968333
TOTAL		81.18	619.06	700.24	185.58	354.68	540.26	158.45	159.98	

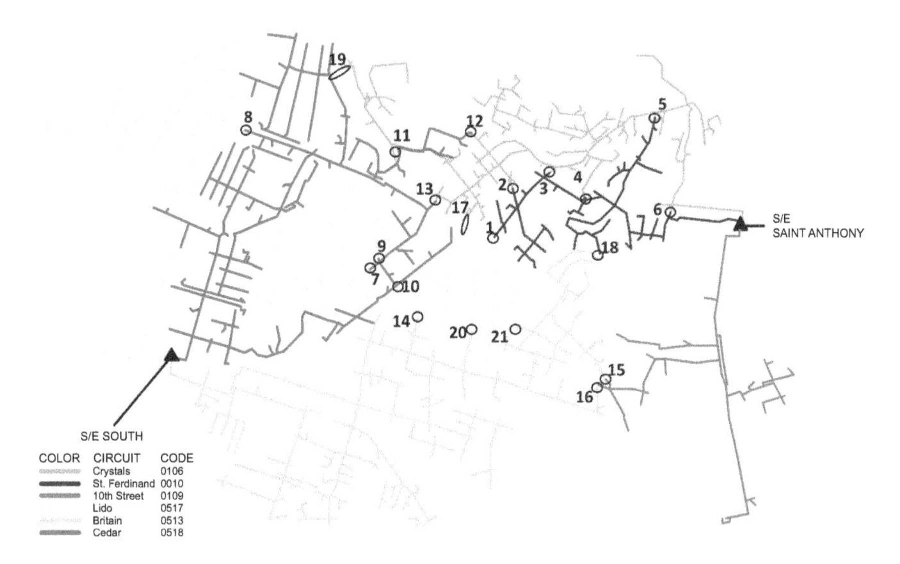

Figure 4.39 Initial topology of case study

Compare the effect from each scenario on the whole system using the reliability values.

Considering the possibility of multiple faults in this system, the calculation becomes difficult since the analysis for each node results from a combination of scenarios/combinatory selection of relationships between different distribution system elements. For that reason, a PSAT, NEPLAN in this case, is used.

The results for the base normal operating scenario, Case (a), are presented in Figure 4.42.

In Case (b) (see Figure 4.43), the circuit breaker just after node N11 was removed. The reliability decreases. This is explained because the circuit breaker removed could have cleared all the faults beyond node N11. Then faults beyond this point can only be cleared when a crew identifies, isolates and restores the

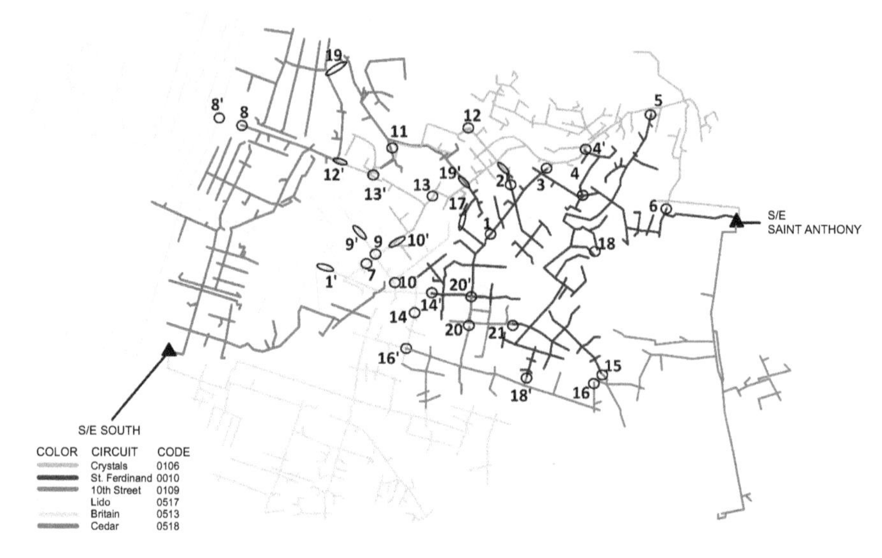

Figure 4.40 Optimal topology considering loss reduction

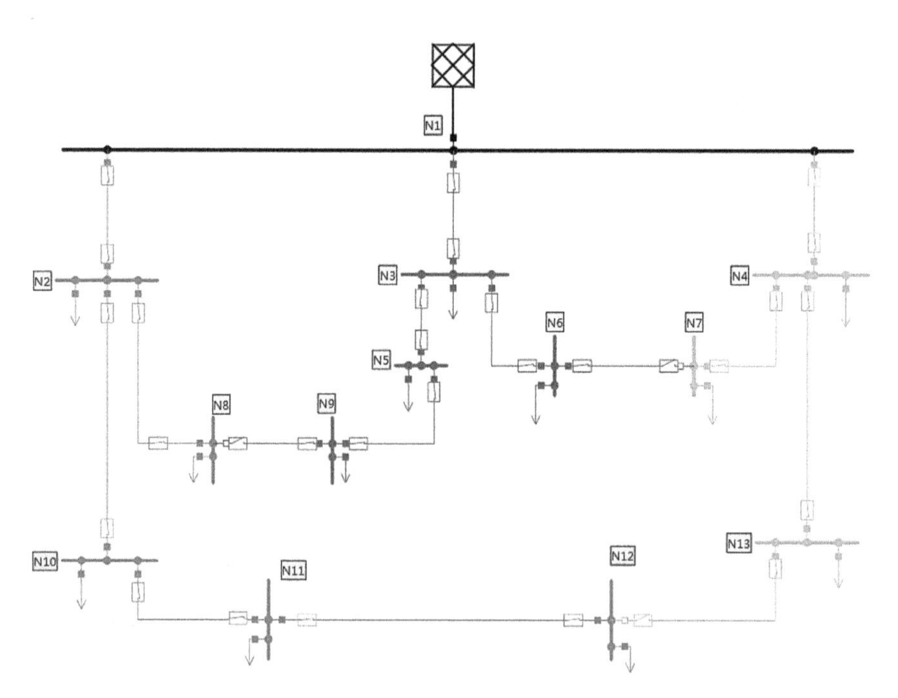

Figure 4.41 System diagram for Example 4.8

Table 4.12 Load data for Example 4.8

Name	P [MW]	Q [MVAR]
LN02	2	1.0
LN03	2	1.0
LN04	2	1.0
LN05	5	2.5
LN06	5	2.5
LN07	5	2.5
LN08	5	2.5
LN09	5	2.5
LN10	7	3.5
LN11	5	2.0
LN12	2	1.0
LN13	7	3.5

Table 4.13 Line data for Example 4.8

Name	Length [km]	T [1/yr*km]	H [h]
$L_{01\text{-}03}$	1.0	1.4	0.5
$L_{06\text{-}07}$	1.2	1.4	0.5
$L_{03\text{-}06}$	1.7	1.4	0.5
$L_{04\text{-}13}$	1.9	1.0	0.75
$L_{01\text{-}04}$	0.8	1.2	0.45
$L_{02\text{-}08}$	2.6	1.2	0.45
$L_{01\text{-}02}$	2.0	1.0	0.75
$L_{09\text{-}08}$	2.2	1.4	0.5
$L_{10\text{-}11}$	3.0	1.4	0.5
$L_{11\text{-}12}$	1.0	1.2	0.45
$L_{03\text{-}04}$	1.5	1.4	0.5
$L_{05\text{-}09}$	2.0	1.2	0.45
$L_{03\text{-}05}$	1.4	1.0	0.75
$L_{02\text{-}10}$	2.1	1.4	0.5
$L_{13\text{-}12}$	1.5	1.4	0.5

service at this point. This increases the probability that node N11 is impaired by a fault for a particular amount of time.

For Case (c) (see Figure 4.44), the circuit breaker just before node N10 was removed. Here, the reliability doesn't change compared to the first scenario. Removing this element from the system doesn't change the ability to clear a fault beyond this node.

For Case (d) (Figure 4.45), the circuit breaker removed is near node N2, between the node N2 and N8. Removing this element from the system impairs the reliability of node N11 in comparison with the first scenario.

Changes in other feeders can also impair reliability in neighbouring feeders. In Case (e) (see Figure 4.46), a circuit breaker from the green feeder was removed.

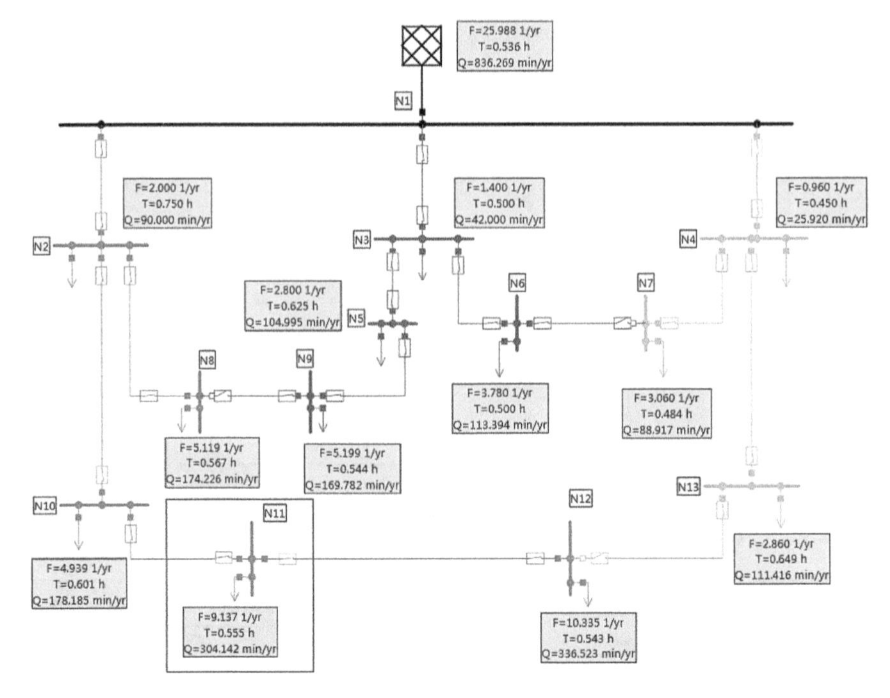

Figure 4.42 Results for scenario (a) of Example 4.8

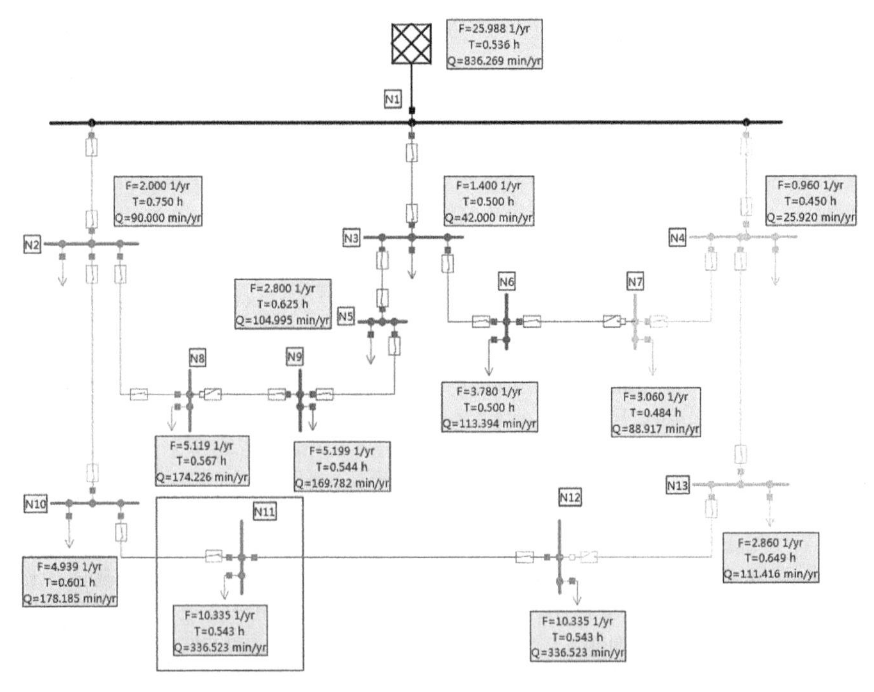

Figure 4.43 Results for scenario (b) of Example 4.8

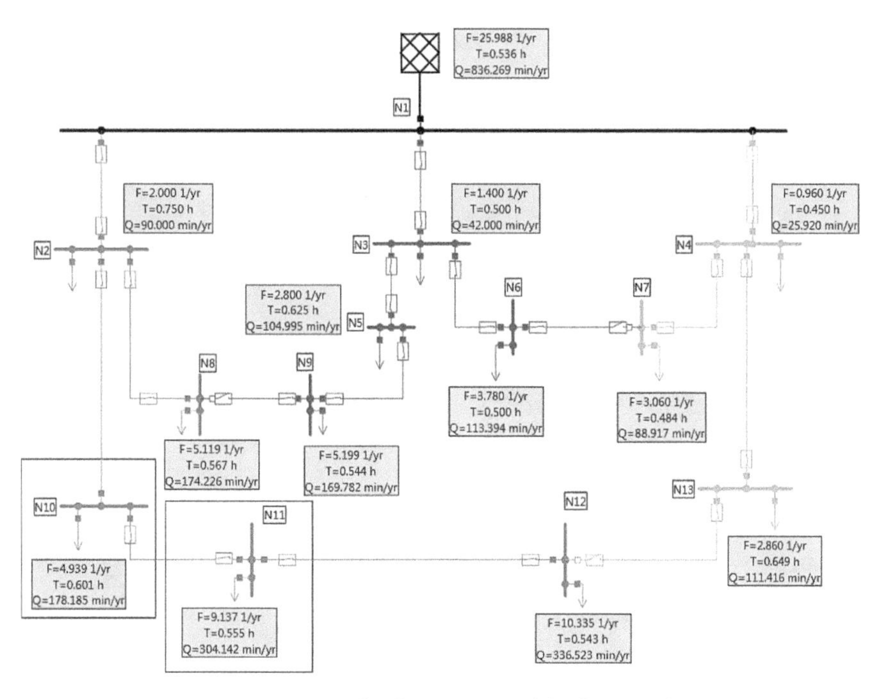

Figure 4.44 Results for scenario (c) of Example 4.8

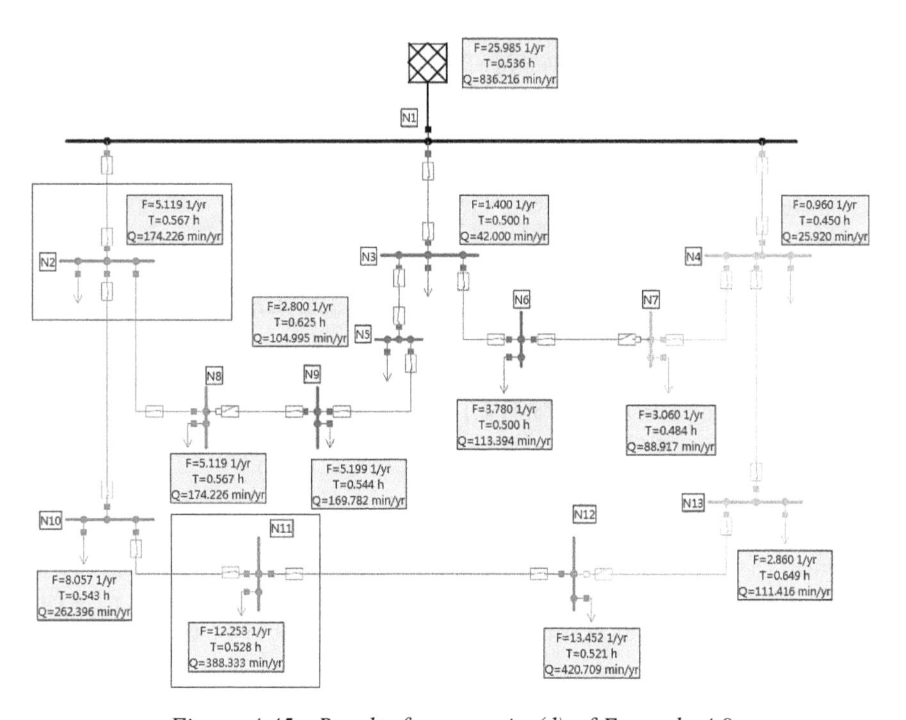

Figure 4.45 Results for scenario (d) of Example 4.8

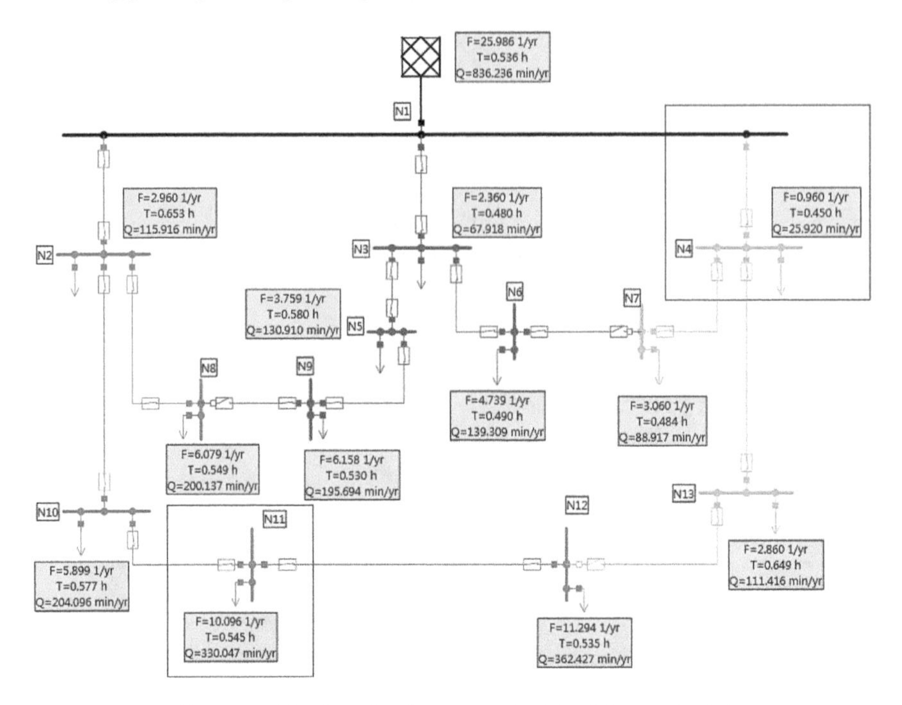

Figure 4.46 Results for scenario (e) of Example 4.8

Table 4.14 Comparison of reliability data for node N11 under different scenarios

	F [1/yr]	T [h]	Q [min/yr]	P [MW/yr]	W [MWh/yr]
Scenario a	9.137	0.555	304.142	45.683	25.345
Scenario b	10.335	0.543	336.523	51.676	28.044
Scenario c	9.137	0.555	304.142	45.683	25.345
Scenario d	12.253	0.528	388.333	61.266	32.361
Scenario e	10.096	0.545	330.047	50.478	27.504

This affects the reliability of node N11 since any fault that takes place at this point disconnects the other feeders (Table 4.14).

Example 4.9

Figure 4.47 shows a distribution system with three feeders coming out from three different substations where switches have been located to increase the flexibility of system operation. The system has in total 13 section switches (NC) and three tie switches (NO) labelled T1, T2 and T3. It can be noted that the tie switches are chosen after the section switches. The five steps for system reconfiguration after fault F_1 are the following:

1. Relay clears fault at Feeder I
2. Switches 1 and 2 open to isolate the fault

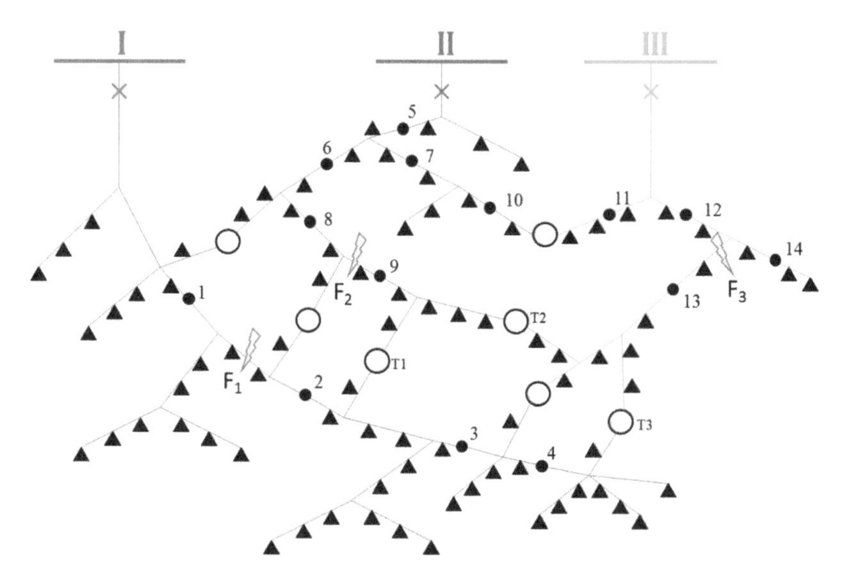

Figure 4.47 Three feeder system illustrating switch location for service restoration

Table 4.15 Reconfiguration options for system of Example 4.9

Fault	Open	Close
F1	1 and 2	T1 or T3
F2	8 and 9	T1 or T2
F3	12 and 13	T2 or T3

3. Relay recloses Feeder I to re-energise up to the location of Switch 1 (upstream restoration)
4. System reconfigures by choosing the best option from the possibilities indicated in Table 4.15 for Fault F_1 (downstream restoration)
5. Fault is fixed and the system is restored to normal configuration.

The same analysis can be conducted for the other two faults, F2 and F3. Table 4.15 shows the operation of the switches for the three different faults indicated in Figure 4.47. Note that for Fault F1, once the section switches 1 and 2 are open, the restoration can be achieved by closing either T1 or T3 or even a combination involving the opening of Switch 3 or 4 and the closing of T1 and T3.

Example 4.10

To illustrate the nature of service restoration, the same system of Figure 4.22 is used. The total loading of each feeder is shown in Figure 4.48 assuming that each section (Z1–Z14) has a loading of 1 p.u. indicating how the operation is affected

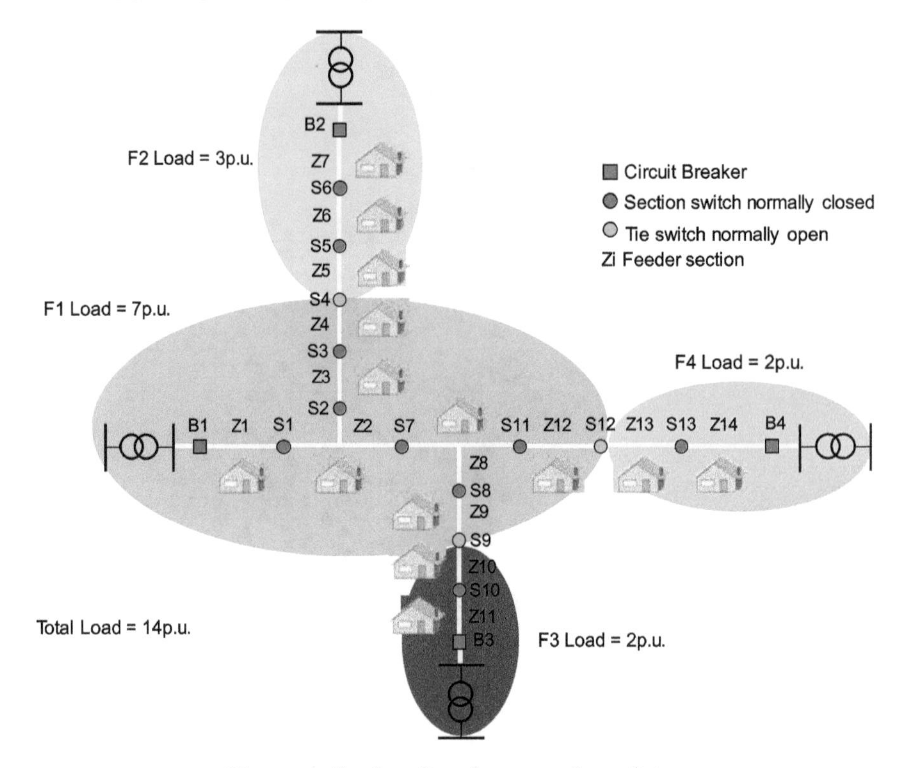

Figure 4.48 Loading for normal condition

when the fault has taken place and the system is reconfigured after step No. 4 that was mentioned previously.

In the scenario of a fault in the Feeder associated with Breaker 1 (B1), the system should clear and isolate the fault and then energise the healthy sections making use of neighbouring circuits, provided that they have sufficient capacity. The options for solutions grow exponentially with an increase in equipment such as section and tie switches. The restoration algorithm should find a configuration that fulfils the operative restrictions. The reconfigured system is shown in Figure 4.25 where all sections are re-energised except section Z2 that needs to be fixed by the maintenance crew. The reconfigured system of Figure 4.49 clearly has different loadings in the Feeders 1–4. Table 4.16 illustrates the section loadings prior to the fault and after the system is reconfigured and the ratio of currents after the reconfiguration and before the fault.

The situation after reconfiguration does not offer any risk to trip Feeder 1. However, it certainly is risky for Feeders 2, 3 and 4 which experience an important increase in loading that might trip the corresponding breakers if the pickup value of the corresponding relays is exceeded. Measures including adaptive relaying must be taken to avoid any nuisance tripping that could worsen the operation of the system.

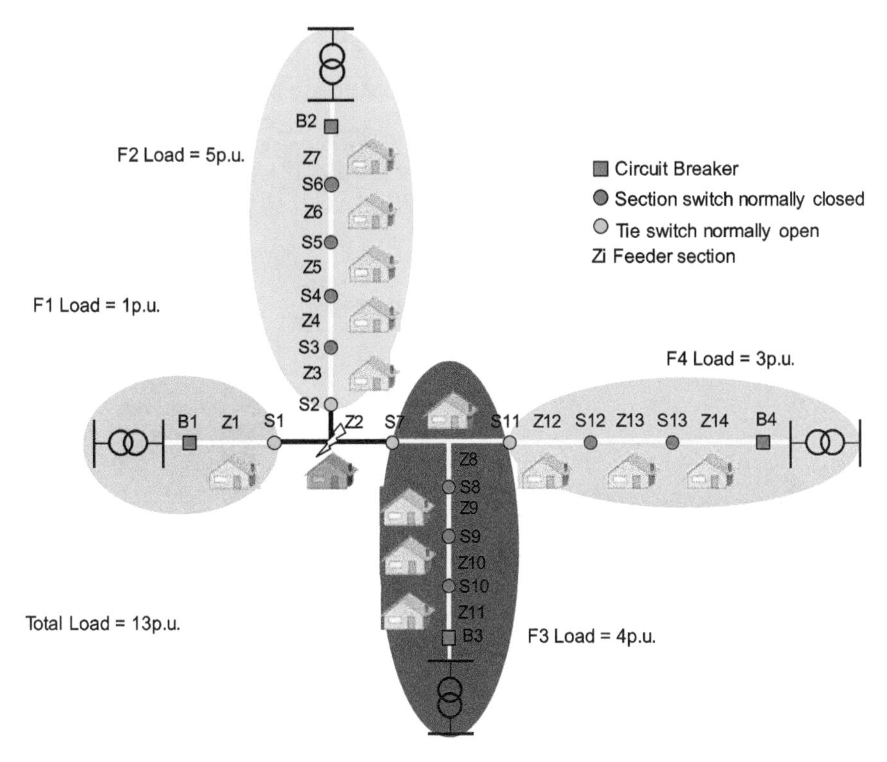

Figure 4.49 Loading after reconfiguration

Table 4.16 Loading values of Example 4.10

Feeder	Sections before	Sections after	I_{after}/I_{before}
B1	7	1	1/7
B2	3	5	5/3
B3	2	4	4/2
B4	2	3	3/2

4.8 Proposed exercises

1. Calculate the equivalent reliability for each system in Figure 4.50, considering each component's respective reliability value.
2. A primary main feeder serves the depicted customers. The annual average fault rates for primary main and branches are 0.068 and 0.13 fault/circuit-km, respectively. The average repair times for each primary main section and for each branch are 4.2 and 1.78 h, respectively. The average time for manual sectionalising of each feeder section is 0.86 h. Assume that at the time of

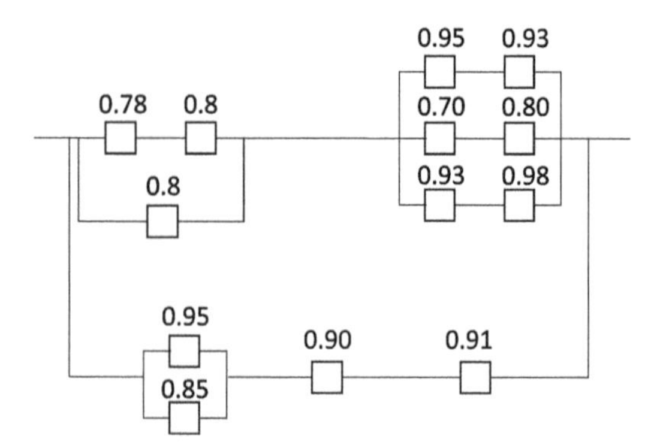

Figure 4.50 Elements of Exercise 2

Table 4.17 Database for Exercise 4.3

Date	Time	Time on	Total time [min]	Number of customers	Load [kVA]	Interruption type
1/3	12:23:34	13:43:49		347	600	
2/2	11:12:50	11:26:54		820	775	
3/10	03:03:15	03:05:08		102	120	
3/15	17:22:28	18:04:05		78	88	
3/15	00:00:49	00:51:30		432	683	
4/22	11:11:52	11:10:55		1700	2480	
4/25	07:17:24	07:18:50		700	1315	
5/02	04:38:15	04:41:27		1300	2765	
6/27	02:53:38	03:36:18		2100	2984	
7/15	16:46:43	16:56:08		2879	3297	
11/15	18:28:21	18:29:52		569	688	

having one of the feeder sections in fault, the other feeder section is manually sectionalised as long as they are not in the mainstream of the fault current, that is, not in between the faulted section and the circuit breaker. Otherwise, they must also be repaired.

Based on the given information, prepare an interruption analysis study for the first contingency only, that is, ignore the possibility of simultaneous outages and determine the following:

a) The total annual sustained interruption rates for customers A, B, C and D.
b) The average annual repair times, that is, downtimes, for customers A, B, C and D (Figure 4.51).

3. Table 4.17 shows a database for a feeder that serves 15,000 customers with a total load of 21.55 MW. Complete the table and calculate the quality indices SAIFI, SAIDI and CAIDI.

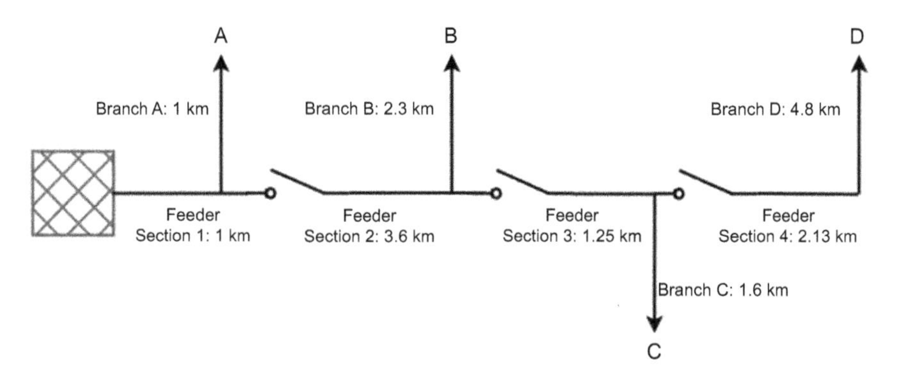

Figure 4.51 Database for Exercise 4.2

4. A substation transformer has a constant hazard rate of 0.0057 per day. Calculate the probability that it will fail during the next 11 years and the probability that it will not fail during the next 9 years. Calculate the MTTF for this transformer and determine the probability of having at least three out of four transformers out of service at any given time.

References

[1] IEEE, 'IEEE Std. 493-2007 – IEEE Recommended Practice for the Design of Reliable Industrial and Commercial Power Systems', 2007.

[2] Lawrence Berkeley National Laboratory (LBNL), 'The Interruption Cost Estimate (ICE) Calculator', 2023. Accessed 28 March 2024. [Online]. Available: https://icecalculator.com/home

[3] IEEE, 'IEEE Std 1366-2022 – IEEE Guide for Electric Power Distribution Reliability Indices', 2022.

[4] Council of European Energy Regulators, *CEER Benchmarking Report 6.1 on the Continuity of Electricity and Gas Supply, Data Update 2015/2016*. Energy Quality of Supply Work Stream (EQS WS), 2018.

[5] S. Civanlar, Grainger, and S. Lee, 'Distribution feeder reconfiguration for loss reduction', *IEEE Trans. Power Deliv.*, vol. 3, pp. 1217–1223, 1988.

[6] D. Shirmohammadi and H. W. Hong, 'Reconfiguration of electric distribution networks for resistive line losses reduction', *IEEE Trans. Power Apparatus Syst.*, vol. 4, pp. 1492–1498, 1989.

[7] Neplan, 'Switching Optimization Module', 2018.

[8] R. W. Uluski, 'Using distribution automation for a self-healing grid', in *PES T&D 2012*, Piscataway, NJ: IEEE, pp. 1–5, 2012, doi:10.1109/TDC.2012.6281582.

[9] J. M. Gers, *Distribution System Analysis and Automation*, 2nd ed. London: The Institution of Engineering and Technology, 2020.

[10] J. H. Eto, K. Lacommare, M. D. Sohn, and H. C. Caswell, 'Evaluating the performance of the IEEE Standard 1366 method for identifying major event days', *IEEE Trans. Power Syst.*, vol. 32, no. 2, pp. 1327–1333, 2017.

[11] M. Karaçelebi, M. Gol, and Ö. Erçin, 'A new index for voltage quality assessment at distribution systems', *53rd International Universities Power Engineering Conference (UPEC)*, Glasgow, UK, 2018, pp. 1–6, doi:10.1109/UPEC.2018.8542073.

[12] T. A. Short, *Electric Power Distribution Handbook*, 2nd ed. Boca Raton, FL: CRC Press, 2014.

[13] Y. P. Agalgaonkar and D. J. Hammerstrom, 'Evaluation of smart grid technologies employed for system reliability improvement: Pacific Northwest smart grid demonstration experience', *IEEE Power Energy Technol. Syst. J.*, vol. 4, no. 2, pp. 24–31, 2017.

[14] R. E. Brown, *Electric Power Distribution Reliability*, 2nd ed. Boca Raton, FL: CRCPress, 2009.

[15] M. R. Behbahani, A. Jalilian, A. Bahmanyar, and D. Ernst, 'Comprehensive review on static and dynamic distribution network reconfiguration methodologies', *IEEE Access*, vol. 12, pp. 9510–9525, 2024, doi:10.1109/ACCESS.2024.3350207.

[16] A. González, F. M. Echavarren, L. Rouco, and T. Gómez, 'A sensitivities computation method for reconfiguration of radial networks', *IEEE Trans. Power Syst.*, vol. 27, no. 3, pp. 1294–1301, 2012.

[17] S. Tatipally, S. Ankeshwarapu, and S. Maheswarapu, 'Swarm intelligence methods for optimal network reconfiguration of distribution system', *IEEE International Power and Renewable Energy Conference (IPRECON), Kollam, India*, pp. 1–6, 2022 doi:10.1109/IPRECON55716.2022.10059540.

[18] R. Billinton and S. Jonnavithula, 'Optimal switching device placement in radial distribution systems', *IEEE Trans. Power Deliv.*, vol. 11, no. 3, 2011.

[19] C. Lee, C. Liu, S. Mehrotra, and Z. Bie, 'Robust distribution network reconfiguration', *IEEE Trans. Smart Grid*, vol. 6, no. 2, pp. 836–842, 2015.

[20] G. Gutiérrez-Alcaraz and J. H. Tovar-Hernández, 'Two-stage heuristic methodology for optimal reconfiguration and Volt/VAr control in the operation of electrical distribution systems', *IET Gener. Transm. Distrib.*, vol. 11, no. 16, pp. 3946–3954, 2017.

[21] G. K. V. Raju and P. R. Bijwe, 'Efficient reconfiguration of balanced and unbalanced distribution systems for loss minimisation', *IET Gener. Transm. Distrib.*, vol. 2, no. 1, p. 7, 2008.

[22] Lin, C. Chen, C. Ku, T. Tsai, and C. C; Ho, 'A multiagent-based distribution automation system for service restoration of fault contingencies', *Eur. Trans. Electr. Power*, vol. 21, pp. 239–253, 2010.

[23] Y. Pradeep, P. Seshuraju, S. A. Khaparde, and R. K. Joshi, 'CIM-based connectivity model for bus-branch topology extraction and exchange', *IEEE Trans. Smart Grid*, vol. 2, no. 2, p. 24, 2011.

[24] S. Saini and M. Mam, 'Reconfiguration of an unbalanced distribution system for loss reduction by software simulation', *Int. J. Eng. Res. Technol.*, vol. 2, no. 6, pp. 3314–3318, 2013.

[25] J. M. Solanki, S. Khushalani, and N. N. Schulz, 'A multi-agent solution to distribution systems restoration', *IEEE Trans. Power Syst.*, vol. 22, no. 3, pp. 1026–1034, 2007.

[26] Y. Takenobu, N. Yasuda, S. Kawano, S. I. Minato, and Y. Hayashi, 'Evaluation of annual energy loss reduction based on reconfiguration scheduling', *IEEE Trans. Smart Grid*, vol. 9, no. 3, pp. 1–11, 2018.

[27] Y.-K. Wu, C.-Y. Lee, L.-C. Liu, and S.-H. Tsai, 'Study of reconfiguration for the distribution system with distributed generators', *IEEE Trans. Power Deliv.*, vol. 25, no. 3, pp. 1678–1685, 2010.

[28] Q. Zhou, D. Shirmohammadi, and W.-H. E. Liu, 'Distribution feeder reconfiguration for operation cost reduction', *IEEE Trans. Power Syst.*, vol. 12, no. 2, pp. 730–735, 1997.

[29] M. B. Ndawula, P. Zhao, and I. Hernando-Gil, 'Smart application of energy management systems for distribution network reliability enhancement', in *IEEE International Conference on Environment and Electrical Engineering and 2018 IEEE Industrial and Commercial Power Systems Europe (EEEIC / I&CPS Europe)*, Piscataway, NJ: IEEE, pp. 1–5, 2018, doi:10.1109/EEEIC.2018.8494478.

[30] L. Guo, Z. Huang, J. Wang, *et al.*, 'Distribution network reconfiguration algorithms: a comparative study', in *IEEE 5th International Conference on Electronics Technology (ICET)*, Piscataway, NJ: IEEE, pp. 472–478, 2022, doi:10.1109/ICET55676.2022.9824709.

[31] K. L. Lo and J. M. Gers, 'Feeder Reconfiguration for Losses Reduction in Distribution Systems', in *Proc. of the UPEC 94 Conference*, University College Galway, Ireland, pp. 290–293, 1994.

[32] K. J. Russell and R. P. Broadwater, 'Model-based automated reconfiguration for fault isolation and restoration', in *IEEE PES Innovative Smart Grid Technologies (ISGT)*, Piscataway, NJ: IEEE, pp. 1–4, 2012.

Chapter 5

Renewable sources, integration and microgrids for cities

Most of the global energy in the world is based on fossil fuels, with only a small percentage of nuclear energy and renewables. However, the generation of electricity with renewable and alternative sources has increased gradually around the world. According to the Renewable Energy Policy Network for the 21st Century (REN21), the global capacity of generation units based on renewables have increased by 348 GW since 2021 [1].

In this chapter, several technologies applied in generation using DERs are presented. The renewable energy resources or technologies that will be discussed in terms of basic operational principles include solar energy, wind generation, small hydro power plants and energy storage.

5.1 Current situation of renewable generation

There is an increased interest in connecting these generation units to the electric grid. The problem with renewable energy is its uncertain nature and the location of the resources. Therefore, it is hard to accurately define where these generation units are going to be connected. This has prompted the possibility of using both transmission and distribution systems in order to properly integrate them.

Distributed energy resource (DER) is the name given to any generation unit which is based on renewable energies of low power capacity and usually connected to a feeder near the user. Larger units known as 'distributed generation' (DG) units are often connected to the transmission system at the point of interconnection (POI).

Only small generators may be connected to the lowest voltage networks, but large installations of hundreds of megawatts are connected to the bus bars of high-voltage distribution systems.

5.2 Inverter-based resources

Inverter-based resources have become a crucial element in power system configurations at various levels, including transmission and distribution. Historically, synchronous generators have been the predominant energy source, with their rotational speeds directly influencing the electrical output frequency. However, with the ongoing displacement of synchronous generation, the proliferation of inverters is significantly impacting system operations.

Despite attempts to enhance system support through rapid frequency response provided by inverters, the behaviour of synchronous generators remains the primary factor determining frequency. This dependency poses a constraint on the effective integration of DERs at higher penetration levels. Therefore, it is imperative to develop solutions that ensure the reliable operation of power systems where inverters play a dominant role.

5.2.1 Grid-forming and grid-following

According to the operational characteristics of inverters, they can be categorised as either grid-following (GFL) or grid-forming (GFM), as presented in Figure 5.1. On one hand, the majority of commercial photovoltaic (PV) installations utilise GFL, in which their power output is regulated by monitoring the phase angle of the grid voltage through a phase-locked loop mechanism. Consequently, these kinds of inverters track the grid's angle and frequency without actively controlling their own frequency output. On the other hand, GFM sources actively regulate both their voltage output and frequency. Nowadays, GFL sources are being replaced by GFM sources, resulting in an increasing reliance on the remaining synchronous generators for frequency control.

The function used for control on each type of inverter is shown in Figure 5.2 in which GFM control are normally power-frequency (P–f) droop and var-voltage (Q–V) droop controls, while GFL uses frequency/watt droop or volt/var droop. GFM inverters can have a positive impact on enhancing the frequency dynamics and stability of power systems dominated by inverters. By actively controlling the frequency, GFM inverters reduce the reliance on mechanical inertia for frequency stability, which contribute to grid stabilisation. Droop control is a commonly employed technique in GFM inverters to derive frequency and voltage commands based on real and reactive power measurements. GFL inverters use the frequency–watt function, in which real power is used as the measured variable and frequency as the controlled variable. Another type of GFM

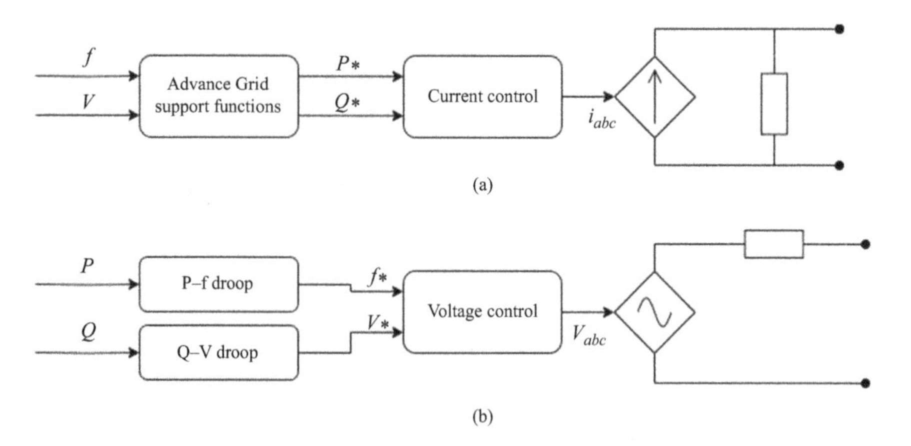

Figure 5.1 Inverter controllers: (a) grid-following (GFL) control with grid support functions; and (b) grid-forming (GFM) droop control

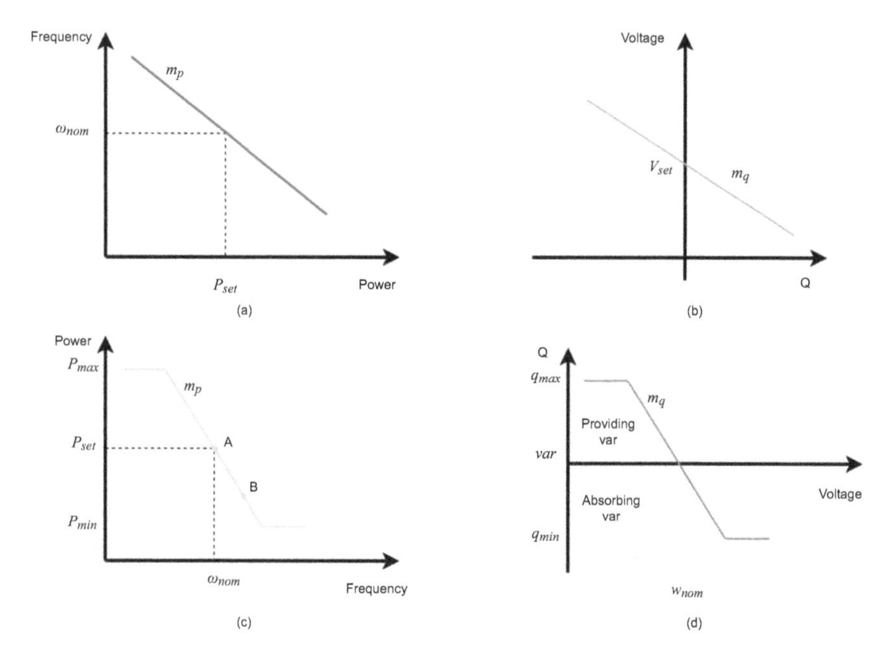

Figure 5.2 Primary frequency and voltage droop control curves of grid-forming and grid-following inverters. (a) Grid forming P-f droop; (b) Grid forming Q-V droop; (c) Grid following freq/watt droop and (d) Grid following volt/var droop.

control described in literature is the emulated/virtual synchronous machine (VSM) [2]. However, this control approach aims to mimic the behaviour of traditional synchronous generators, thereby not fully exploiting the faster response capabilities of inverters to enhance system dynamics. Additionally, the relative simplicity of droop-based control algorithms compared to VSM approaches makes them an attractive choice.

In the case of a GFL inverter, the dynamics of the current controller are considerably faster compared to the power response and can be disregarded. The power response is approximated as a first-order system, and the phase-locked loop is represented as a low-pass filter. Nonlinearities associated with the frequency-watt function are neglected, and a simple droop gain is used to represent this component, as shown in Figure 5.3. The voltage controller dynamics of GFM inverters are also relatively fast and can be neglected. The only dynamics accounted for in the GFM inverter model are those of the low-pass filter used for power measurement. The dynamics of the power-limit controller are not included in this initial system model. The network equations are formulated as algebraic constraints for each specific case.

The presence of GFL inverters in power systems often has a detrimental effect on the system's eigenvalues, leading to underdamped responses. The dominant electromechanical modes, which are influenced by system inertia, persist at all penetration levels. Conversely, the incorporation of GFM inverters has a positive influence on eigenvalues, resulting in faster modes and improved damping of slower modes due to active frequency control.

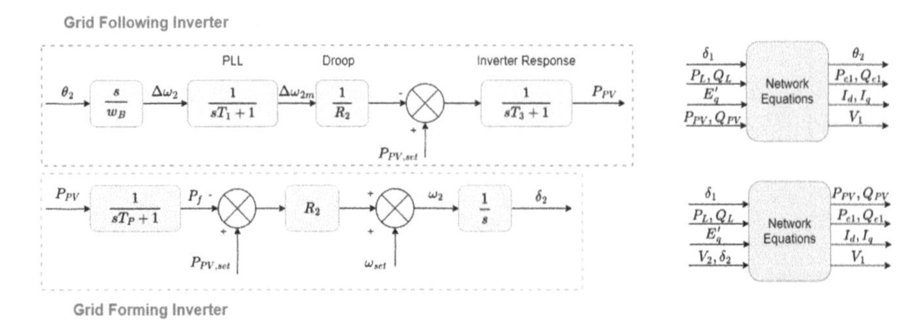

Figure 5.3 Grid-following and grid-forming inverter models

GFM inverters have certain limitations as they operate as voltage sources rather than current sources. In a GFM control scheme, the inverter current is not tightly regulated. This poses challenges in limiting overcurrent during load changes, as well as abnormal conditions like faults and grid transients. Advanced control schemes such as virtual impedance current limiting may be necessary to achieve current limiting, but they can impact the overall system stability. Instantaneous events such as grid voltage sag and swell can cause significant current flow from the inverter.

Addressing anti-islanding for distributed inverters is a major challenge to be implemented. Since GFM inverters can operate in islanded mode, implementing effective anti-islanding protection schemes becomes crucial. All grid-tied inverters operate as GFL sources. Therefore, if future installations adopt GFM inverters, a thorough examination of the dynamics of power systems with mixed source types (GFM-GFL-Generator) will be essential. While most research on GFM inverters has focused on microgrid configurations, their application to larger power systems requires more comprehensive investigation.

5.2.2 Solar plants

The total energy that the earth receives from the sun represents an important amount that can be used to supply energy demand requirements. The total amount of energy received from the sun can vary between locations and the atmospheric conditions. For instance, solar radiation is higher in locations near the equator than near poles in an environment with low cloudiness.

The solar energy reaching 1 square meter (1 m^2) at the earth's surface may vary between 1000 and 2000 kWh/m^2 [3]. It is required to optimise the performance of the system that takes advantage of this sunlight energy and uses it into other useful forms of energy. This process needs several data points about the sunlight properties, atmospheric effects, sun position, among others.

There are many applications in which the energy from the sun can be transformed. For instance, large solar facilities are based on concentrating sunlight to boil water. This high-pressure vapour powers a turbine, in a similar way as done in a conventional thermal power station [4].

The most common way of using solar radiation is by using solar panels, which are installed on a roof or in large PV solar farms [3]. PV systems use the sunlight and convert it into electricity. This type of technology is designed around the PV cell. The connection of these cells is called PV module, which is connected into different arrays depending on the desired voltage or current at the system output. An advantage of this technology is the power produced close to the load because the electric losses of transporting the energy around the system are reduced considerably. In the following sections, more details about this technology and its components are developed.

A PV cell is formed by a *p–n* junction, which is a layer of combined *p*-type and *n*-type semiconductor materials. When the photon in the sunlight reaches the PV cell, an electric potential difference across the junction terminals is produced. This excites the electrons in both layers. The incident light creates electron-hole pairs in the junction, which produce a diffusion of electrons through the cell and the outer contacts become electrodes (the *n*-type side becomes positive electrode, while the *p*-type side becomes negative). A current will flow around the electric circuit once a load is connected between both electrodes [3,5,6], as shown in Figure 5.4.

The ideal electrical behaviour of a PV cell is described by the *I–V* characteristic curve of a diode [7], where both current and voltage depends on the cell irradiation level and temperature of exposure.

$$I = I_{Ph} - I_0 \cdot \left(e^{\frac{qV}{kT}} - 1\right) \tag{5.1}$$

$$V = \frac{kT}{q} \ln\left(1 - \frac{I - I_{Ph}}{I_0}\right) \tag{5.2}$$

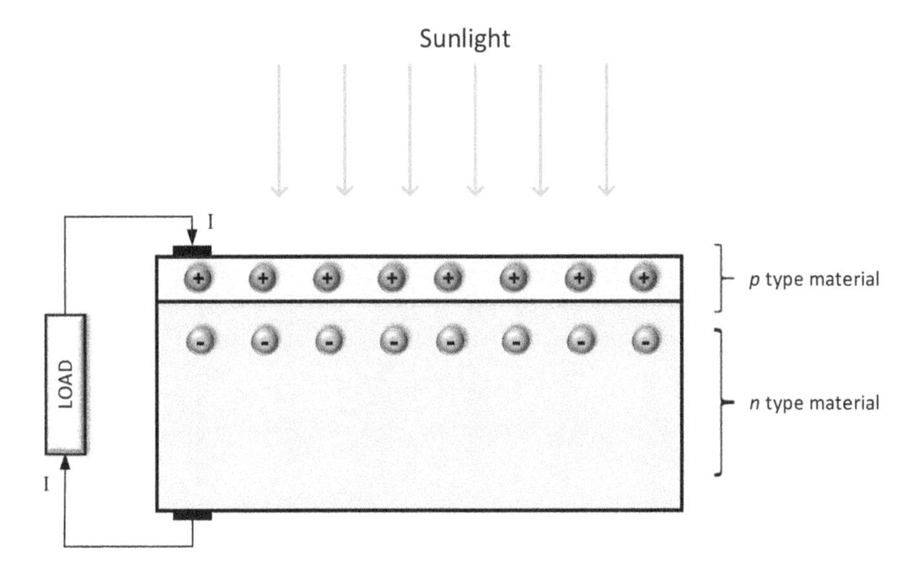

Figure 5.4 A typical PV cell

Then,

$$V_{OC} = \frac{kT}{q} \ln\left(\frac{I_{Ph}}{I_0}\right)$$

I : total current flowing through the cell terminals in amps, A
V : the cell voltage across terminals in volts, V
V_{OC} : is the open-circuit voltage across terminals
I_{Ph} : cell current due to the incident photons or photocurrent
I_0 : is the saturation current of the diode
q : the electron electric charge in coulombs, $q = 1.6021 \times 10^{-19}$ C
k : Boltzmann constant, 1.3806×10^{-23} J/K
T : the cell temperature in Kelvin, K

In order to determine the short-circuit current of a PV cell, the voltage should be set on $V = 0$ for the exponent. This leads to $I_{SC} = I_{Ph}$. Typically, a PV cell produces an open-circuit voltage around 0.6 V at 25 °C. Figure 5.5 shows the *I–V* curve of a PV cell for different values of short-circuit currents and cell temperatures.

Figure 5.6 shows the power output of a PV cell, which corresponds to a non-linear expression that relates the current and voltage across the cell terminals. The maximum power point can be determined as the peak of the curve, which is located around the inflection point in the *I–V* curve. This point can be expressed in terms of the short-circuit current, the open-circuit voltage and a fill factor *FF*, which indicates the overall behaviour or quality of the PV cell, as indicated in the following expression:

$$P_{max} = I_{max} \cdot V_{max} = FF \cdot I_{SC} V_{OC} \tag{5.3}$$

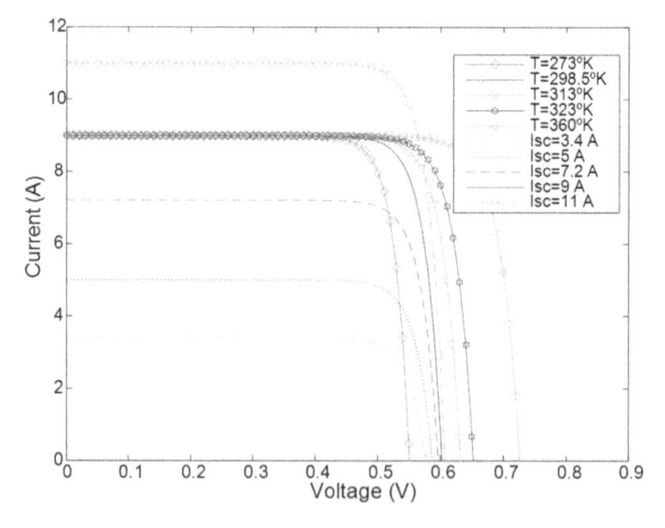

Figure 5.5 I–V characteristic of a PV cell

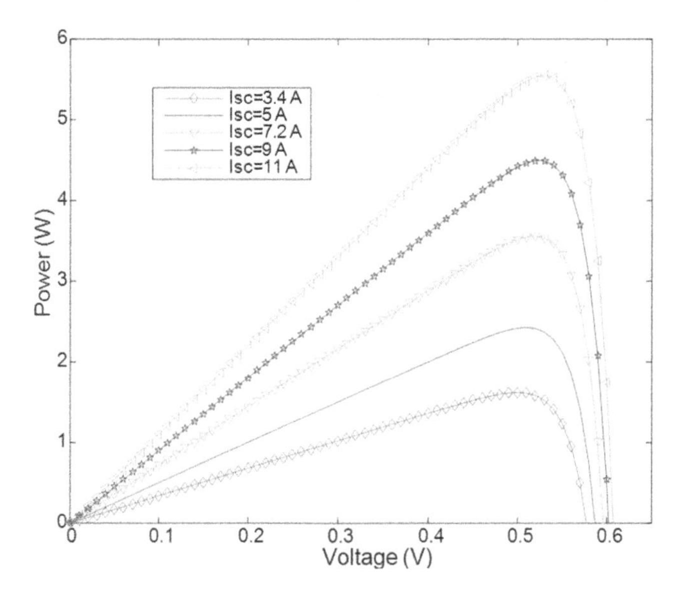

Figure 5.6 P–V characteristic of a PV cell

A large internal resistance and higher saturation will turn into smaller fill factors and vice versa. Fill factor for PV cells may vary from 0.5 to 0.82 [7,8]. From (5.3), the fill factor can be determined as follows:

$$FF = \frac{P_{\max}}{I_{SC} \cdot V_{OC}} = \frac{I_{\max} \cdot V_{\max}}{I_{SC} \cdot V_{OC}} \tag{5.4}$$

PV cells can be connected in series configuration to obtain more voltage at the output and in parallel configuration for more current. These connections together produce a PV module, which provides more power. Several modules can achieve open-circuit voltages around 600 V DC. They are connected to an inverter, which has a maximum power point tracking (MPPT) algorithm to extract power [9].

Figure 5.7 illustrates the equivalent circuit of a PV cell. The dependency and proportionality of the surface area and the incident irradiance are represented by a current source. The shunt resistors represent the losses associated with leakage currents through the edges and structure of the cell, while the series correspond to the resistance of terminals, temperature effects and semiconductor materials.

The electric signals that come out the PV system are DC signals, and they must be converted into AC. A power inverter is a single-phase or three-phase solid-state electronic device formed by transistors or thyristors that operate as switches, converting the DC input to an AC symmetrical output with a desired amplitude and frequency. The basic operation of an inverter is shown in Figure 5.8.

The simplest inverter configuration is a single-phase square wave inverter shown in Figure 5.9. This configuration has a significant amount of harmonic content and may cause the electric loads to overheat or malfunction. With

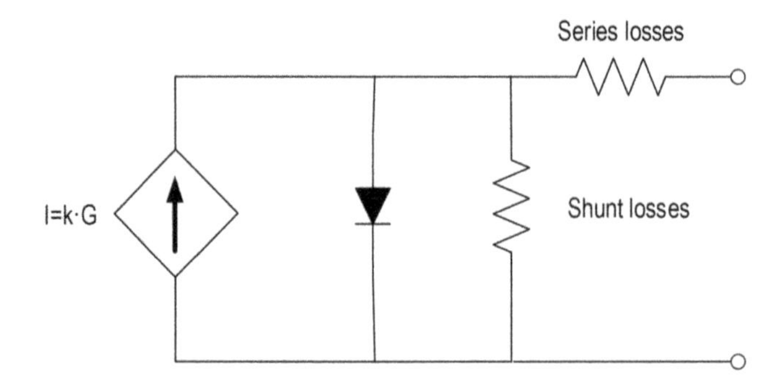

Figure 5.7 Real PV cell equivalent circuit [3]

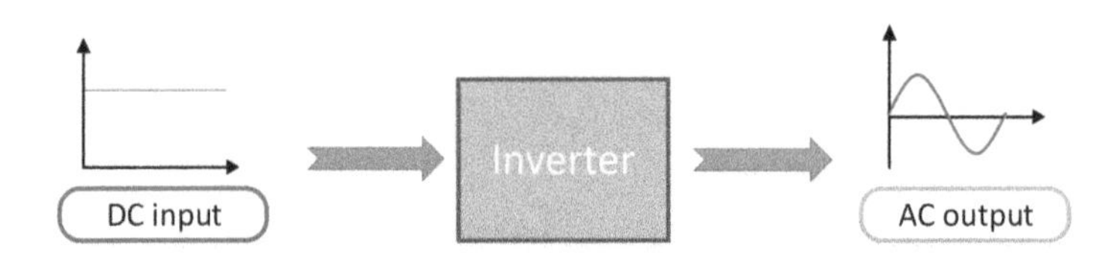

Figure 5.8 Power inverter operation scheme

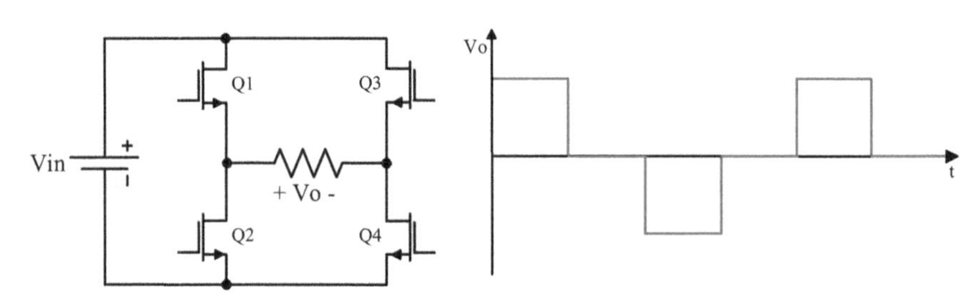

Figure 5.9 One-phase square wave inverter

pulse-width-modulation (PWM), a sine wave can be obtained by controlling the on-and-off switching of a pulse waveform as seen in Figure 5.10. This produces less harmonic content and more efficiency in the resultant waveform.

Figure 5.11 illustrates a common configuration for a PV system as a DG unit, which is formed by a PV array or PV system, a MPPT algorithm, an energy storage device, DC–DC converter, DC–AC inverter, an isolation transformer and an output filter to avoid DC and harmonic injection.

There are several commercial manufacturers such as SMA and Siemens (Germany), ABB (Switzerland), Fronius (Austria), among others. They offer different power ranges up to 3000 kW at unit power factor (pf = 1.0) for solar string applications [12].

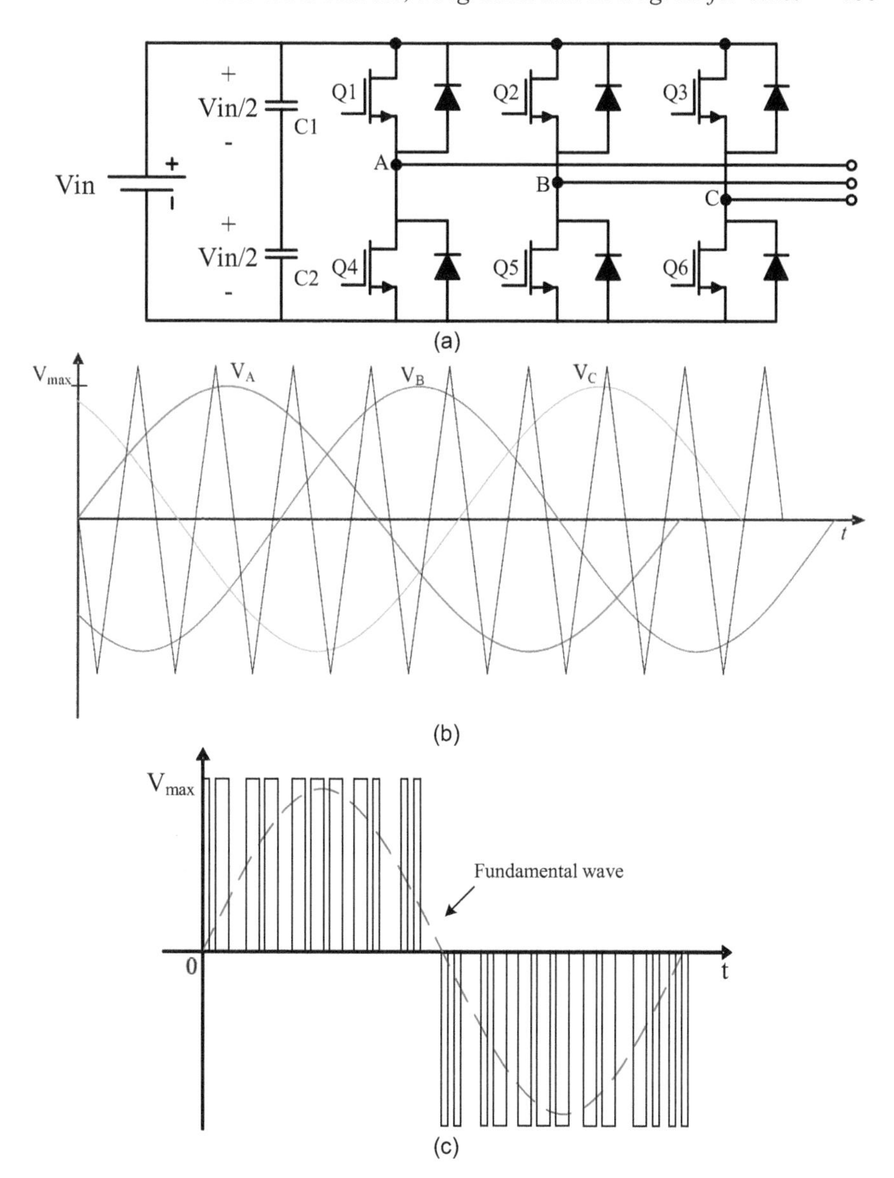

Figure 5.10 Three-phase sine wave PWM inverter: (a) equivalent circuit; (b) resultant waveform and (c) fundamental wave modulation [10,11]

A common application of the PV technologies includes exporting energy to the main grid, also referred to as grid-connected systems. There is also stand-alone operation, in which the system is disconnected from the electric system. This system is equipped with an energy storage system that stores the energy surplus and releases it at times of low power generation. Finally, there are hybrid systems that

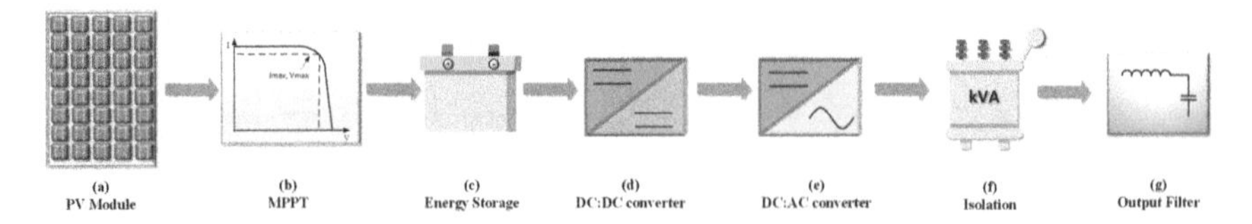

Figure 5.11 Inverter system for a PV module as a DG unit

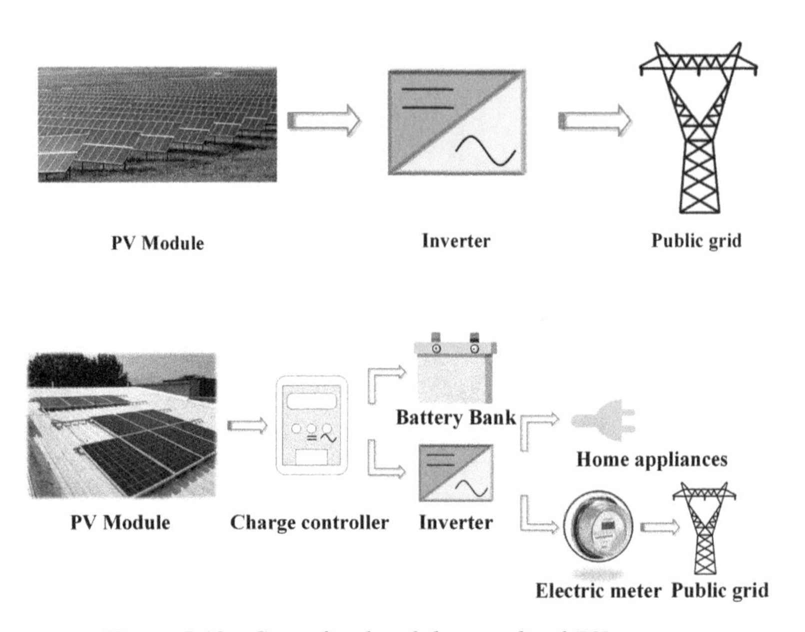

Figure 5.12 Centralised and decentralised PV systems

are coupled to loads and surplus is exported to the grid or stored if a storage system is installed.

PV integration into the power system can be either centralised or decentralised [13]. A centralised system is a large-scale installation that produces a significant amount of power. In order to dispatch this power, electrical substations and transmission lines are required. Decentralised systems produce energy on-site or close to the consumers, which are owners of the system. The energy surplus can be sold to the local utility according to the market regulations. Decentralised configuration has more freedom to be operated. Figure 5.12 shows the typical components of each configuration.

Example 5.1

To illustrate the performance of a grid-connected PV system on a distribution network, a 2 MVA PV system is connected to a 0.4 kV bus as shown below. The power consumed by the load is 1.5 MVA with unity power factor. Both the load and the PV system are modelled with a load profile, which indicates the variation of the load and sunlight through a day. The example has been modelled and run with a PSAT and the results are depicted in Figures 5.13 and 5.14. By running the load

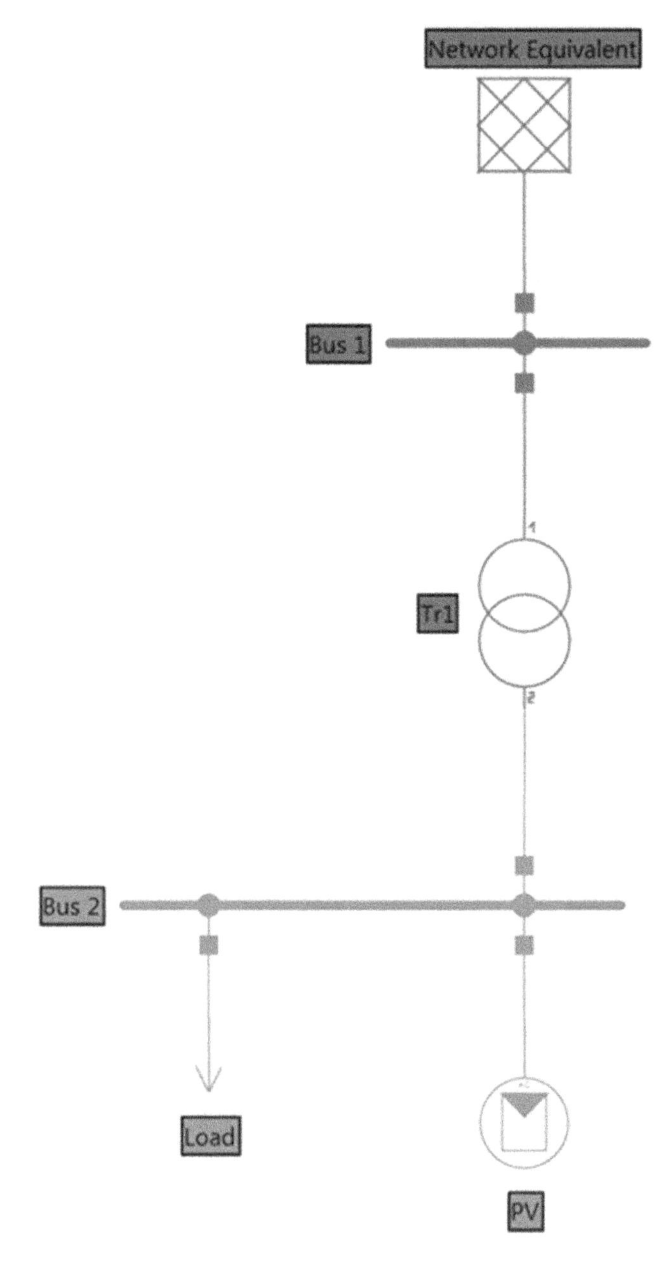

Figure 5.13 Grid-connected PV system example

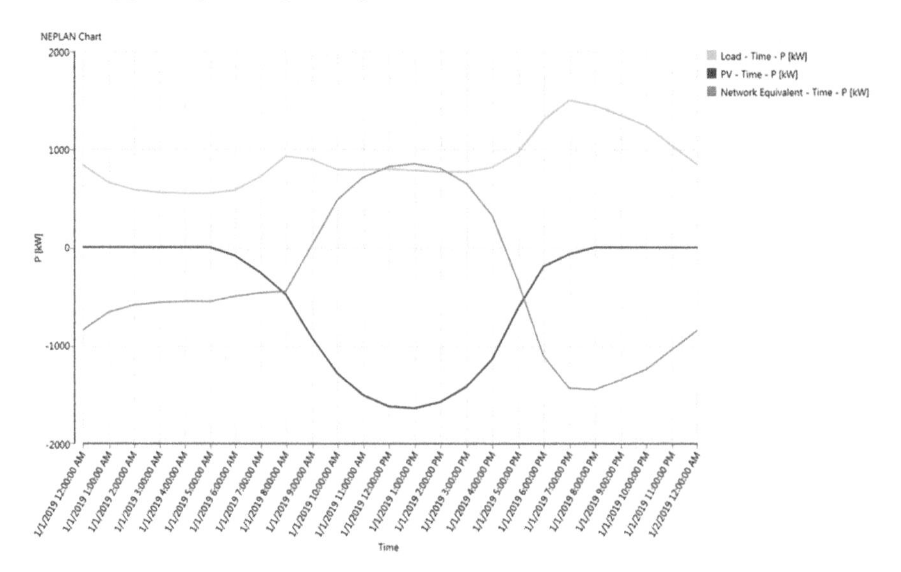

Figure 5.14 Load flow of the PV system through one day

flow through one day it can be seen how the surplus energy, that is, the energy that is not being consumed by the load, is delivered to the system.

Example 5.2

In order to size a PV system for a stand-alone (off-grid) application, some basic requirements must be met.

1. Energy demand estimation
2. Solar resource availability
3. PV system sizing
4. Charge regulator
5. Battery bank
6. Solar inverter

A brief explanation is given in the next sections to size a PV system correctly.

Step 1: Energy demand estimation

As the PV system is thought to supply 100% of the load demand, the total energy demand must be estimated. Table 5.1 indicates some common appliances with their respective power demand and average hours of usage per day.

By computing the total energy per appliance (Table 5.2):

$$Total\ demand\left(\frac{Wh}{day}\right) = (P \cdot time \cdot \#units)$$

Table 5.1 Basic home appliances and typical power consumption for Example 5.2

Description	Type	Units	Power (W)	hrs/day	Energy (Wh/day)
Lightning	AC	8	10	5	
Fridge	DC	1	80	12	
TV	AC	1	120	5	
Laptop	AC	1	75	4	
Fan	AC	1	40	12	
Stereo system	AC	1	100	6	
Mobile charger	AC	4	5	3	
Washing machine	AC	1	500	1.5	

Table 5.2 Energy demand per home appliance for Example 5.2

Description	Type	Units	Power (W)	hrs/day	Energy (Wh/day)
Lightning	AC	8	10	5	400
Fridge	DC	1	80	12	960
TV	AC	1	120	5	600
Laptop	AC	1	75	4	300
Fan	AC	1	40	12	480
Stereo system	AC	1	100	6	257
Mobile charger	AC	4	5	3	60
Washing machine	AC	1	500	1.5	107

The total demand is:

$$Total\ demand\left(\frac{Wh}{day}\right) = \sum_{i=1}^{n=8}(P_i \cdot time_i \cdot \#units_i)$$

$$Total\ demand\left(\frac{Wh}{day}\right) = (400 + 960 + 600 + 300 + 480 + 257 + 60 + 107)\frac{Wh}{day}$$

$$Total\ demand = 3164\frac{Wh}{day}$$

The demand per type of appliance, considering a safety factor of 15% and a power factor of 0.95 lagging:

$$L_{DC} = E_{DC} \cdot 1.15$$

$$L_{DC} = 960\frac{Wh}{day} \cdot 1.15 = 1104\frac{Wh}{day}$$

$$L_{AC} = \frac{E_{AC} \cdot 1.15}{pf}$$

$$L_{AC} = \frac{2204\frac{Wh}{day} \cdot 1.15}{0.95} = 2668\frac{Wh}{day}$$

The total load of the PV system is then:

$$L_T = L_{DC} + L_{AC} = 3772 \frac{Wh}{day}$$

$$PVS_{demand} = 3772 \frac{Wh}{day}$$

Step 2: Solar resource availability

It is also important to consider the availability of the energy resource and its variability through the year according to the facility location. Such information can be obtained from websites with meteorology and solar energy data. The main relevant information for this example is the daily solar radiation and earth temperature of each month as shown in Table 5.3.

As the PV system performance is dependent on the environmental conditions, months with the higher and lower temperatures must be identified. Also, demand can vary with months and seasons; thus, months with higher demand/irradiation ratio are classified as critical months (Table 5.4).

In this case, November is the most critical month as the ratio demand/irradiation is higher.

Step 3: PV system design

Table 5.5 shows the technical specifications of a PV poly-crystalline module. Each module has a maximum power capacity of 270 W and open-circuit voltage of 38.8 V.

From the modules datasheet, information such as operating temperature conditions, short-circuit and open voltages characteristics are very useful for the system specification.

Table 5.3 Daily solar radiation per month for Example 5.2

Month	**Daily solar radiation horizontal (kWh/m²/d)**	**Earth temperature (°C)**
January	4.06	20.9
February	4.28	21.4
March	4.37	21.6
April	4.21	21.6
May	4.10	21.4
June	4.05	21.3
July	4.34	22.3
August	4.31	23.6
September	4.26	23.4
October	3.99	22.1
November	3.89	20.9
December	3.82	20.7
Annual	4.14	21.8

Table 5.4 Demand/irradiation ratio per month for Example 5.2

Month	Irradiation (kWh/m^2/day)	PSH (hours)	Demand		Demand/ Irradiation
			(%)	(kWh/day)	
January	4.06	4.06	45	1.7	0.42
February	4.28	4.28	100	3.7	0.88
March	4.37	4.37	100	3.7	0.86
April	4.21	4.21	100	3.7	0.90
May	4.10	4.10	100	3.7	0.92
June	4.05	4.05	45	1.7	0.42
July	4.34	4.34	45	1.7	0.39
August	4.31	4.31	100	3.7	0.88
September	4.26	4.26	100	3.7	0.89
October	3.99	3.99	45	1.7	0.43
November	3.89	**3.89**	100	**3.7**	**0.97**
December	3.82	3.82	45	1.7	0.44

Note: Bold values represent the selected values to be used in the exercise.

Table 5.5 Typical PV module datasheet for Example 5.2

	STC		NOCT
Maximum power (*P*max)	270 Wp		202 Wp
Maximum power voltage (*V*mp)	31.7 V		29.0 V
Maximum power current (*I*mp)	8.52 A		6.97 A
Open-circuit voltage (*V*oc)	38.8 V		35.6 V
Short-circuit current (*I*sc)	9.09 A		7.35 A
Module efficiency STC (%)	16.50%		
Operating temperature (°C)		−40 °C + 85 °C	
Maximum system voltage		1000 VDC (UL)	
Maximum series fuse rating		15 A	
Temperature coefficients of *P*max		−0.40 %/°C	
Temperature coefficients of *V*oc		−0.30 %/°C	
Temperature coefficients of *I*sc		0.06 %/°C	
Nominal operating cell temperature (NOCT)		45±2 °C	

The number of modules is determined as the ratio of the energy demand from the PV system and the energy considering the power output of each module, the peak sun-hours of the critical month and a module performance factor, assumed as 75%.

For this example, five modules are needed to fulfil the energy requirements, the total power of the PV system is equal to 1.35 kW. These modules can be arranged in series and parallel configuration depending on the voltage and current limitations of the inverter and regulator involved.

$$\#Modules = \frac{L_T}{P_{module} \cdot PSH_{critical\ month} \cdot PR}$$

$$\#Modules = \frac{3772\,\frac{Wh}{day}}{270W \cdot 3.89\frac{h}{day} \cdot 0.75} = 4.78 \approx 5\ modules$$

$$P_{PVS} = 5\ modules \cdot 270\ W = 1.35\ kW$$

As mentioned, temperature plays an important role on the PV modules performance. The maximum and minimum temperatures at which the modules are exposed to define the number of series and parallel elements of the array according to the voltage and current limits, respectively. This number of units also depends on the charge regulator parameter ranges and the operating cell temperature conditions. Figure 5.15 shows the variation of series and parallel modules with temperature.

For the maximum number of elements connected in series, the minimum operating cell temperature must be determined. If the temperature is reduced, the open-circuit voltage is then increased.

For the lower temperature scenario from the site data, this is 20.7 °C, a maximum number of three series units are obtained. As can be seen, the open-circuit voltage of the cell rises from 38.8 V to 41.21 V due to a reduction in the operating temperature.

$$T_{p\ \min} = T_{a\ \min} + \frac{T_{NOC} - 20}{800\frac{W}{m^2}} * I_a$$

$$T_{p\ \min} = 20.7\ °\mathrm{C} + \frac{45\ °\mathrm{C} - 20}{800\frac{W}{m^2}} * \frac{3820\frac{Wh}{m^2}}{4.5\ hours} = 47.23\ °\mathrm{C}$$

$$V_{oc(T_{p\ \min})} = \left(1 + \frac{(T_{P\ \min} - 25\ °\mathrm{C}) * \Delta V_{oc}\%}{100}\right) * V_{oc(STC)}$$

$$V_{oc(T_{p\ \min})} = \left(1 + \frac{(47.23\ °\mathrm{C} - 25\ °\mathrm{C}) * 0.3}{100}\right) * 38.8\ Vdc = 41.39\ Vdc$$

Hence,

$$n_{series\_Max} = \frac{V_{Max-Equip}}{V_{oc(T_{p\ \min})}}$$

$$n_{series\_Max} = \frac{145\ Vdc}{41.2\ Vdc} = 3.503\ \approx 3\ series\ units,\ maximum$$

For the minimum number of elements connected in series, the maximum operating cell temperature must be determined. If the temperature is increased, the open-circuit voltage then decreases.

On the other hand, for the higher temperature scenario, this is 23.6 °C, a minimum number of two series units are obtained. The open-circuit voltage of the cell falls from 38.8 V to 35.04 V due to an increase in the cell operating temperature.

$$T_{p\ \max} = T_{a\ \max} + \frac{T_{NOC} - 20}{800\frac{W}{m^2}} * I_a$$

$$T_{p\ \max} = 23.6\ °\mathrm{C} + \frac{45\ °\mathrm{C} - 20}{800\frac{W}{m^2}} * \frac{4310\frac{Wh}{m^2}}{4.0\ hours} = 57.27\ °\mathrm{C}$$

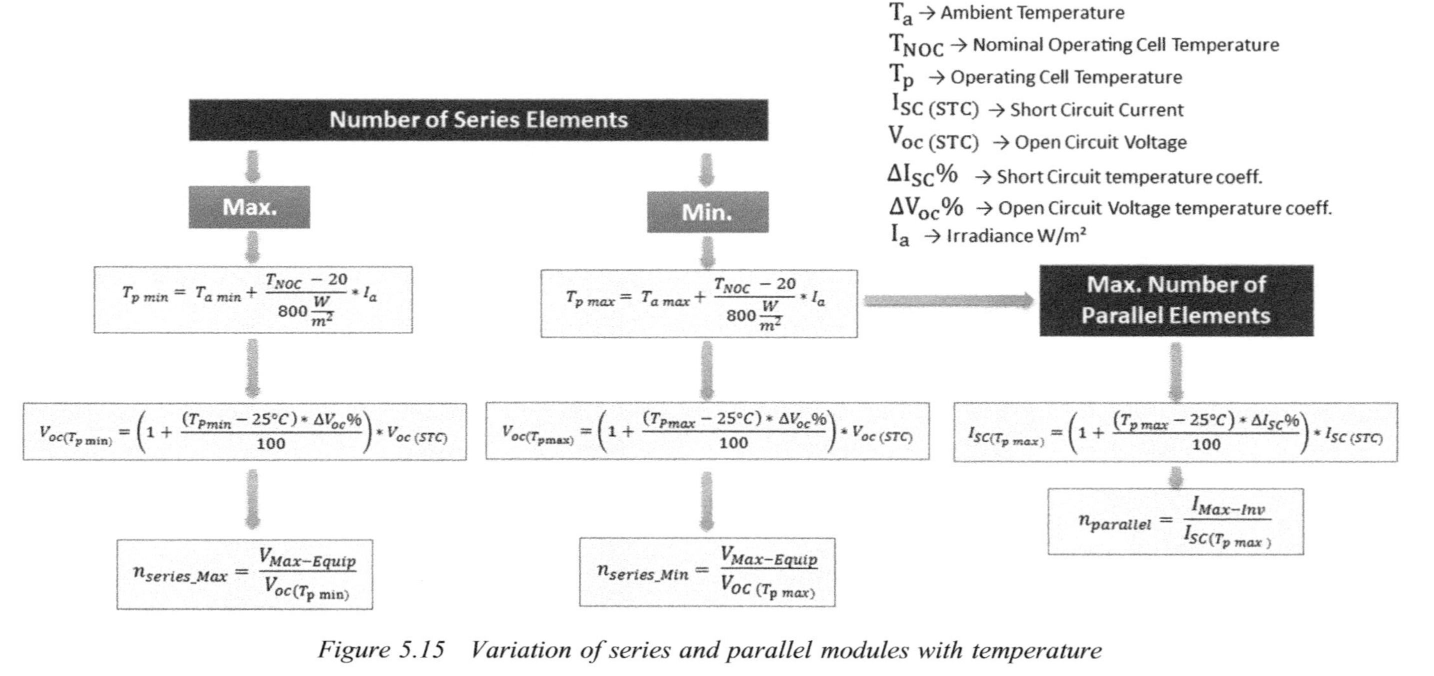

Figure 5.15 Variation of series and parallel modules with temperature

$$V_{oc(T_{p\,\max})} = \left(1 + \frac{(T_{P\max} - 25\,^{\circ}C) * \Delta V_{oc}\%}{100}\right) * V_{oc(STC)}$$

$$V_{oc(T_{p\,\max})} = \left(1 + \frac{(57.27\,^{\circ}\text{C} - 25\,^{\circ}\text{C}) * 0.3}{100}\right) * 38.8\ Vdc = 35.04\ Vdc$$

$$n_{series\_Min} = \frac{V_{Min-Equip}}{V_{oc(T_{p\,\max})}}$$

$$n_{series\_Min} = \frac{60\ Vdc}{35.04\ Vdc} = 1.71\ \approx 2\ series\ units,\ minimum$$

Real PV cells or modules can be modelled as dependent current sources, representing the dependency and proportionality of the cell to the incident irradiance. Therefore, as the irradiance increases, the short-circuit current also increases. As can be noticed, the short-circuit current increases from 9.09 A to 9.27 A due to the irradiance.

This current capacity yields a maximum of five parallel units connected to the regulator.

$$T_{p\,\max} = T_{a\,\max} + \frac{T_{NOC}\ - 20}{800\,\frac{W}{m^2}} * I_a$$

$$T_{p\,\max} = 23.6\,^{\circ}\text{C} + \frac{45\,^{\circ}\text{C}\ - 20}{800\,\frac{W}{m^2}} * \frac{4310\,\frac{Wh}{m^2}}{4.0\ hours} = 57.27\,^{\circ}\text{C}$$

$$I_{sc(T_{p\,\max})} = \left(1 + \frac{(T_{P\,\max} - 25\,^{\circ}C) * \Delta I_{sc}\%}{100}\right) * I_{sc(STC)}$$

$$I_{sc(T_{p\,\max})} = \left(1 + \frac{(57.27\,^{\circ}\text{C} - 25\,^{\circ}\text{C}) * 0.06}{100}\right) * 9.09\ A_{dc} = 9.27\ A_{dc}$$

$$n_{parallel} = \frac{I_{Max-Equip}}{I_{sc(T_{p\,\max})}}$$

$$n_{parallel} = \frac{50\ A_{dc}}{9.27\ A_{dc}} = 5.39\ \approx 5\ Parallel\ strings$$

Step 4: Charge regulator

A charge regulator is chosen to supply a battery energy storage system if needed; also, the regulator is useful for DC/DC voltage control. Table 5.6 shows the charge regulator datasheet for this example. The 1.6 kW regulator has DC voltage ranges from 60 to 115 V with a maximum PV array open-circuit voltage of 145 V and a maximum operating current of 50 A. Hence, the PV system design must be in accordance with these parameters.

Table 5.6 Charge regulator datasheet for Example 5.2

Input	**Values**
Operating voltage	60 V_{DC}–115 V_{DC}
Maximum PV array open-circuit voltage	145 V_{DC}
Maximum PV array power	1600 W
Maximum current	50 A
Output	**Values**
Nominal battery voltage	48 V_{DC}
Maximum charging current	60 A

For this, a brief verification must be done:

$$P_{PVS} = 5\ Modules * 270\ W_p = 1.35\ kW_p$$

$$I_{in} = I_{SC\_Panel}\ * \#Parallel\ Strings[A]$$

$$I_{in} = 9.09\ A_{dc} * 2 = 18.18\ A$$

$$I_{out} = \frac{P_{DC} + \frac{P_{AC}}{0.95}}{V_{bus\_DC}}\ [A]$$

$$I_{out} = \frac{80\ W + \frac{850\ W}{0.95}}{48\ V_{dc}} = 20.31\ A$$

It is important to validate that the array design accomplishes the regulator requirements. A configuration of two parallel strings is chosen. The total regulator input current from the PV system is 18.18 A and the output current is equal to 20.31 A, which is in accordance with the regulator specifications as shown in the datasheet.

Step 5: Battery bank

The charge of the battery storage system is calculated with a safety factor of 10%.

$$Q_{Ah} = \frac{L_t * D_{aut}}{V_{bus\_DC} * P_d} * 1.1\ [Ah]$$

$$Q_{Ah} = \frac{3772\ \frac{Wh}{day} * 2}{48V_{DC}\ * 0.5} * 1.1\ [Ah]$$

$$Q_{Ah} = 346\ Ah;48\ Vdc$$

where:

L_T is the total demand
D_{aut} is the days of autonomy
V_{bus_DC} is the DC bus voltage
P_d is the discharge depth

Step 6: Solar inverter

The solar inverter is chosen considering the voltage level and the power output oversized by 25% of the PV system capacity.

$$P_{Inverter} = 1.25 * P_T \; [VA]$$

$$P_{Inverter} = 1.25 * (850/0.95 + 80) \; [VA]$$

$$P_{Inverter} = 1218 \; VA$$

Table 5.7 shows the selected solar inverter.

The final arrangement of the PV system is shown in Figure 5.16.

5.2.3 Wind generation

The energy available from the wind drafts is obtained from kinetic energy due to large masses of air moving over the earth atmosphere [14]. The blades of a wind

Table 5.7 Inverter datasheet for Example 5.2

Output	**Values**
Output voltage	120 V_{AC}±3%
Maximum output current	12.5 A
Output frequency	60 Hz±1%
Total harmonic distortion	<3%
Continuous output power (at pf = 1)	1500 W
Input	**Values**
Nominal DC input voltage	48 V_{DC}
Maximum input current	50 A

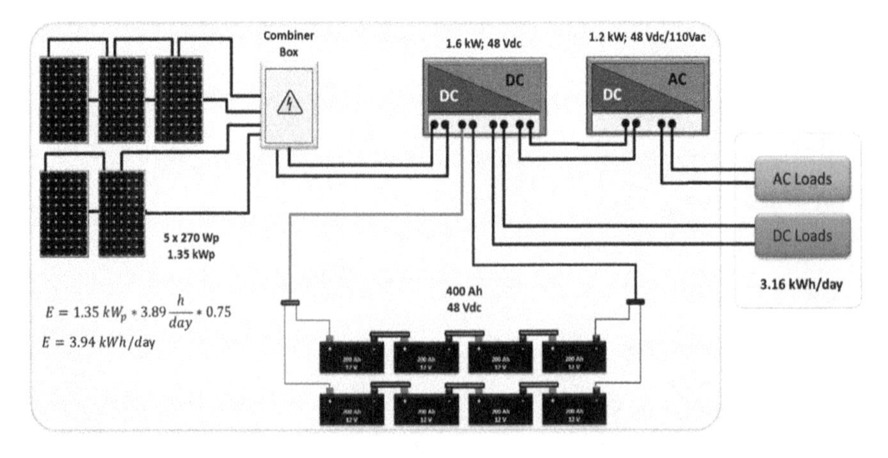

Figure 5.16 PV arrangement for Example 5.2

turbine allow the transformation of this kinetic energy into mechanical energy and then electrical energy through a system of gearboxes, a turbine and the generator. The total system efficiency depends greatly on the efficiency with which the rotor interacts with the wind stream.

The energy of a wind stream can be estimated with the total air mass (m) and the moving velocity (V) as follows:

$$E_K = \frac{1}{2} m V^2 \tag{5.5}$$

The total amount of kinetic energy available for a wind turbine can be determined considering the wind rotor cross sectional area A exposed to the wind stream or parcel. This is illustrated in Figure 5.17.

The total air interacting with the rotor per unit time can be defined in terms of its cross-sectional area, which is equal to the area covered by the rotor (A_T) and thickness in function of the wind velocity (V). The air density and volume available to the rotor are denoted as ρ_a and v, respectively. Therefore, the expression for kinetic energy can be defined as follows:

$$E_K = \frac{1}{2} \rho_a v \cdot V^2 \tag{5.6}$$

Hence, the power, which is energy per unit time, is expressed as shown as follows:

$$P = \frac{1}{2} \rho_a A_T V^3 \tag{5.7}$$

These expressions show that power of the turbines is defined by air density, the area of the wind rotor and the wind velocity. The geographical location and the rotor size are relevant to the design. Wind velocity is the most relevant design factor, since this term has a cubic effect in the previous equation, and this helps to reduce the cost of the wind turbine. Therefore, rotors can be reduced eight times to obtain the same power.

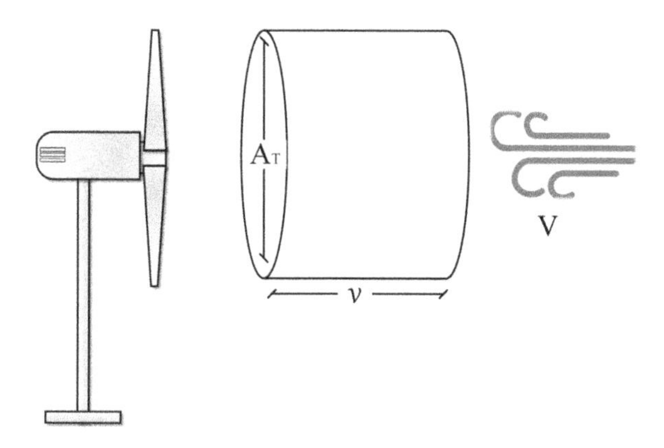

Figure 5.17 Air parcel interacting with a wind turbine rotor [13]

The efficiency of a wind turbine is defined in terms of the power coefficient (C_P), which represents the part of the kinetic energy from the air mass that is transferred to the rotor and turned into electric energy (the rest of the air mass does not interact with the blades). Therefore, the power coefficient is the ratio of actual power developed by the rotor to the theoretical power available in the wind parcel, as indicated in the following expression:

$$C_P = \frac{2P_T}{\rho_a A_T V^3} \tag{5.8}$$

where P_T is the total power developed by the turbine. The power coefficient of a turbine depends on many factors such as the profile of the rotor blades, blade arrangement and setting [15].

The power developed by a rotor depends on the relative speed between the blade and the wind speed. For instance, if the blades are moving slow, a portion of the air parcel may pass without interacting with the turbine. Similarly, if the rotor is rotating fast and the wind is very low, the wind may be deflected from the turbine and the turbulence will produce a poor energy conversion. This is known as the tip-speed ratio (TSR or λ), which is the ratio between the velocity of the rotor tip and the wind velocity, as shown in the following expression:

$$\lambda = \frac{R\Omega}{V} = \frac{2\pi NR}{V} \tag{5.9}$$

where R radius, N is the rotational speed and Ω the angular velocity of the rotor.

The power coefficient can be obtained through experimental tests and simulations of the wind turbine performance. Nonetheless, some numerical approximations can be done as shown in [5] by expressing the C_P as a nonlinear function of the TSR and pitch angle of the blades, both dependent on the wind speed. This is shown in the following equations:

$$C_P(\lambda, \beta) = \frac{1}{2}\left(\frac{98}{\lambda_i} - 0.4\beta - 5\right) \cdot e^{\frac{-16.5}{\lambda_i}} \tag{5.10}$$

$$\lambda_i = \left[\frac{1}{\lambda + 0.089} - \frac{0.035}{(\beta^3 + 1)}\right]^{-1} \tag{5.11}$$

where:

β is the blade pitch angle in degrees
λ is the tip-speed ratio

Typical modern wind turbines operate at TSR ranges between 5 and 10, close to the maximum value of C_P. The maximum limit is known as the 'Betz's limit', a theoretical maximum that states that only wind power up to 59.3% can be achieved.

Figure 5.18 shows several C_P curves considering the pitch angle for a three-blade wind turbine. As can be seen, the maximum C_P obtained is near 0.47.

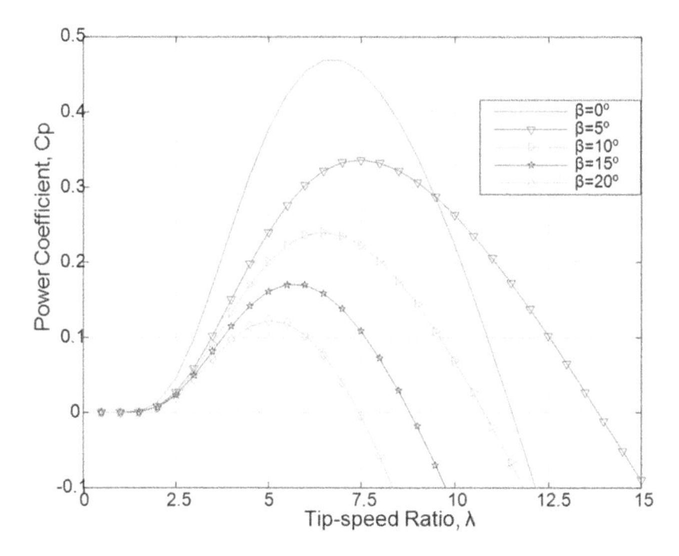

Figure 5.18 Power coefficient as a function of the TSR

Wind turbines are categorised according to the way the blades interact with the wind, the orientation of the rotor axis and the type of generator. The following sections discuss in more detail their characteristics [14,16].

Drag blades are characterised by the air pushing the blade itself, forcing the rotor to turn on its axis. This type of blade is very limited as its efficiency since the blades cannot turn faster than the wind speed.

Lift-bladed turbines use airfoils, which has the advantage of moving faster than the wind speed and being more efficient due to its aerodynamic construction.

Wind turbines can be also classified as horizontal axis or vertical axis turbines. Horizontal axis wind turbines (HAWT) rotate with their axis parallel to the wind stream and to the ground. This is the most common and commercial type of wind turbine. The components of this type of turbine are shown in Figure 5.19.

The main advantages of the HAWT include a low cut-in wind speed, access to high wind streams because of their tower height, relatively high-power coefficient and high efficiency. Their disadvantages include a complex design and maintenance, massive structure and complex control systems. Also, their generators, gearboxes and power transformers are placed over the tower. In some new applications, the power transformer is located outside the nacelle to reduce space and material.

The other turbine type according to its axis orientations is the vertical axis wind turbines (VAWT). These turbines rotate with their axis perpendicular to the wind stream and to the ground. The advantages of the VAWT include receiving wind from any direction, simpler and more economical designs, maintenance can be done at ground level and complex yaw mechanisms are no longer needed. Disadvantages of the VAWT include a lower wind speed because the height of

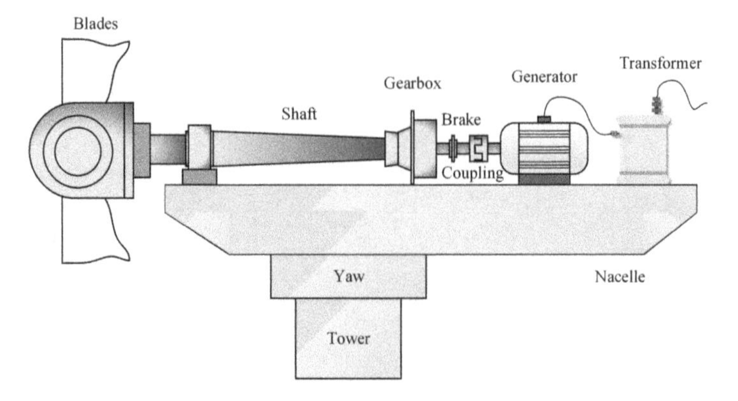

Figure 5.19 Mechanical structure of a HAWT [15]

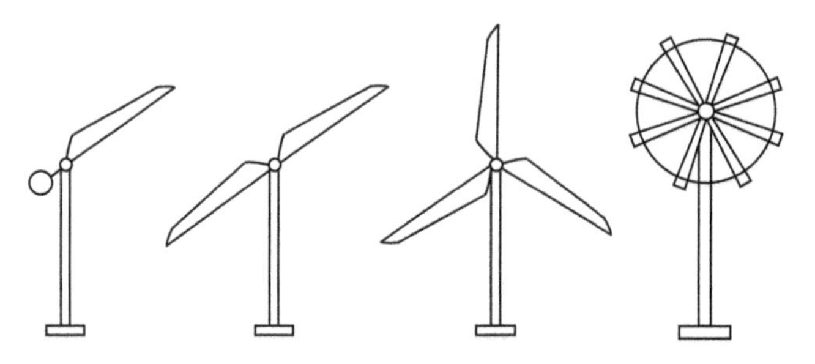

Figure 5.20 Single bladed, two bladed, three bladed and multi-bladed wind turbines [14]

towers, which makes less power available and air flow around the ground can create turbulent flows. Also, their inertia is higher, which requires more power to start up the turbine.

Horizontal axis turbines can also be classified considering the number of blades. Construction of single-, two-, three- and multi-bladed systems can be implemented. These configurations are illustrated in Figure 5.20.

Single-bladed turbines use less materials, which makes them cheaper. However, a counterweight is needed to balance the blade. Two-bladed turbines also need a counterweight and drag losses can be minimal. Most of the commercial wind turbines used for electricity generation are three-bladed. Multi-bladed systems are often used for water pumping applications because of the need for high starting torque.

According to the speed of rotation of the rotor, wind turbines are classified into fixed-speed wind turbines (FSWT) or variable-speed wind turbines (VSWT). FSWT are simpler to operate because they generate electricity when the wind speed makes the generator shaft speed higher than the synchronous speed. These turbines are limited in the generated power, although require less maintenance and are cheaper in comparison with the other type. FSWT operates almost linearly, as illustrated in Figure 5.21. When the speed is outside the linear region, the turbine enters an unstable nonlinear region. This type is known as 'constant' or 'fixed speed' because of the narrow speed range.

VSWT systems are more complex than FSWT because of the power converters that allow them to generate electricity at a wide range of turbine speeds. A voltage is injected in the rotor circuit of the generator to reach different torque–speed characteristics.

Wind turbine generators may have different types according to the generator configurations [17,18].

The type 1 configuration corresponds to the squirrel cage induction generator (SCIG). This type of wing turbine generator uses a fixed speed turbine with a SCIG. The induction generator generates electricity when it is driven above synchronous speed. A negative slip indicates that the wind turbine operates in generating mode, in which the operating slips for an induction generator are normally between 0% and −1%. Figure 5.22 shows the simplified single-phase equivalent circuit of a squirrel-cage induction machine and Figure 5.23 shows an example single-line connection diagram.

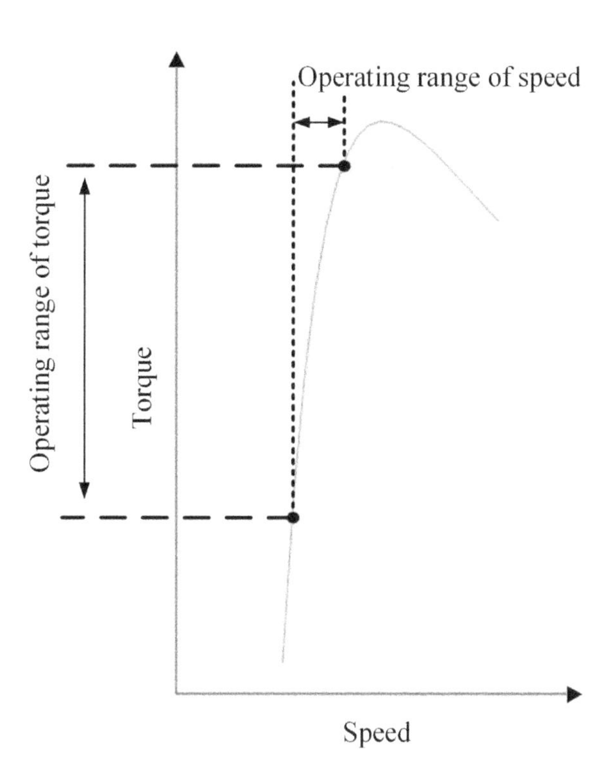

Figure 5.21 Torque–speed characteristic of a FSWT [15]

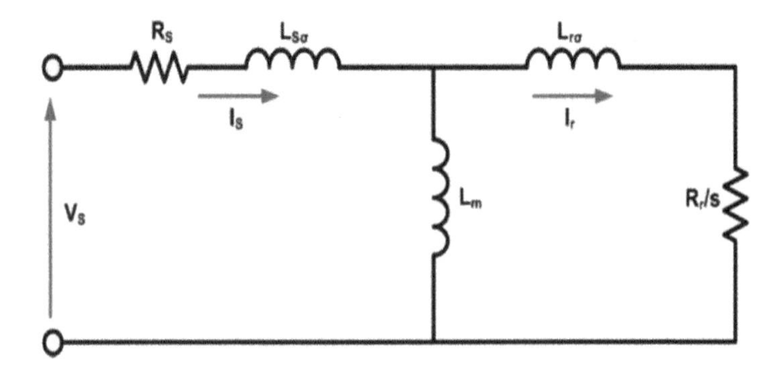

Figure 5.22 Equivalent circuit of a type 1 generator [17]

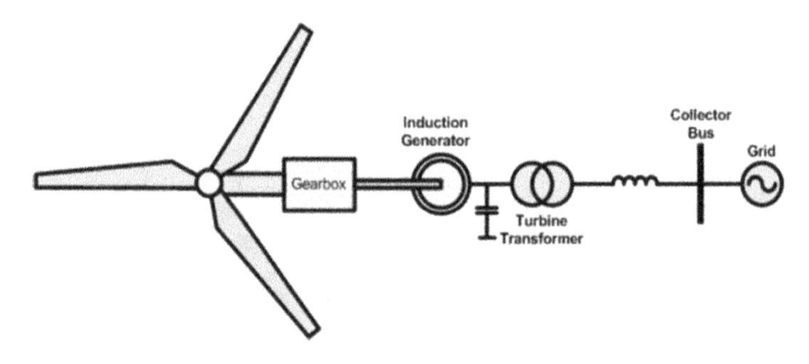

Figure 5.23 Circuit diagram of a type 1 generator [17]

Figure 5.22 is referred to the stator where R_S and R_r are stator and rotor resistances, $L_{S\sigma}$ and $L_{r\sigma}$ are stator and rotor leakage inductances, L_m is magnetising reactance and S is rotor slip. In the case of voltage fault, the inertia of the wind rotor drives the generator after the voltage drops at the generator terminals. The rotor flux may not change instantaneously right after the voltage drop due to fault. Therefore, voltage is produced at the generator terminals causing fault current flow into the fault until the rotor flux decays to zero. This process takes a few electrical cycles.

The type 2 configuration corresponds to a wound-rotor induction generator with a variable external rotor resistance. A three-phase rotor winding is connected to a power electronic component and three-phase external resistance. The external rotor-resistance controller (ERRC) is a very fast controller that allows the effective rotor resistance to vary; thus, the torque speed characteristic of this type of generator can be shaped accordingly. The equivalent circuit for a type 2 generator is illustrated in Figures 5.24 and 5.25 shows an example single-line connection diagram.

The type 3 configuration is called doubly fed induction generator. This type of wind turbine generator is implemented by a doubly fed induction generator. The

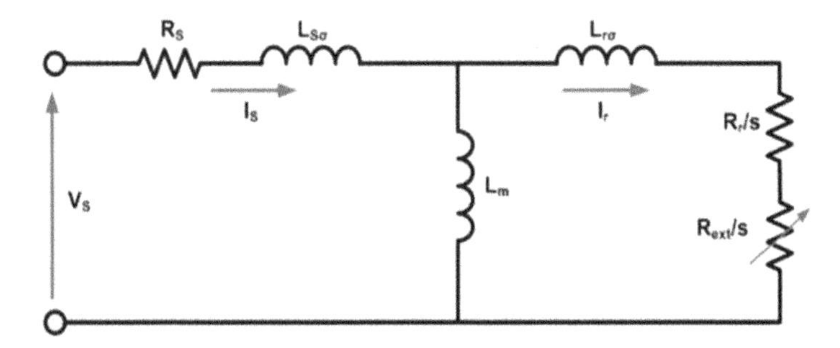

Figure 5.24 Equivalent circuit of a type 2 generator [17]

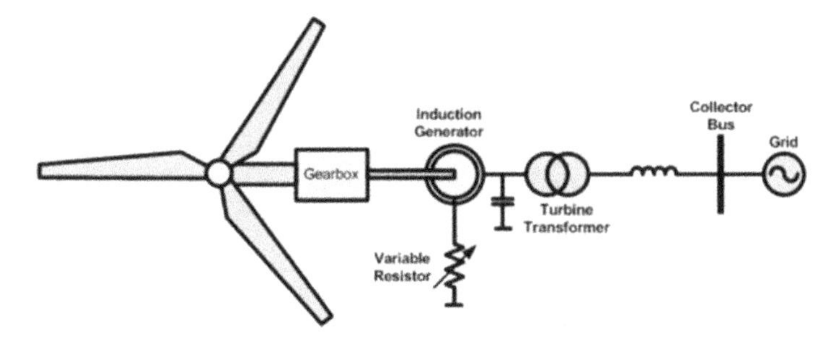

Figure 5.25 Circuit diagram of a type 2 generator [17]

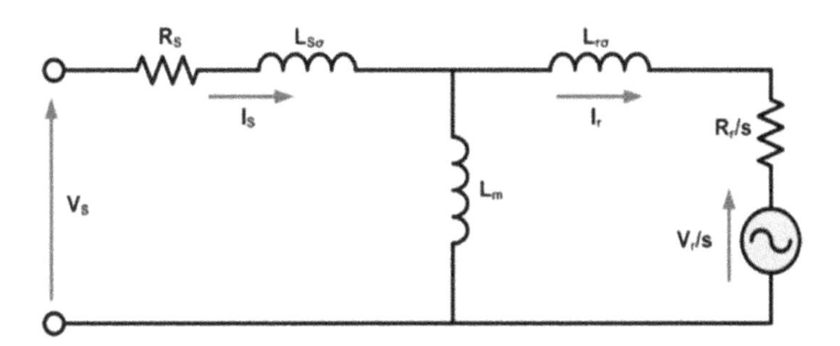

Figure 5.26 Equivalent circuit of a type 3 generator [17]

rotor speed is allowed to vary between 0.3 slip and −0.3 slip; thus, the power converter can be sized to about 30% of rated power (partial rating). Maximum energy yield is accomplished for low to medium wind speeds. Above rated wind speeds, the aerodynamic power is controlled by pitch to limit rotor speed and minimise mechanical loads. The equivalent circuit for a type 3 generator is illustrated in Figures 5.26 and 5.27 shows an example single-line connection diagram.

The type 4 wind turbine corresponds to a full-converter wind turbine generator. This is a variable-speed wind turbine generator implemented with full power

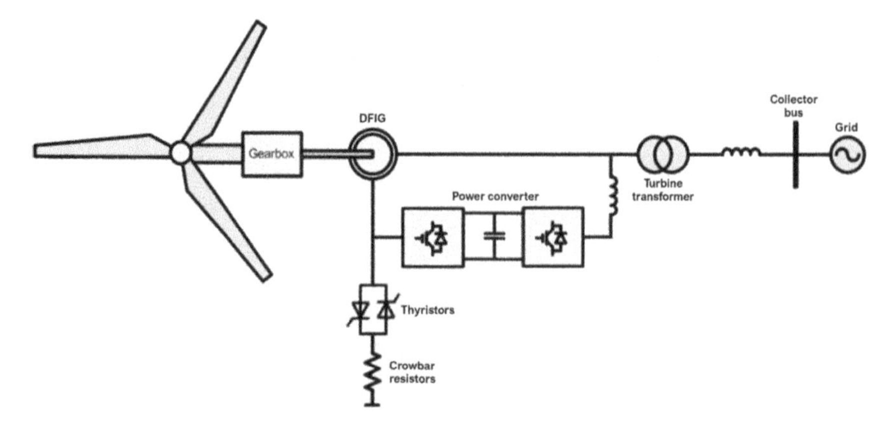

Figure 5.27 Circuit diagram of a type 3 generator [17]

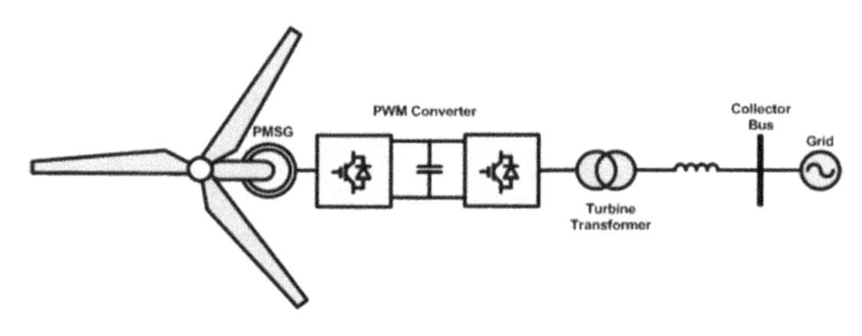

Figure 5.28 Circuit diagram of a type 4 generator [17]

conversion. Recent advances and lower cost of power electronics make it feasible to build variable-speed wind turbines with power converters with the same rating as the turbines. Maximum energy yield is accomplished for low to medium wind speeds. Above rated wind speeds, the aerodynamic power is controlled by pitch to limit rotor speed and to minimise mechanical loads. With the use of a power converter, the real and reactive power can be controlled independently and instantaneously within design limits. Figure 5.28 shows an example of a single-line connection diagram.

The short-circuit current contribution for a three-phase fault is limited to its rated current or a little above its rated current. An overload capability of 10% above rated power is a common practice to design a power converter for a type 4 wind turbine. The generator stays connected to the power converter in any fault condition and is isolated from the faulted lines on the grid. Although there is a fault on the grid, the generator output current is controlled to stay within the current limit (e.g., 1.1 p.u.).

Wind turbines have several types of control mechanisms [5,15,19]. The most common are soft starting, generation control, pitch control and yaw control. Other control strategies also implemented are feathering, reactive power control, stability, low and high voltage ride through and ramp control. The former will be discussed in this section.

The soft starting of a wind turbine is an electrical device used in fixed-speed turbines during the grid connection process, which reduces the inrush-current and to avoid voltage disturbances into the grid.

Most of the modern turbines implement this type of control to regulate the output power. The blades are pitched as the wind speed varies in order to obtain a better performance. As the wind speed increases, the blades turn into the opposite direction of the wind to increase the attack angle and to control its power at higher wind speeds without any damage. This type of control depends on the operation region of the wind turbine by modifying the angle of the blades. This is done to achieve a higher aerodynamic efficiency.

The operation region is shown in Figure 5.29. Phase I (wind speed below 5 m/s) indicates that the generator is not generating any power since the wind speed is not high enough. In Phase II, the output power is increased gradually with the wind speed up to its rated power (wind speed around 12 m/s). In this phase, the pitch angle maximises the output power by tracking the maximum C_P, until the output power remains constant at its rated value at Phase III. Control mechanisms are applied to avoid overcurrent and/or overloading. When the wind speed exceeds the cut-out speed in Phase IV (wind speed around 25 m/s), the blades are adjusted to limit the output power until they are stopped.

Yaw control is useful to face the wind turbine to the wind stream. It is used to turn the rotor partly away at higher wind speeds. When the wind speed exceeds the cut-off limit of the wind turbine, the axis position is adjusted to a nearly perpendicular position with respect to the wind; thus the generated power is reduced.

When large amounts of power are required, wind turbines can be gathered in a cluster, forming what is known as a wind farm [15]. Some of the advantages of clustering wind turbines are in terms of an easier installation than several scattered units, operation and maintenance (O&M) and power transmission efficiency.

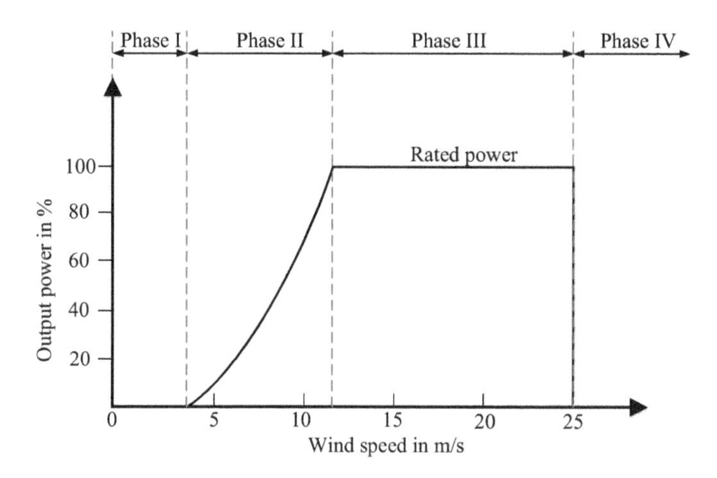

Figure 5.29 Phases of wind power generation [15]

The wind turbines in a wind farm are arranged in rows near each other. Factors like wind direction, topographic and environmental conditions, available area and existing electrical network must be considered.

To avoid windshadow and wind turbulence due to nearby units' safety distances are needed. In [14], a spacing of 3DT to 4DT within the rows is suggested, where DT is the rotor diameter. Between rows, a spacing of 10DT is recommended to restore the wind stream before it interacts with the turbines in the next row. This is illustrated in Figure 5.30.

Alternatively, the number of turbines per row (N_{TR}) and the number of row (N_R) can be calculated as follows:

$$N_{TR} = \frac{L_R}{S_R} + 1 \tag{5.12}$$

where:

L_R is the row length
S_R is the row spacing.

Considering:

$$N_T = \frac{P_F}{P_T} \tag{5.13}$$

where:

N_T is the total number of turbines in the farm
P_F is the total capacity of the wind farm
P_T is the rated power of a single turbine

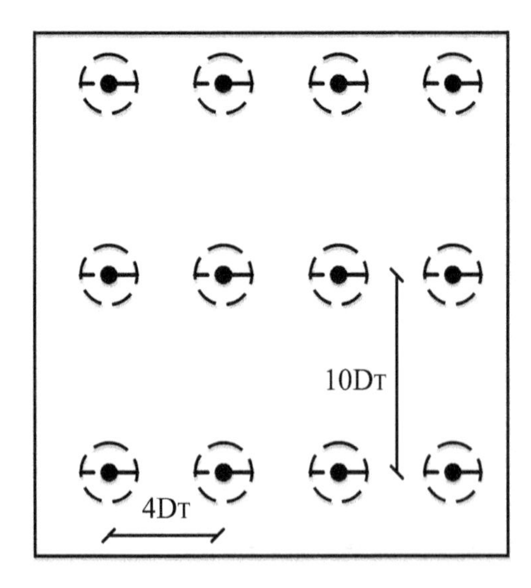

Figure 5.30 Wind farm layout

Hence, the total number of rows is:

$$N_R = \frac{N_T}{N_{TR}} \tag{5.14}$$

In some cases, areas with high wind speeds are located far from the habitable zones and existing electrical networks. Therefore, a small network must be installed to take advantage of the wind resource. Offshore wind farms are an example of the latter. Offshore farms often offer higher efficiency and productivity than onshore wind farms.

5.3 Small hydroelectric plants

Hydro power plants represent a mature technology with a well-known operational scheme [3,20]. A dam is constructed to collect water that powers the turbines, as shown in Figure 5.31. Some hydro schemes can experience large variations of water according to the dam's capacities and seasonal rainfalls. Small hydro power plants become an interesting option for customers close to mountain areas, in which it is not possible to construct large hydro power plants. They can be constructed fast with a technology that offers over 80% of efficiency. They have automatic operating systems and low operational and maintenance costs.

Most of these plants operate in grid connected mode, but several units operate in isolated operation mode, which requires an exact match between the demand for and the supply of electrical energy. The generation can be controlled considering that criteria, in which the dam is designed to match the required power demand. In some scenarios in which the water capacity surpasses the required amount, it must be bypassed by the turbines. The available power from a hydro plant is calculated by using the following expression:

$$P_H = \eta \cdot g \cdot Q \cdot H \, [kW] \tag{5.15}$$

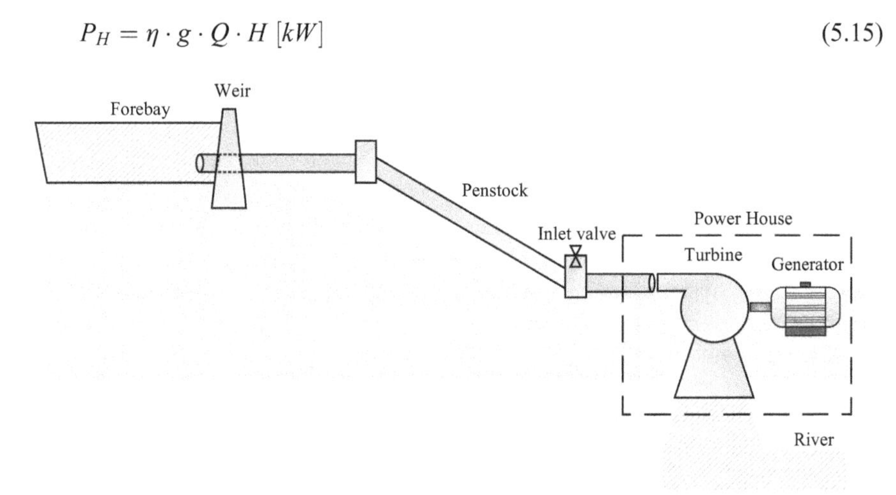

Figure 5.31 Small-scale hydropower plant

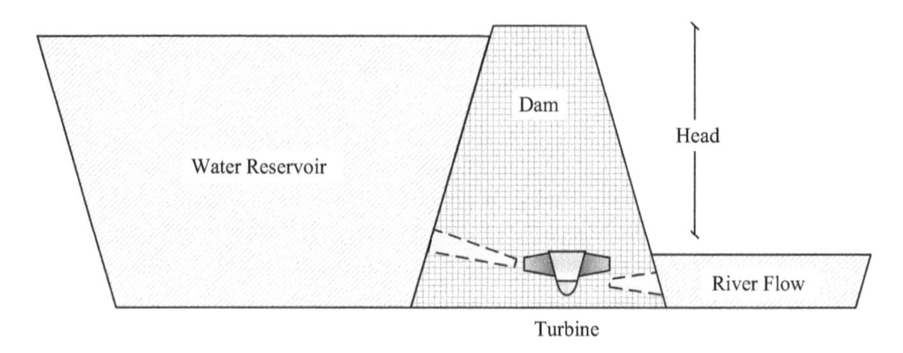

Figure 5.32 Schematic diagram of a hydroelectric power plant with water reservoir

where η is the efficiency of the turbine, g the gravity (9.81 m/s^2), Q the total flow through the turbine and H is the head [8].

Hydro power plants are classified according to the head of the installation, which is the vertical distance through which the water falls and by linking it with the flow. These are low head, middle head and high head hydro power plants. Low head hydro plants are rivers and channels with a small downward gradient. They have the water supplied to the turbine practically without storage. Middle head hydro power plants are built exclusively in connection with dams (control works), which have heights up to 200 m. The sole of the dam uses pressure pipelines or penstocks to supply water to the powerhouse. High head hydro power plants are in central and high mountains with small flow rates up to 2000 m, from where the energy comes rather than from the flow rate.

In some applications, small hydro plants with water reservoirs can be used as a storage system for wind energy systems, which have the storage capacity able to support the total wind energy supply. Figure 5.32 shows the schematic diagram of a hydroelectric power plant with a water reservoir.

5.4 Energy storage systems

Due to the uncertainty nature associated with some sources such as wind or solar energy, it is required to support their generation capacity by storage units. The energy storage systems are becoming very important in the increased utilisation of the intermittent and periodic alternative energy sources such as wind, solar, tidal, wave energy, among others [21–24]. These storage systems get the energy when there is surplus after attending the local demand.

Some of the most relevant advantages of using energy storage devices on a distribution network are the following [21,25–28]:

- Capacity firming: Storage units can help to maintain a committed level of supplied energy for a period of time. They smooth the output and control the ramp rate (MW/min) to eliminate rapid voltage and power swings on the electrical grid.

- Peak shaving: This helps to respond to high variability load demands, which stress system requirements and makes the operation more expensive.
- Voltage support: These units help to maintain the grid voltage by injecting or absorbing both active and reactive power when the system presents operational voltage values outside the allowed ranges.
- Frequency regulation: This approach for regulating frequency is an attractive option due to its rapid response time, which is charged or discharged according to the system demand.
- Power quality: In power quality applications, an energy storage system helps protect downstream loads against short-duration events that affect the quality indices.

Example 5.3

An energy storage device is connected to the PV system shown in Figure 5.33. The system is the same as Figure 5.13 where the energy storage device has been added. The example has been modelled and run with a PSAT and the results are depicted in Figure 5.34 that shows the load flow profile when the power output from the PV system is reduced. No power from the grid is then needed.

5.4.1 Type of energy storage systems

The most frequently used systems for the storage of energy are presented in this section [8,22,24]. There are some based on electromechanical principles such as pump systems or flywheels, and based on thermal and chemical systems as batteries, fuel cells and supercapacitors.

5.4.1.1 Electromechanical storage

There are two types of energy storage that result from the application of forces upon materials systems. One of these involves changes in potential energy (such as pumped hydro storage, compressed-air) and the other involves changes in the kinetic energy (such as flywheel energy storage).

5.4.1.2 Electrochemical storage

Chemical storage is a very desirable method for energy storage, which has a very high energy density (in kJ/kg) and is easy to be used for combustion or direct conversion to electricity. Additionally, an energy storage device with desirable properties must charge and discharge quickly. This allows providing large quantities of energy and high power when it is required. Some examples of the electrochemical energy storage devices are fuel cells, supercapacitors, superconducting magnetic energy storage and batteries.

One of the most common technologies used for power system applications are batteries, which can be used for a very wide range of applications, from assisting the very large-scale electrical grid down to tiny portable devices used for many

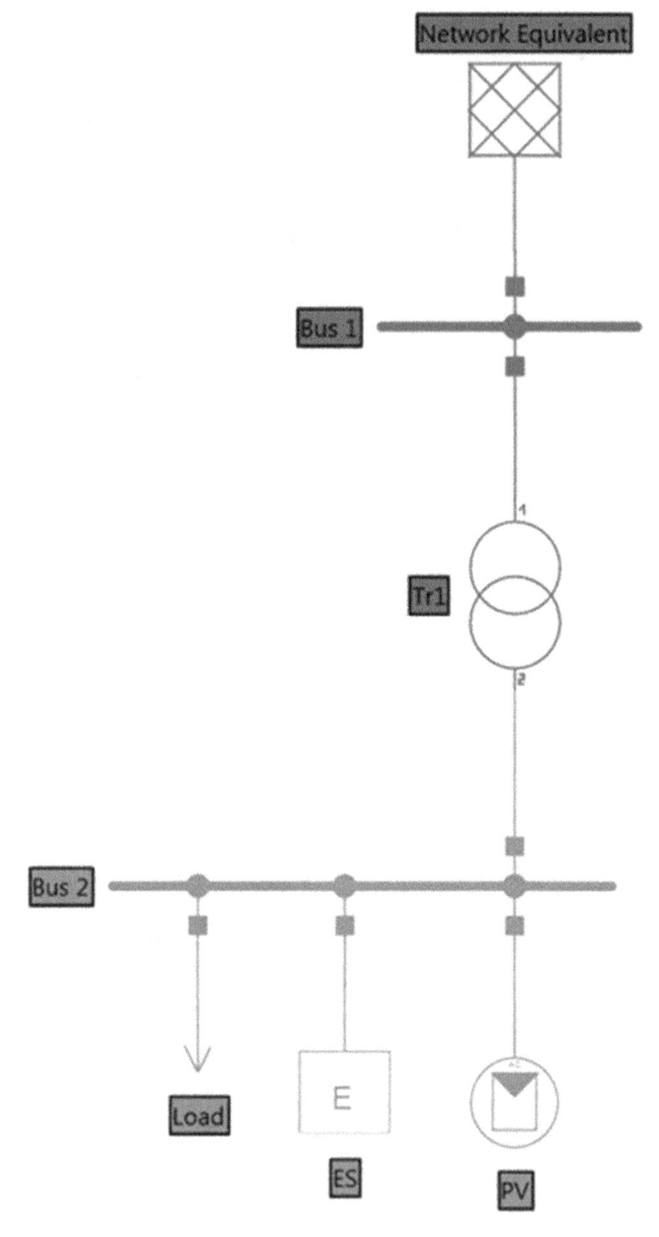

Figure 5.33 Integration of a PV system with energy storage in the power system

purposes. There are standard batteries such as lead acid or alkaline batteries, lithium batteries. Also, there are high temperature or thermal batteries, which are activated at high temperatures by the application of heat from an external source to liquefy the electrolyte.

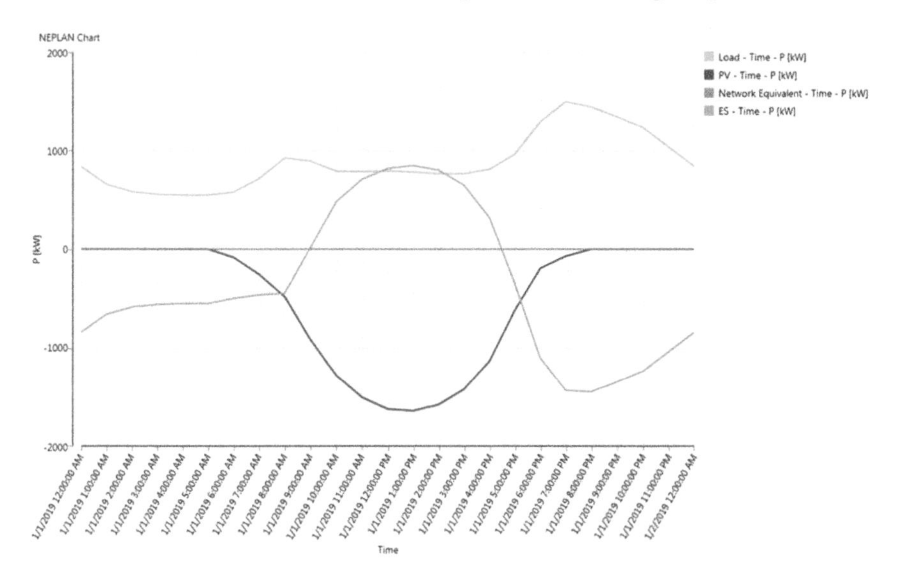

Figure 5.34 Example of a load flow profile through one typical day

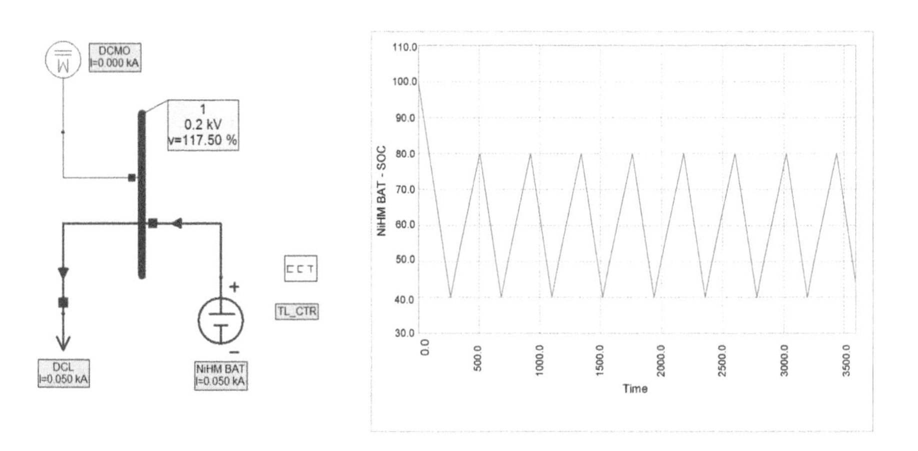

Figure 5.35 NiMH battery state of charge operation

The system illustrated in Figure 5.35 is taken from NEPLAN [29] and represents a 200 V-6.5 Ah NiMH battery connected to a constant DC load of 50 A. The DC machine (in shunt excitation) is connected in parallel with the load and operates freely at no load torque. When the state of charge (SOC) of the battery goes below 40%, a negative load torque of 200 Nm is applied to the machine, so it acts as a generator, recharging the battery and supplying the load. When the SOC goes above 80%, the load torque is removed so only the battery supplies the load. This control scheme is implemented in the control circuits block (CCT) of the corresponding battery that has been modelled in the referred software.

As illustrated above, the SOC determines the available capacity (in Ah) as a percentage of its rated capacity. Accurate SOC estimation is one of the main tasks of battery management systems, which will help improve the system performance and reliability, and will also lengthen the lifetime of the battery. In fact, precise SOC estimation of the battery can avoid unpredicted system interruption and prevent the batteries from being overcharged and over discharged [30].

In this chapter, the fundamentals of microgrid will be presented. It is important to understand the main characteristics of a microgrid as a subsystem and active component of the distribution network, as well as its operation modes.

Some general concepts of control strategies and protection solutions in microgrids are also explained, considering the impacts of high penetration of DG units and energy resources into the distribution network.

5.5 Introduction to microgrids

The incorporation of microgrids in the electric system has taken place because of the growing concern regarding energy availability and cost of kWh in some portions of electrical systems. Also, some infrastructure of current electrical transmission and distribution systems have been deteriorated by ageing. One efficient way to improve the performance of energy distribution is through the use of microgrids [31].

A microgrid can be described as a modern component of the electric network in which different components such as loads, DER and energy storage systems interact together through control devices. This makes microgrids seen as a single entity by the main grid. An illustration of this concept is shown in Figure 5.36 [32].

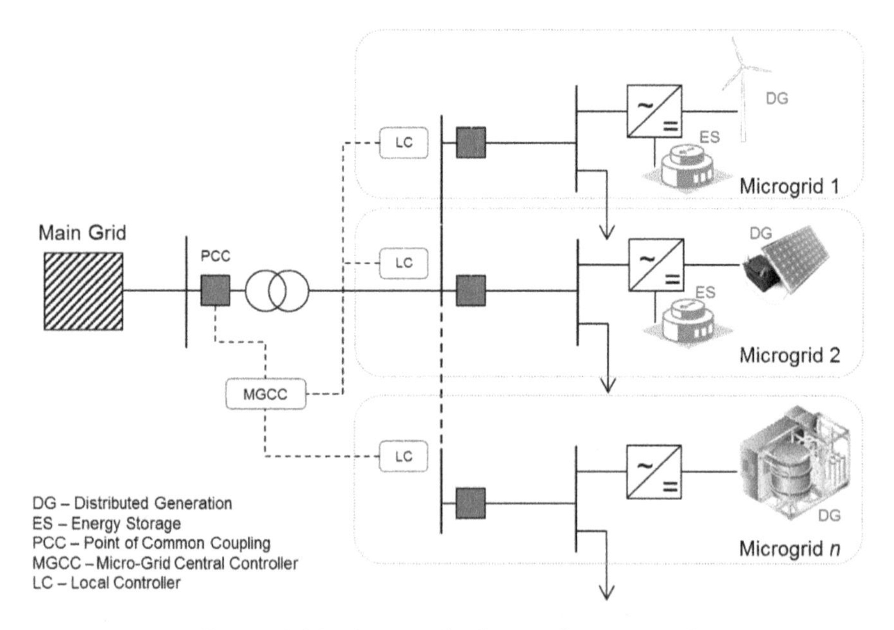

Figure 5.36 A general scheme of a microgrid

Generation technologies such as conventional DG units (micro turbines, fuel cells, diesel), renewable energy units (wind energy, solar PV and others), energy storage (thermal storage, batteries) and combined heat and power units are common components of microgrids.

The main motivations to build microgrids can be summarised as follows:

- Improve reliability since the quality indices values are becoming more stringent every day, and particularly those of SAIDI, SAIFI and CAIDI.
- Increase resiliency: which is the ability to withstand challenges and continue in operation. A clear example is the value of microgrids during storms by maintaining the power supply.
- Improve economic and infrastructure cost savings by avoiding investments for replacement and/or expansion of old power plants.
- Promote sustainability/emissions reduction.

There is no standard term to define a microgrid however; there are some international standards have incorporated related specifications for microgrids including IEEE 446-1995, *Recommended Practice for Emergency and Stand-by Power Systems for Industrial and Commercial Applications* [33]; IEEE 519-2022 [34], 'Recommended Practice and Requirements for Harmonic Control in Electric Power Systems', IEEE 1547-2018, 'Standard for Interconnection and Interoperability of Distributed Energy Resources with Associated Electric Power Systems Interfaces' [35]; IEC TS 62257 series, 'Recommendations for renewable energy and hybrid systems for rural electrification' [36]; and IEC TS 62898, 'Microgrids parts 1 & 2' [37,38].

5.6 Microgrid components

According to the microgrid definitions given above, the main components as suggested are presented as follows [5,31,39]:

- Distributed generation (DG)
- Energy storage (ES)
- Distributed energy resources
- Loads (traditional loads, plug-in electric vehicles, charging stations)
- Metering System (AMR/AMI)
- Control devices
- Point of common coupling (PCC)
- Protection devices

Additionally, the information and communication technology (ICT) plays a relevant role in modern electric power networks. Several technologies have been used or tested with distribution systems and microgrids such as advanced metering infrastructure (AMI) and smart meters, which can use wireless, optical communication and power line communication. Some common requirements of the communication system are defined to achieve bidirectional, real-time and efficient communication. The communication system should be established based on open and common standards, which may support high-speed and accurate communication between sensors, advanced electronics and applications [39].

5.7 Classification of microgrids

The previous definition of a microgrid described a group of interconnected loads with DERs and control devices. Nevertheless, it does not describe how these components are related or how they interact with the main grid or among themselves. Different types of microgrids should be established according to their configuration, capacity and type of DG technologies. Therefore, microgrids can be classified by their configuration, AC/DC type of connection, their main feeder location and operation mode [31,39,40].

5.7.1 Classification by configuration

Microgrids can be classified according to the type of configuration and the load associated. These classifications are consumer microgrids, community microgrids and utility microgrids [39,41].

- Consumer microgrids are composed of few DGs, simple control functions and design for a single consumer. Normally, they are managed by the same customers.
- Community microgrids require a more complex design and operation than the consumer microgrid. This type has multiple consumer sides, and it is normally consumer owned. Some loads are defined as 'sheddable' in case of a frequency issue to maintain power balance.
- Finally, utility microgrids are those whose loads are prioritised based on user's requirements on reliability, and high-priority loads will be powered preferentially in an emergency. The supply resources are provided on the utility side with consumer interactions. They are organised according to the utility objectives.

5.7.2 Classification by AC/DC type

According to the type of connection, microgrids are divided into DC, AC and AC/DC or hybrid microgrids [39]. In a DC microgrid, the generation and consumption units are connected to a DC bus by using converters. An AC microgrid is connected to the distribution network via an AC bus, which is connected to the distribution system through the PCC. A hybrid microgrid combines AC buses and a DC bus according to the nature of loads connected. This integrates the advantages of both AC and DC microgrids.

On one hand, DC microgrids depend on DG voltage, which results in a coordinated operation of the DGs. Also, a DC microgrid does not involve any synchronisation process with the electric grid, which simplifies the control process. However, most of the loads in the current system are AC loads, which implies the use of more converters.

AC microgrids require no inverter for power supply to AC loads, which makes this type of microgrid the most commonly used around the world. Since it operates with the electric grid, it requires a synchronisation process in the control process.

5.7.3 Classification by modes of operation

A microgrid, in general, can operate either in grid-connected mode and islanded mode [39]. The first mode of operation indicates that the microgrid is connected to the main power grid, while the other refers to a completely independent operation from the rest of the electric grid. There is an intermediate operation mode stage, in which the microgrid changes from the grid-connected mode to the islanded mode of operation and vice versa. This is done considering a series of steps to achieve the transition of operational modes correctly and without affecting the microgrids components.

Some of the advantages of grid-connected mode include economic optimal dispatch by using resources from the distribution network, automatic voltage control and reactive power control. Also, it makes use of the energy storage to minimise total feeder maximum demand and the ratio between total feeder maximum and minimum demand.

The islanded mode can be used intentionally or unintentionally. For the latter, predetermined control strategies and unintentional islanding strategies are defined to prevent unplanned or uncontrollable behaviours.

5.7.4 Classification by feeder location

Based on the characteristics and properties of feeders, a microgrid can be categorised in urban, rural and off-grid microgrids [40].

Some of the urban microgrid characteristics include a high-density balance load structure around the system. The short-circuit ratio between the system grid short-circuit capacity at the PCC and the total DER generation capacity of the microgrid is regularly above 25.

Another type of microgrids is the rural microgrids, in which load is scattered around the system. The short-circuit ratio is not as high as it is for the urban microgrid. However, it presents voltage imbalance and fluctuation. Therefore, DER units can be used to regulate the feeder voltage.

Finally, an off grid microgrid operates in islanded mode since it is located far from the main grid or is surrounded by a difficult environment for transmission line connection. This means it cannot strictly be considered a 'microgrid'. Nevertheless, strategies for off-grid microgrids are fully applicable to islanded mode.

5.8 Microgrid control

As stated in the previous section, there are many possible configurations for microgrids where the main objective is the coordinated operation of both energy supply and demand in order to guarantee voltage and frequency stability. Therefore, microgrid control has an important role in guaranteeing operational conditions. There is no general structure of a control architecture for microgrids because it depends on the type of microgrid or the existing infrastructure. Typical control actions are network reconfigurations by switching operation and voltage control via capacitor switching or transformers tap changing.

A typical distribution network with high penetration of DGs is shown in Figure 5.37 [31]. The DMS defines how the system is controlled and managed, which monitors the HV/MV sections. This control system is not in charge of controlling the DGs or loads. However, it can control large installations in certain specific scenarios. Another required element for controlling is the AMR/AMI system, which is responsible for the collection of electronic meter readings. When there are integrated smart meters, the data management capabilities are increased, which also enables two-way communication between utilities and customers. The distribution system operator is the entity that manages and controls the distribution system and collects the energy metering data. It sends the measured data to the Energy Service Company, which is responsible for billing.

The previous structure just mentioned one layer is required for microgrid control. Local control levels at DGs and loads are required in order to enable an advanced market participation. Also, scalability capacities are expected to integrate a large number of users. Finally, plug-and-play characteristics are expected to enable new participants and integrate new functionalities and business cases.

Two control structures can be identified for microgrid control: centralised and decentralised structures.

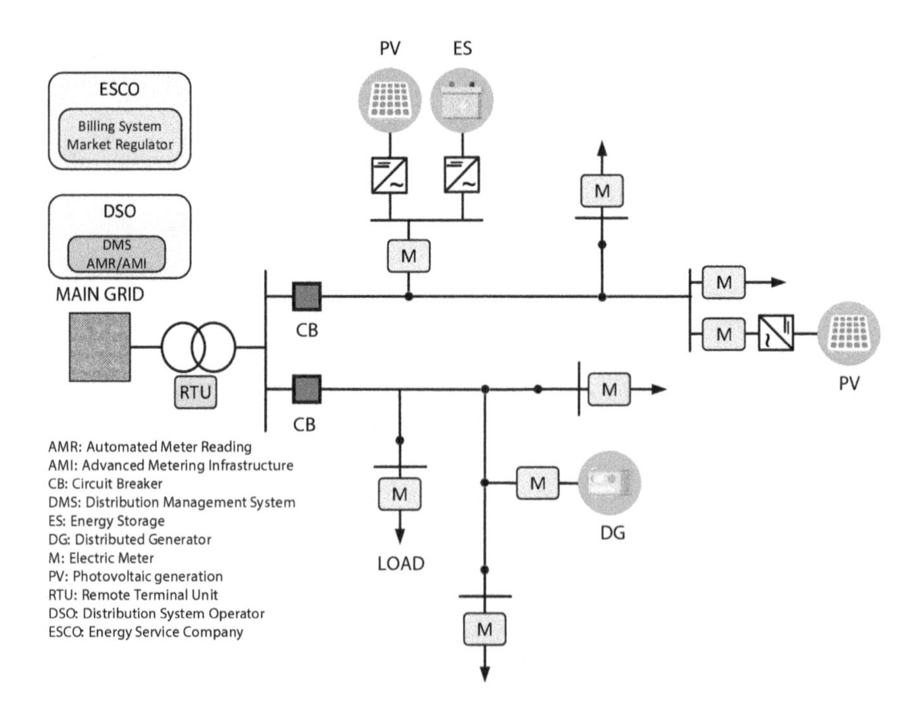

Figure 5.37 Typical distribution system management structure

5.8.1 Centralised control

In centralised control, the microgrid has a three-layer control scheme as indicated in Figure 5.38 [31,39,42,43]. The first layer called *local control layer* is the bottom layer composed of a local controller, which is in charge of primary voltage and frequency regulations. It also has a local protection, which provides fault protection for the microgrid. The middle layer called the *centralised control layer* is the microgrid control centre (MGCC) and the core of the microgrid control system. During grid-connected operation, this control layer tries to bring the performance of the microgrid to the best results. When the microgrid operates in islanded mode, it gives priority to stability and safe operation of the microgrid. Finally, the *distribution network layer* is the last layer, which is coordinated by the distribution

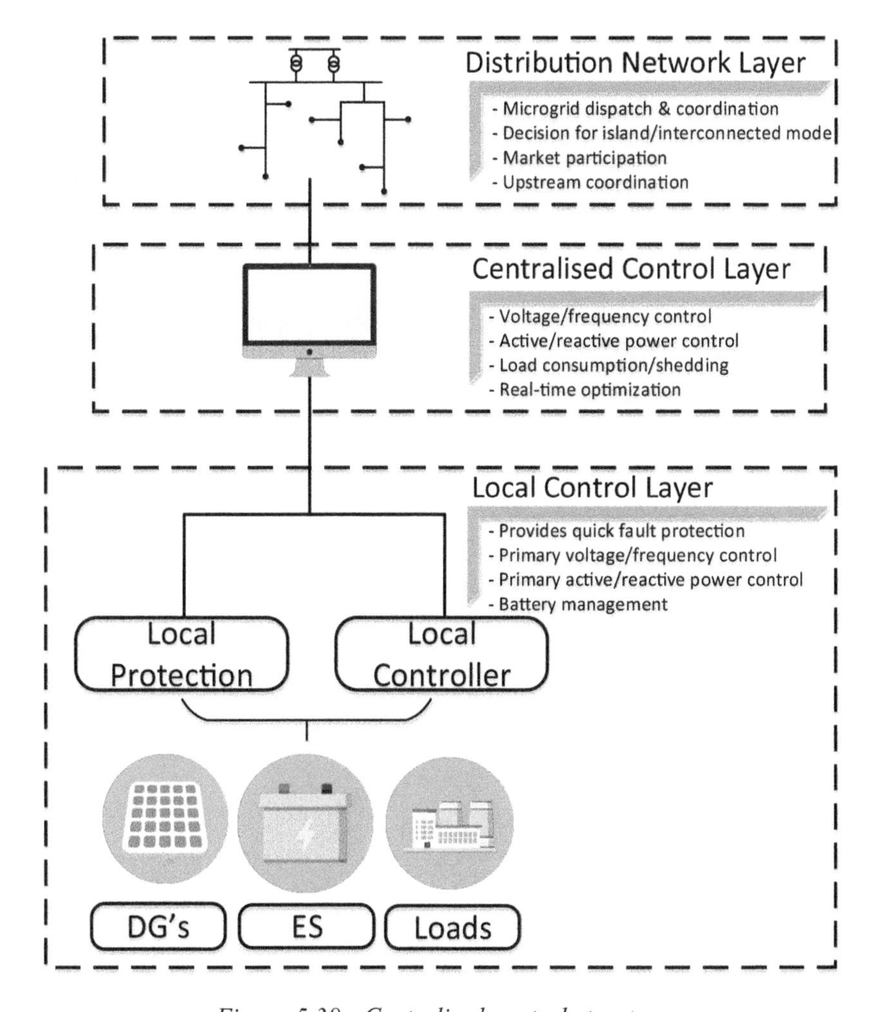

Figure 5.38 Centralised control structure

network, and it oversees dispatching the microgrid to maximise its profit and provide ancillary services to the main grid. In order to achieve this, this layer relates microgrids to the market according to their participation.

The advantage of a centralised control system is that all data is centralised which allows the defining of global objectives in the control approach, which results in high controllability of the system. However, the computational burden is high as the optimisation is computed based on a large amount of information. Moreover, there is a single point of failure and redundancy of the central controller is expensive. The loss of communication may cause a shutdown of the overall system.

5.8.2 Decentralised control

In a decentralised control, the microgrid demand is defined in the control goal by the collaboration of different controllers [31,42]. Aspects like the market prices and the status of microgrid elements are considered in the definition of the control approach, but the control divides the problem into local controllers or agents. They achieve an optimal state through a hierarchical system coordination of all controllers. There is less information required at the central controller of different microgrid DG units, which results in a more scalable system. A central agent communicates global constraints to the other agents such as transfer power limits or voltage limits. The information can be flowing around different agents when a single communication channel fails [44–46].

Table 5.8 compares the characteristics of both types of control.

Table 5.8 Microgrid control methods comparison [31]

	Centralised control	**Decentralised control**
DG ownership	Single owner	Multiple owners
Goals	A clear, single task, e.g. minimisation of energy costs	Uncertainty over what each owner wants at any particular moment
Availability of operating personnel (monitoring, low-level management, special switching operations, etc.)	Available	Not available
Market participation	Implementation of complicated algorithms	Owners unlikely to use complex algorithms
Installation of new equipment	Requirements of specialised software	Should be plug-and-play
Optimality	Optimal solutions	Mostly suboptimal solutions
Communication requirements	High	Modest
Market participation	All units collaborate	Some units may be competitive
Microgrid operation is attached to a larger and more critical operation	Possible	Not possible

Example 5.4

To illustrate the above-mentioned technical benefits, the following calculations were performed for the microgrid illustrated in Figure 5.39 with a PSAT:

(a) Total losses in the system.
(b) The best topology considering loss reduction and a radial configuration.
(c) List of the lines whose ends must change status.
(d) Comparison of the total losses results with those obtained for a mesh network.

(a) Total losses in the system:

The total losses and losses by feeder for the given system were calculated running a load flow analysis. The results are indicated in Table 5.9.

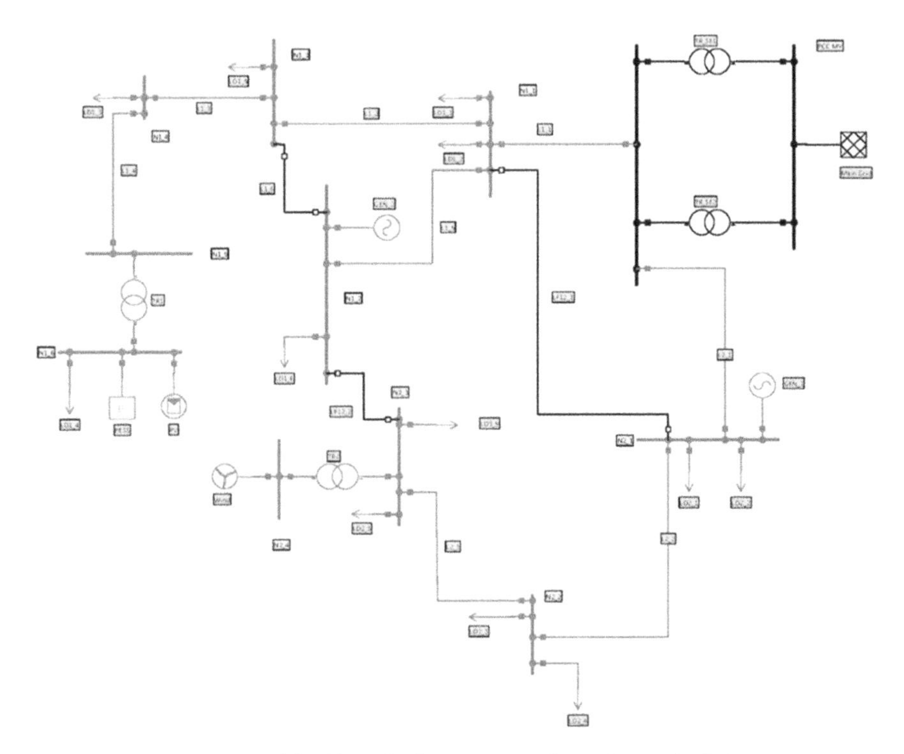

Figure 5.39 Two-feeder microgrid for Example 5.4

Table 5.9 Total losses for the given system in Example 5.4

	P loss MW	Q loss MVAr
Network	0.201	0.343
Feeder_01	0.068	0.077
Feeder_02	0.127	0.125

Table 5.10 indicates the total losses for the elements in the given system at the different voltage levels.

(b) The best topology for the system considering loss reduction and a radial configuration:

The best topology was determined with a switching optimisation for a system reconfiguration using a PSAT. Figure 5.40 illustrates the best topology considering loss reduction and a radial configuration.

The total losses and losses by feeder for the new topology were calculated running a load flow analysis. The results are indicated in Table 5.11.

Table 5.12 indicates the total losses of the elements for the reconfigured system at different voltage levels.

Table 5.10 Element losses for the given system in Example 5.4

Un	P loss line	Q loss line	P loss transformer	Q loss transformer
kV	MW	MVAr	MW	MVAr
20	0.193	0.097	0.002	0.106
110	0	0	0.006	0.14

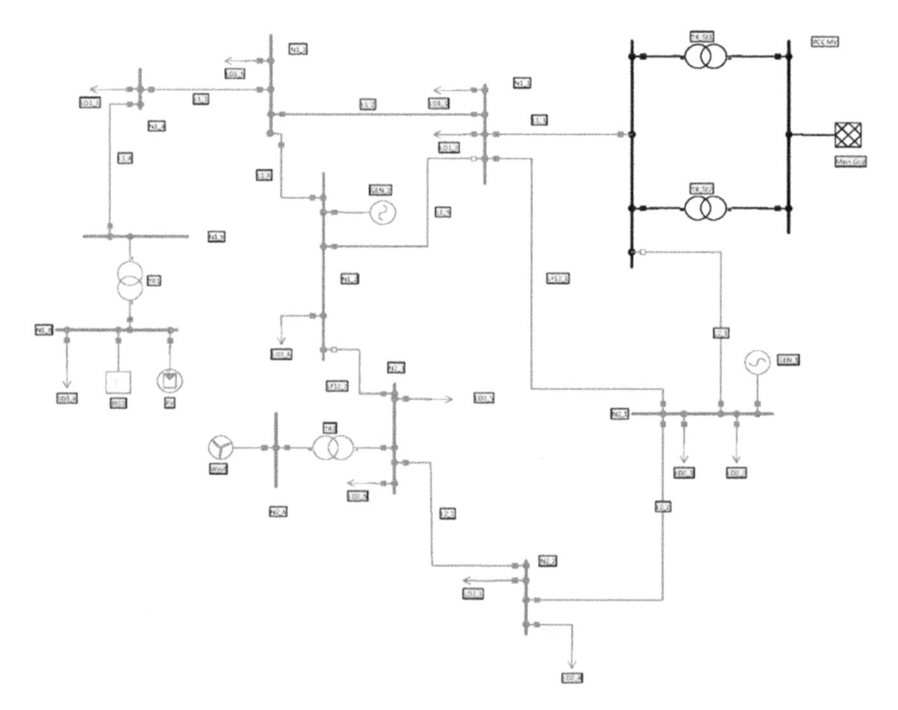

Figure 5.40 Reconfigured microgrid for Example 5.4

Table 5.11 Total losses for the new topology for Example 5.4

From area/zone	P loss MW	Q loss MVAr
Network	0.125	0.318
Feeder_01	0.119	0.177
Feeder_02	0	0

Table 5.12 Line losses for the new topology for Example 5.4

Un kV	P loss line MW	Q loss line MVAr	P loss transformer MW	Q loss transformer MVAr
20	0.117	0.071	0.002	0.106
110	0	0	0.006	0.14

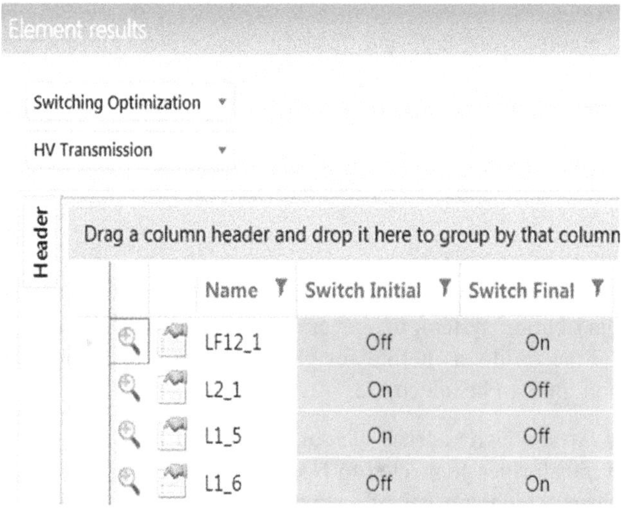

Element results

Switching Optimization

HV Transmission

Header

Drag a column header and drop it here to group by that column

Name	Switch Initial	Switch Final
LF12_1	Off	On
L2_1	On	Off
L1_5	On	Off
L1_6	Off	On

Figure 5.41 Lines whose ends must change their status

(c) List of the lines whose ends must change status:

Figure 5.41 illustrates the lines whose ends must change their status to allow the system's reconfiguration for optimal topology.

(d) Comparing the total losses results with those obtained for a mesh network:

Table 5.13 indicates the total losses obtained for the given system, the best topology for the system in mesh configuration. As can be seen, the mesh configuration allows lower losses in the system.

Environmental benefits of a microgrid can be expected from two aspects: a shift towards renewable or low-emission fuels and the adoption of more energy-efficient energy supply solutions.

Table 5.13 Total losses comparison for the given system, the new topology and mesh configuration for Example 5.4

	P loss (MW)			**Q loss (MVAr)**		
	Initial network	**New topology**	**Mesh network**	**Initial network**	**New topology**	**Mesh network**
Network	0.201	0.125	0.089	0.343	0.318	0.292
Feeder_01	0.068	0.119	0.064	0.077	0.177	0.14
Feeder_02	0.127	0	0.019	0.125	0	0.012

With supportive policies for distributed renewable resources, the fuel-switching credit of microgrid is expected to grow as renewable energy sources costs decrease over the years.

Social benefits of microgrids includes the following:

- Large land use impacts are avoided.
- Raising public awareness and fostering incentives for energy saving and greenhouse gas (GHG) emission reduction.
- Microgrids can provide electric service to regions and communities that are currently unserved.
- Creation of new research and job opportunities.

Example 5.5

Figure 5.42 illustrates a load flow simulation for a distribution system without DG. A summary of the results is presented in Table 5.14.

In this distribution system, the generation units shown below will be added to produce the same results as before for grid-connected mode and islanded mode. The generation units to be added are:

a. A wind farm, which consists of four sets of wind turbine generators and one step up transformer connected to N3_3 Busbar.
b. A PV farm, which consists of a single unit connected to N2_4 Busbar.
c. Two synchronous generators, Generator 1 connected to N3_1 Busbar and Generator 2 connected to N2_2 Busbar.

The characteristics of all units are shown in Tables 5.15–5.18.

Figure 5.43 presents the new results (with distributed generators) and the previous ones in the base case (without distributed generators). It is shown that DG supports the loads and improves the voltage levels. In addition, load flow results in the main transformers are having a significant reduction due to the generation systems next to the loads.

Also, Table 5.19 presents that distributed generation decreases the losses around the system.

Finally, this new modification creates a microgrid with multiple uses of different kinds of distributed generation. Figure 5.44 and Table 5.20 present the load flow results in islanded mode opening the main transformers.

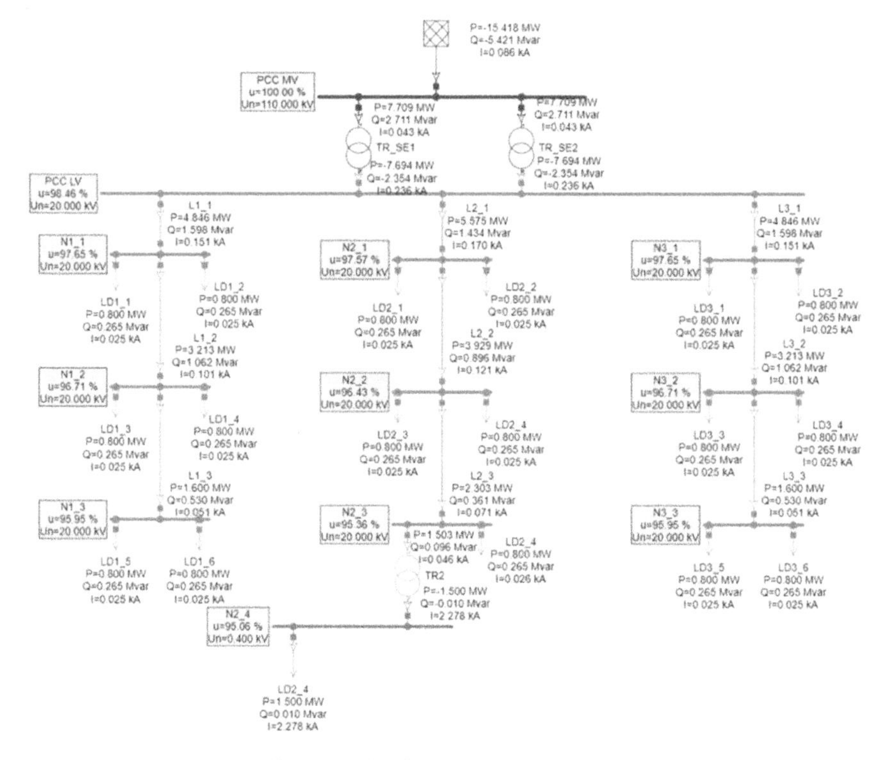

Figure 5.42 Load flow results for a distribution system illustration case

Table 5.14 Summary of load flow results

P Loss MW	Q Loss MVAr	P Imp MW	Q Imp MVAr	P Gen MW	Q Gen MVAr	P Load MW	Q Load MVAr
0.318	0.906	15.418	5.421	15.418	5.421	15.1	4.515

Table 5.15 Synchronous generator 1 data

Synchronous generator 1			
Symbol:	Synchronous machine	Sr (MVA)	7
Vr (kV)	20	Cos(phi)	0.9
LF Analysis Parameters			
LF Type	PQ (grid-connected) Slack (islanded mode)		
PGen (MW)	6.3		
QGen (MVAr)	−3.05		

Table 5.16 Synchronous generator 2 data

Synchronous generator 2			
Symbol:	Synchronous Machine	Sr (MVA)	3
Vr (kV)	20	Cos(phi)	0.9
LF analysis parameters			
LF type	PQ (grid-connected and islanded mode)		
PGen (MW)	2.7		
QGen (MVAr)	−1.307		
Scaling factors:	GEN		

Table 5.17 PV system data

PV system			
Symbol	Disperse generation	Sr (MVA)	2
Vr (kV)	0.4	Cos(phi)	1
LF analysis parameters			
P (MW)	2		
Q (MVAr)	0		
Scaling factors	Solar		
Wind turbines			
Symbol	Disperse generation		
No. units	4	Sr (MVA)	0.53
Vr (kV)	0.69	Cos(phi)	0.95
LF analysis parameters			
P (MW)	0.503		
Q (MVAr)	0.165		
Scaling factors	Wind		

Table 5.18 Wind turbines data

Transformer (only 1 is required)			
Sr: 3 MVA			
HV: 20 kV	LV: 0.69 kV		
R(1)%	0.43	kW	12.9
Zcc(1)%	5.357		
R(0)%	0.43	kW	12.9
Zcc(0)%	5.357		

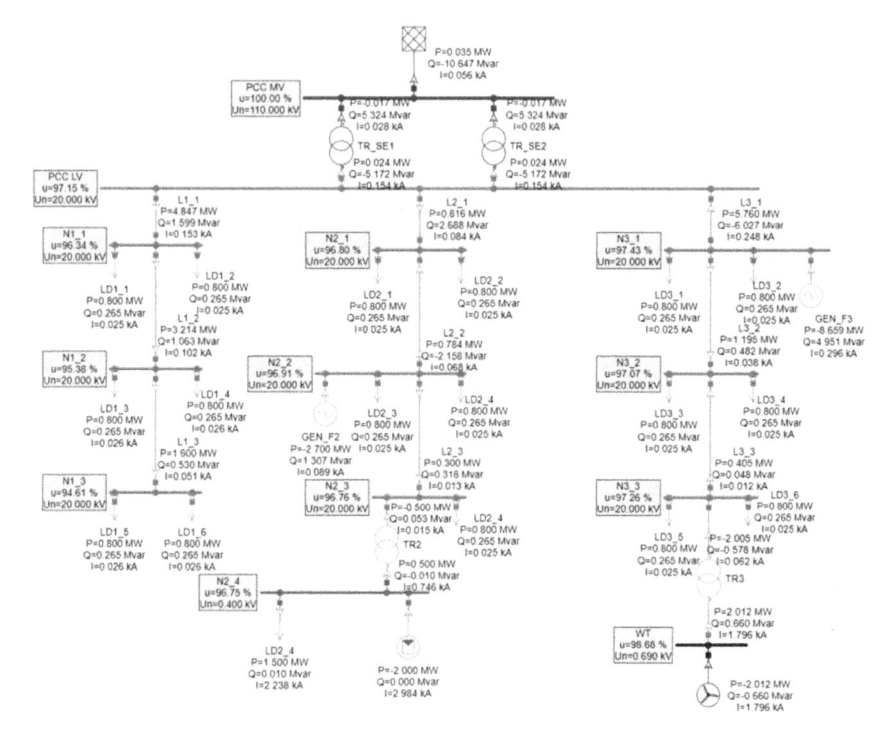

Figure 5.43 Comparison of load flow results with DG Vs Base Case

Table 5.19 Summary of load flow results with distributed generation

P Loss MW	Q Loss MVAr	P Imp MW	Q Imp MVAr	P Gen MW	Q Gen MVAr	P Load MW	Q Load MVAr
0.183	0.205	8.571	5.367	15.371	4.304	15.187	4.099

5.9 Proposed exercises

1. Write a script in MATLAB® that plots the *I–V* curve for a PV cell as shown in Figure 5.6. The inputs for the script are the saturation current *Io*, the photocurrent *I*ph and the temperature *T* and only valid values can be assigned by the user.
2. Write a script in MATLAB that plots the power coefficient curve as shown in Figure 5.18. The user should be able to set the pitch angle and the TSR as inputs.
3. Based on Figure 5.45, find the optimal angular velocity for a wind turbine with a 60 m blade length and wind speed of 11 m/s. Assume the wind turbine is operating with the maximum performance coefficient.

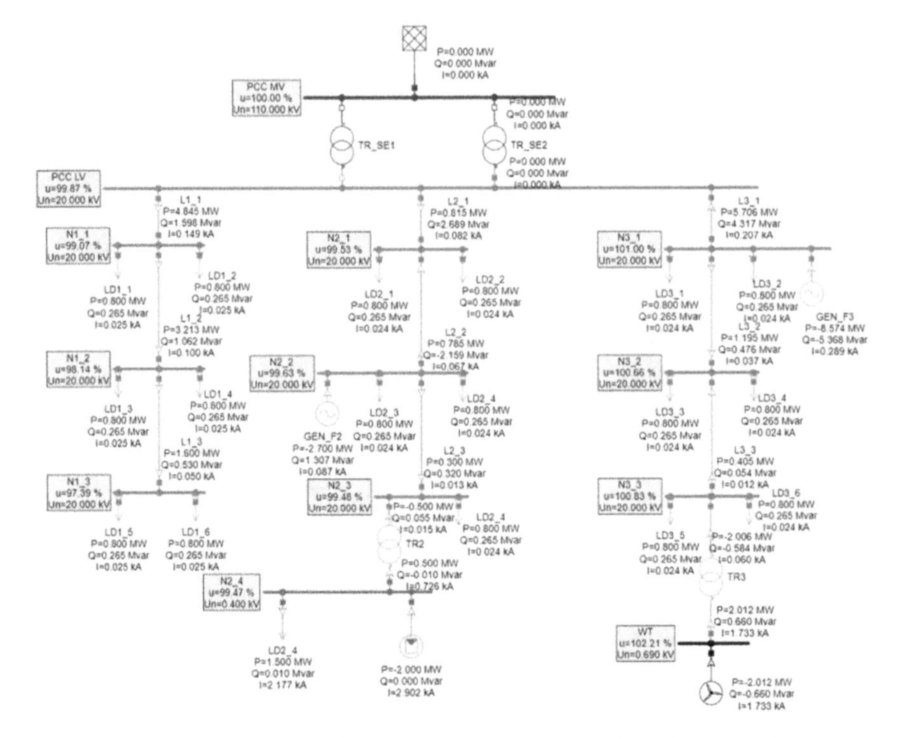

Figure 5.44 Comparison of load flow results islanded mode versus base case

Table 5.20 Summary of load flow results in islanded mode

P Loss MW	Q Loss MVAr	P Imp MW	Q Imp MVAr	P Gen MW	Q Gen MVAr	P Load MW	Q Load MVAr
0.186	0.206	8.574	5.368	15.286	4.721	15.1	4.515

4. Determine the total power delivered by a pumped-hydro storage plant when the water flow rate is 115 m^3/s and the conversion efficiency is 0.78. Assume an elevation change of 30 m and the density of water as 1000 kg/m^3.

$$E_{stored} = g\rho_w V\left(H_{upper} - H_{lower}\right)$$

5. Determine the total amount of charge in Ah of a battery bank in order to supply the total demand of an off-grid PV system, considering that the total demand of energy from the customer is 8760 Wh/day, the battery bank has an autonomy of 1.5 days, the bus voltage is 36 V and the battery bank discharge depth is 70%.
6. For the microgrid shown in Figure 5.46, determine if the generated energy meets the energy that is being consumed when the microgrid operates

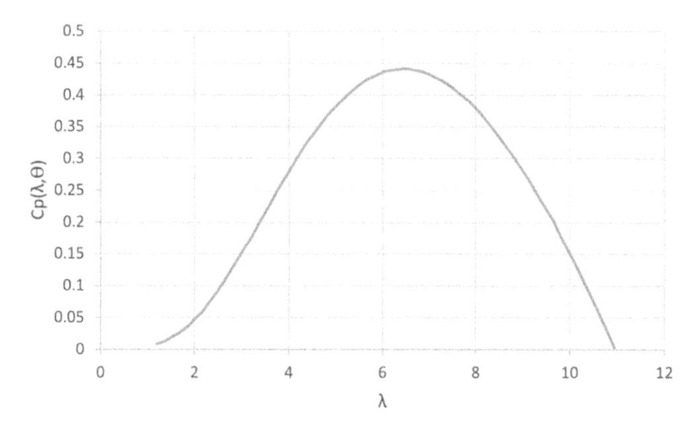

Figure 5.45 Cp versus TSR curve for Exercise 3

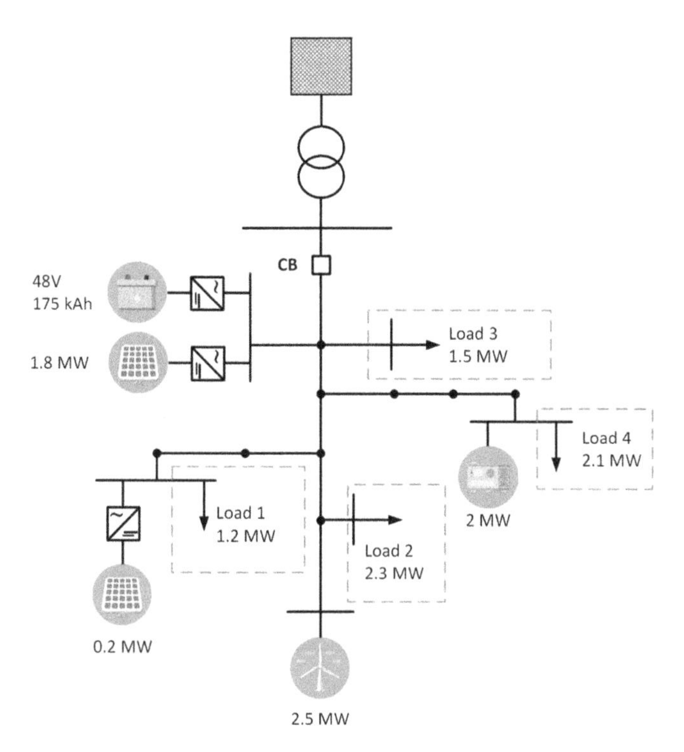

Figure 5.46 Microgrid for Exercise 6

disconnected from the main grid. The reference load and generation profiles are shown in Table 5.21. Consider the following:

- Use the load profile for all loads in the microgrid. This profile provides the consumed power in p.u.

Table 5.21 Load and generation profiles data for Exercise 6

Hours (h)	Factors in p.u.				Hours (h)	Factors in p.u.			
	Solar	Wind	Diesel	Load		Solar	Wind	Diesel	Load
0:00	0.000	0.150	0.000	0.300	12:00	0.750	0.660	1.000	0.800
1:00	0.000	0.220	0.000	0.200	13:00	0.890	0.690	1.000	0.710
2:00	0.000	0.280	0.000	0.150	14:00	0.630	0.710	1.000	0.660
3:00	0.000	0.260	0.000	0.200	15:00	0.600	0.630	1.000	0.610
4:00	0.000	0.310	0.000	0.250	16:00	0.520	0.660	1.000	0.530
5:00	0.000	0.380	0.000	0.270	17:00	0.360	0.760	1.000	0.470
6:00	0.030	0.360	0.000	0.380	18:00	0.150	0.720	1.000	0.440
7:00	0.070	0.430	1.000	0.470	19:00	0.000	0.510	1.000	0.550
8:00	0.075	0.440	1.000	0.580	20:00	0.000	0.350	1.000	0.590
9:00	0.095	0.490	1.000	0.640	21:00	0.000	0.380	0.000	0.700
10:00	0.114	0.550	1.000	0.690	22:00	0.000	0.330	0.000	0.450
11:00	0.350	0.580	1.000	0.740	23:00	0.000	0.200	0.000	0.350

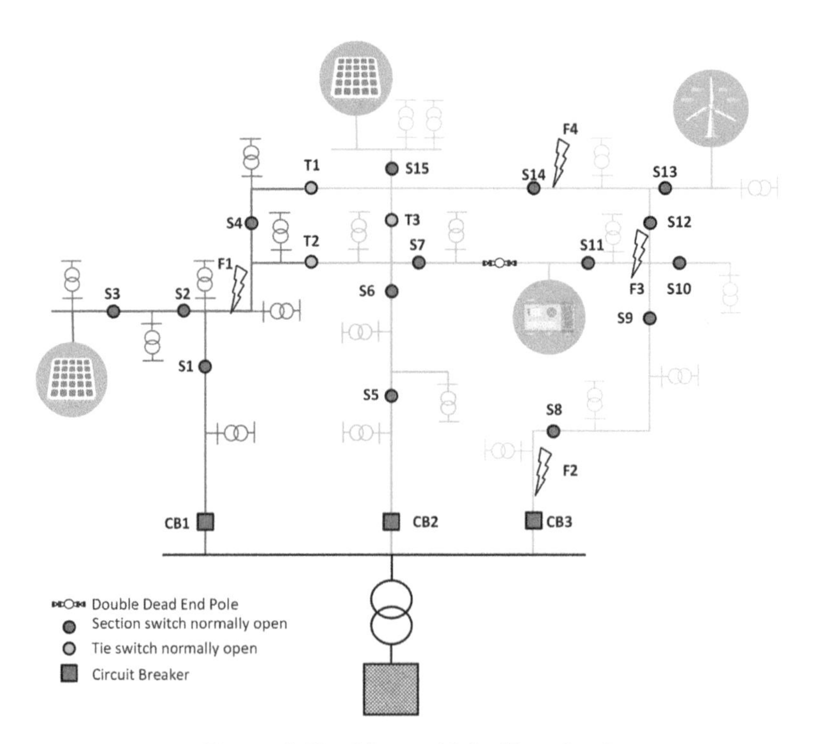

Figure 5.47 Microgrid for Exercise 7

Table 5.22 Switches operation sequence for Exercise 7

FAULT	OPEN	CLOSE	Total affected customers
F_1			
F_2			
F_3			
F_4			

- Use the generation profile for the generation units in the microgrid. This profile provides the output power in p.u.
- Compute the produced and consumed energy for a period between 6:00 am and 8:00 pm of one day.
- Calculate the amount of energy that can be provided by the energy storage system to the unsupplied load.

7. Consider the microgrid that is shown in Figure 5.47 with the tie and section already switches located. For each one of the faults F1, F2, F3 and F4, indicate the operating sequence of the tie and section switches in order to restore service to the highest possible number of customers, as requested in Table 5.22.

References

[1] Renewable Energy Policy Network for the 21st Century-REN21, 'Renewables 2023 Global Status Report', 2023.
[2] D. Pattabiraman, R. H. Lasseter, and T. M. Jahns, 'Comparison of grid following and grid forming control for a high inverter penetration power system', in *2018 IEEE Power & Energy Society General Meeting (PESGM)*, Piscataway, NJ: IEEE, pp. 1–5, 2018, doi:10.1109/PESGM.2018.8586162.
[3] N. Jenkins, J. Ekanayake, and S. Goran, *Distributed Generation*. London: IET, 2010.
[4] Z. Sen, *Solar Energy Fundamentals and Modeling Techniques*. London: Springer, 2008. doi:10.1007/978-1-84800-134-3.
[5] A. Z. de Souza and M. Castilla, *Microgrids Design and Implementation*. Berlin: Springer, 2019.
[6] Z. Sen, *Solar Energy Fundamentals and Modeling Techniques*. Berlin: Springer, 2008.
[7] R. Messenger and A. Abtahi, *Photovoltaic Systems Engineering*, 4th ed. Boca Raton, FL: CRC Press, 2017.
[8] M. Kaltschmitt, W. Streicher, and A. Wiese, *Renewable Energy*. Berlin: Springer, 2007.
[9] M. Ahmed, I. Harbi, R. Kennel, J. Rodriguez, and M. Abdelrahem, 'Model-based maximum power point tracking algorithm with constant power generation capability and fast DC-link dynamics for two-stage PV systems', *IEEE Access*, vol. 10, pp. 48551–48568, 2022.

[10] M. Rashid, *Power Electronics: Circuits, Devices and Applications*, 4th ed. Englewood Cliffs, NJ: Prentice-Hall, 2013.

[11] N. Mohan, T. Undeland, and W. Robbins, *Power Electronics: Converters, Applications and Design*, 3rd ed. New York: Wiley, 2002.

[12] Clean Energy Reviews (CER), 'Best Solar Inverters 2023', 2024 [Online]. Available: https://www.cleanenergyreviews.info/blog/best-grid-connect-solar-inverters-sma-fronius-solaredge-abb [accessed 19 May 2024].

[13] F. Mustafa, I. Ahmed, A. Malek, and A.-Y. MobasSharin, *Centralized and Decentralized Solar Power Control and Distribution: A Comparative Analysis of Operational and Economic Aspects*. Dhaka: BRAC University, 2017.

[14] S. Mathew, *Wind Energy Fundamentals*. Berlin: Springer, 2006.

[15] M. El-Sharkawi, *Wind Energy: An Introduction*. Boca Raton, FL: CRC Press, 2015.

[16] V. Nelson and K. Starcher, *Introduction to Renewable Energy*, 2nd ed. Boca Raton, FL: CRC Press, 2016.

[17] V. Gevorgian and E. Muljadi, 'Wind power plant short circuit current contribution for different fault and wind turbine topologies', in *9th Annual International Workshop on Large-Scale Integration of Wind Power into Power Systems as well as on Transmission Networks for Offshore Wind Power Plants*, pp. 570–578, 2010.

[18] J. M. Gers and C. Viggiano, 'Protective relay setting criteria considering DERs and distributed automation', *13th International Conference on Development in Power System Protection 2016 (DPSP)*, Edinburgh, UK, 2016, pp. 1–9, doi: 10.1049/cp.2016.0069.

[19] F. Gharedaghi, M. Deysi, H. Jamali, and A. Khalili, 'Soft starter investigation on grid connection of wind turbines', *Aust. J. Basic Appl. Sci.*, vol. 5, no. 10, pp. 1146–1153, 2011.

[20] M. Bollen and F. Hassan, *Integration of Distributed Generation in the Power System*. New York: Wiley, 2011.

[21] E. Efstathios, *Alternative Energy Sources*. Berlin: Springer, 2012.

[22] D. O. Akinyele and R. K. Rayudu, 'Review of energy storage technologies for sustainable power networks', *Sustain. Energy Technol. Assessments*, vol. 8, pp. 74–91, 2014.

[23] A. Oberhofer and P. Meisen, 'Energy-Storage Technologies', 2012.

[24] R. Huggins, *Energy Storage: Fundamentals, Materials and Applications*. 2nd ed. Cham, Switzerland: Springer International Publishing, 2016.

[25] Hitachi Energy, 'Frequency Regulation', Accessed 19 March 2024 [Online]. Available: https://www.hitachienergy.com/products-and-solutions/grid-edge-solutions/applications/energy-storage-applications#tab-tabs-fb331494aa-item-ca9e5a9255

[26] Hitachi Energy, 'Spinning Reserve' [Online]. Accessed 19 March 2024 [Online]. Available: https://www.hitachienergy.com/products-and-solutions/grid-edge-solutions/applications/energy-storage-applications#tab-tabs-fb331494aa-item-76875011c6

[27] Hitachi Energy, 'Load Leveling' [Online]. Accessed: 19 March 2024 [Online]. Available: https://www.hitachienergy.com/products-and-solutions/grid-edge-solutions/applications/energy-storage-applications#tab-tabs-fb331494aa-item-be2620143f
[28] Hitachi Energy, 'Peak Shaving' [Online]. Accessed: 19 March 2024 [Online]. Available: https://www.hitachienergy.com/products-and-solutions/grid-edge-solutions/applications/energy-storage-applications#tab-tabs-fb331494aa-item-ec8904db6d
[29] Neplan, 'NiHM Battery with DC Machine Shunt Excited', 2018.
[30] M. Murnane and A. Ghazel, *A Closer Look at State of Charge (SOC) and State of Health (SOH) Estimation Techniques for Batteries*. Wilmington, MA: Analog Devices, Inc., 2017.
[31] N. D. Hatziargyriou, *Microgrids*. New York: Wiley, 2014.
[32] Office of Electricity Delivery and Energy Reliability Smart Grid R&D Program, *DOE Microgrid Workshop Report*. San Diego, CA, USA, 2011.
[33] IEEE, 'IEEE Std 446 – 1995, IEEE Recommended Practice for Emergency and Standby Power Systems for Industrial and Commercial Applications', Piscataway, NJ: IEEE, pp. 1–320, 1996.
[34] IEEE, 'IEEE Std. 519-2022. IEEE Standard for Harmonic Control in Electric Power Systems', 2022.
[35] Institute of Electrical and Electronics Engineers, *IEEE Std. 1547-2018. Standard for Interconnection and Interoperability of Distributed Energy Resources with Associated Electric Power Systems Interfaces*, 2018. doi: 10.1109/IEEESTD.2018.8332112.
[36] International Electrotechnical Commission, 'IEC TS 62257-100:2022. Renewable Energy Off-Grid Systems', 2022. [Online]. Available: https://webstore.iec.ch/publication/64175 [accessed 19 March 2024].
[37] International Electrotechnical Commission, 'IEC TS 62898-1:2017+AMD1:2023. Microgrids – Part 1: Guidelines for microgrid projects planning and specification', 2023.
[38] International Electrotechnical Commission, 'IEC TS 62898-2:2018. Microgrids – Part 2: Guidelines for Operation', 2018..
[39] L. Fusheng, L. Ruisheng, and Z. Fengquan, *Microgrid Technology and Engineering Application*. Amsterdam: Elsevier, 2016.
[40] A. Hooshyar and R. Iravani, 'Microgrid protection', *Proceedings of the IEEE*, vol. 105, no. 7, pp. 1332–1353, 2017, doi:10.1109/JPROC.2017.2669342.
[41] J. Miller, 'Power system optimization: smart grid, demand dispatch and microgrids', presented at the Smart Grid Implementation Strategy Team Lead, U.S. Department of Energy, 2011.
[42] H. Almasalma, J. Engels, and G. Deconinck, 'Peer-to-Peer Control of Microgrids', 2017. Available: https://doi.org/10.48550/arXiv.1711.04070.
[43] J. Gers and E. Holmes, *Protection of Electricity Distribution Networks*, 4th ed. London: Institution of Engineering and Technology, 2021.

[44] IEEE, 'IEEE Std 1547-2018. IEEE Standard for Interconnection and Interoperability of Distributed Energy Resources with Associated Electric Power Systems Interfaces', 2018.

[45] A. Singhal, T. L. Vu, and W. Du, 'Consensus control for coordinating grid-forming and grid-following inverters in microgrids', *IEEE Trans. Smart Grid*, vol. 13, no. 5, pp. 4123–4133, 2022, doi:10.1109/TSG.2022.3158254.

[46] J. Araia and Y. Taguchi, 'Coordinated control between a grid forming inverter and grid following inverters suppling power in a standalone micro-grid', *Global Energy Interconnect.*, vol. 5, no. 3, pp. 2096–5117, 2022, doi:10.1016/j.gloei.2022.06.002.

Chapter 6
Basics of design, protection, automation and control for urban grids

Urban power grids are having a significant transformation that includes changes in the load management as well as the new generation systems along the feeders. Due to these changes, conventional protection and control systems are incorporating new applications to adjust the system response to the different operation conditions.

Most typical protections in distribution systems include reclosers and fuses. These protections are based in time–current characteristics for overcurrent protection principles. However, under the framework of modern urban power grids, traditional protections like fuses are being complemented by protection relays with sophisticated features including directional protection, voltage protection and communications.

On the other hand, with the dynamic nature of modern urban grids, different strategies involve real-time monitoring and change of the protection settings for the different operating conditions. For this reason, adaptive protection and control strategies are becoming essential to meet the evolving needs of urban power grids.

6.1 Fundamentals of overcurrent protection

Overcurrent relays are the most common form of protection used to operate only under fault conditions. They should not be installed purely as a means of protecting systems against overloads. The relay settings that are selected are often a compromise in order to cope with both overload and overcurrent conditions.

Overcurrent relays can be classified as definite current, definite time and inverse time as shown in Figure 6.1(a)–(c). The time delay units can work in conjunction with the instantaneous units as shown in Figure 6.1(d).

6.1.1 Protection coordination principles

Relay coordination is the process of selecting settings that will assure that the relays will operate in a reliable and selective way. Overcurrent relays can have different protection units and setting groups. These units can be enabled at the same time in order to set the $i(t)$ characteristic of the relay. In overcurrent relays,

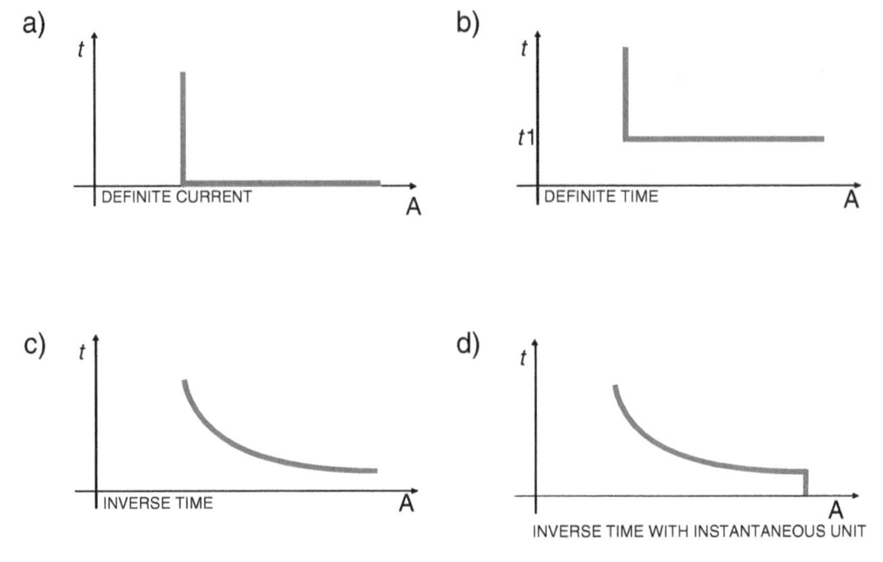

Figure 6.1 Time–current operating characteristics of overcurrent relays: (a) definite current, (b) definite time, (c) inverse time, and (d) inverse time with instantaneous unit

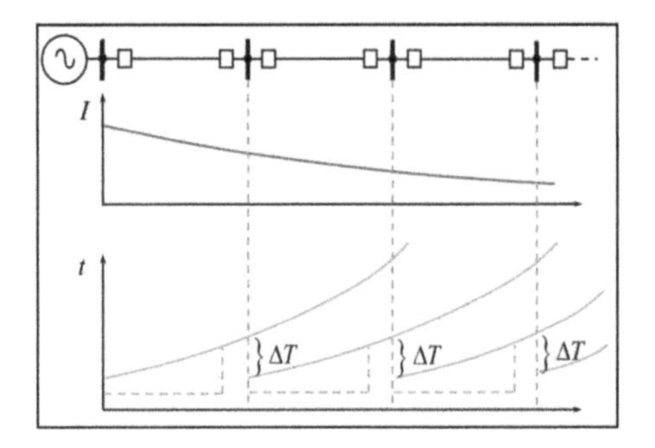

Figure 6.2 Overcurrent relay coordination procedure in a distribution system

the coordination is based on the relay time–current characteristics of instantaneous and/or time delay units.

1. Instantaneous units should be set so they do not trip for fault levels equal or lower to those at busbars or elements protected by downstream instantaneous relays.
2. Time delay units should be set to clear faults in a selective and reliable way, assuring the proper coverage of the thermal limits of the elements protected.

Figure 6.2 illustrates for a typical distribution substation, the process to carry out the coordination, starting from the relays associated with the downstream breakers to the main substation.

6.1.2 Criteria for setting instantaneous units

For these protection units, it is expected the lowest tripping time of the relay where the fault usually corresponds to the higher currents due to short circuits. Instantaneous units are set by adjusting the pickup level current at which the relays operate. Most numerical relays now have the possibility of setting an operating time that will change the instantaneous trip but allows the relay to behave as a definite time unit.

Some typical criteria are presented below, depending on the primary element:

i. Distribution lines
   Between six and ten times, the maximum circuit rating
   50% of the maximum short circuit at the point of connection of the relay
ii. Lines between substations
   125%–150% of the short-circuit current existing on the next substation
iii. Transformer units
   125%–150% of the short-circuit current existing on the LV side

The units at the LV side are overridden unless there is communication with the relays protecting the feeders.

6.1.3 Setting time-delay relays

These relays have different standard curves and can be categorised as inverse, very inverse, extremely inverse and others defined in IEEE and IEC standards that will be presented in Table 6.1. Different manufacturers offer the availability of some additional curves, typically owner curves. But in general, the selection of the curve defines the time-characteristic and the relays have the availability to select different units with different curves. Time delay units are set by selecting the time-curve characteristic that is defined by two parameters:

1. Tap or pick-up value: A value that defines the pick-up current of the relay. Current values are expressed as multiples of this value in the time–current characteristic curves.
2. Dial: Defines the time curve at which the relay operates for any tap value. Higher dial values represent higher operating times.

Table 6.1 IEEE and IEC constants for standards, overcurrent relays

IDMT curve description	Standard	α	β	L
Moderately inverse	IEEE	0.02	0.0515	0.114
Very inverse	IEEE	2	19.61	0.491
Extremely inverse	IEEE	2	28.2	0.1217
Inverse	IEEE	2	5.95	0.18
Short-time inverse	IEEE	0.02	0.02394	0.01694
Standard inverse	IEC	0.02	0.14	
Very inverse	IEC	1	13.5	
Extremely inverse	IEC	2	80.0	
Long-time inverse	IEC	1	120	

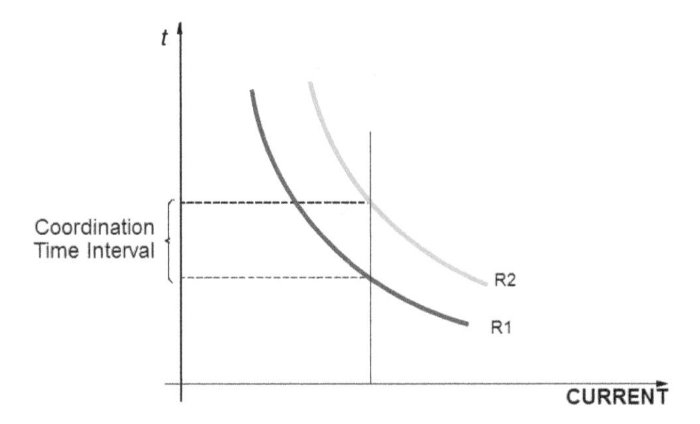

Figure 6.3 Overcurrent inverse-time relay curves associated with two breakers on the same feeder

Figure 6.3 shows overcurrent inverse time relay curves associated with two breakers on the same feeder.

Regarding coordination times, a margin between two successive devices in the order of 0.2–0.4 s should be used to avoid losing selectivity due to one or more of the following reasons:

1. the breaker opening time,
2. the overrun time after the fault has been cleared,
3. variations in fault levels, deviations from the characteristic curves of the relays and errors in the current transformers.

For phase relays, the tap or pick-up value is determined by:

$$TAP = (OLF \cdot I_{nom}) / CTR$$

For ground fault relays, the tap value is determined, with the maximum unbalance, typically around 20%:

$$TAP = ((0.2)\ x\ Inom) / CTR$$

The overload factor recommended is as follows:

- Motors = 1.05
- HV lines, transformers and generators = 1.25–1.5
- Distribution feeders = 2.0

For phase relays, three phase faults and maximum short time overload should be considered. For ground relays, line to ground faults and max $3I_o$ should be considered.

The procedure to determine the time dial settings is based on operating time targets corresponding to the multiples of pick-up or tap values at the instantaneous values.

The process starts at the furthest downstream relay and finishes with the furthest up relay, as illustrated in Figure 6.2.

For the furthest downstream relays, the lowest Time Dial is chosen or that considering cold load pick-up conditions.

Normally the settings are first carried out for phase relays and then for ground (neutral) relays. For the latter, the lowest time dial is selected whenever an open ground circuit is established, like that through Dy transformers.

The process to determine the time dial setting is a rather elaborate and is summarised in the following steps:

1. Calculate the multiple of pickup value for the secondary short-circuit current corresponding to the instantaneous setting of the relay where the process starts. If the instantaneous unit is overridden, the calculation is carried out with the total secondary short-circuit current at the relay location.
2. With the value calculated above, determine the operating time t_1 of the relay for the given time dial.
3. Determine the operating time t_{2a} of the upstream relay with the expression $t_{2a} = t_1 + t_{margin}$.
4. Calculate the multiple of pick-up value of the upstream relay using the same short-circuit current used in the first relay (Step 1).
5. Knowing t_{2a} and having calculated the multiple of pick-up value of the upstream relay, select the above nearest time dial for that relay.

The process follows the same steps for the next upstream relay and is repeated until the settings of the furthest up relay are calculated.

Operating times defined by IEC 60255 [1] and IEEE C37.112 [2] are:

$$t = \frac{k \cdot \beta}{\left(\frac{I}{I_s}\right)^{\alpha} - 1} + L$$

t is the relay operating time in seconds
k is the time dial, or time multiplier setting
I is the fault current level in seconds amps
I_S is the tap or pick-up current selected
L is the constant
α is the slope constant
β is the slope constant

Table 6.1 shows the constant values of the parameters for curves defined by IEEE C 37.112 and IEC 60255, and Figure 6.4 illustrates these curves.

6.1.4 *Setting overcurrent relays using software techniques*

Overcurrent protections coordination is usually supported by PSAT. These studies need the power system model for short-circuit calculations and the relays model, with the corresponding relays and circuit breakers. Finally, using selectivity diagrams is essential to avoid overlapping of the curves and an appropriate zone of protection for all the relays.

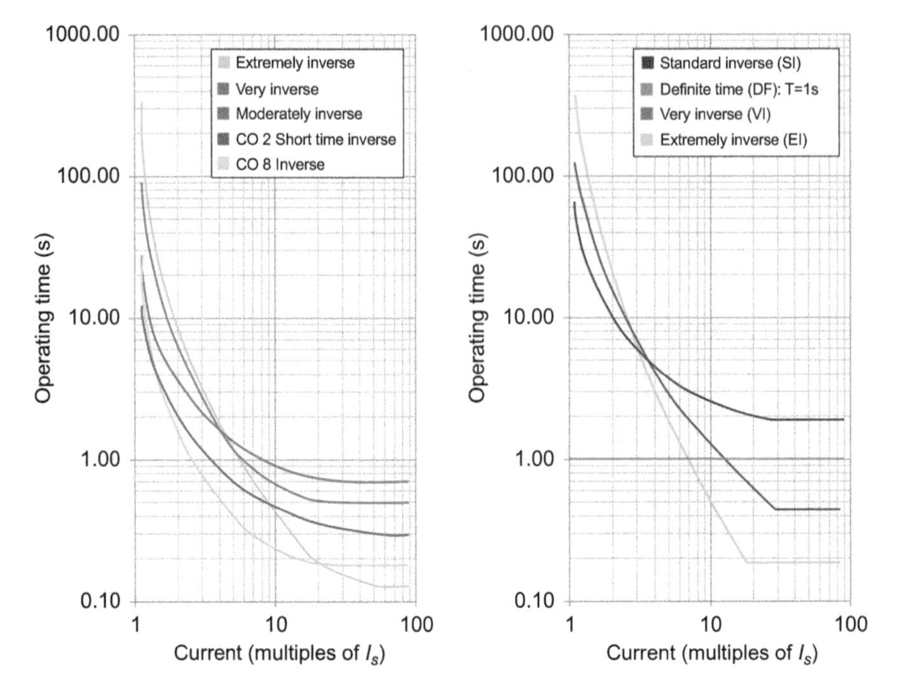

Figure 6.4 Typical ANSI/IEEE and IEC overcurrent relay curves

The procedure can follow the following recommendations:

1. Locate the fault and obtain the current for setting the relays.
2. Identify the pairs of relays to be set, first determining which one is farther away from the source.
3. Verify that the requirement's thermal capabilities are protected and devices operate for minimum short-circuit levels.

6.2 Coordination across Dy transformers

The transformers vector group plays an important role for the overcurrent protections coordination since fault current is affected by the winding connections. One of the most significant impacts is associated to Dy transformers where the secondary carries a current less than the current flowing through the primary relays for phase-to-phase faults.

The results of the three cases are summarised in Table 6.2. Analysing the results, it is shown that the critical case for the coordination of overcurrent relays is the phase-to-phase fault. In this case, the relays installed in the secondary carry a current less than the equivalent current flowing through the primary relays that could lead to a situation where the selectivity between the two relays is at risk. For this reason, the discrimination margin between the relays is based on the operating

Table 6.2 Summary of fault conditions

Fault	I_{primary}	$I_{\text{secondary}}$
Three phase	I	I
Phase-to-phase	I	$\sqrt{3}\cdot I/2$
Phase-to-earth	I	$\sqrt{3}\cdot I$

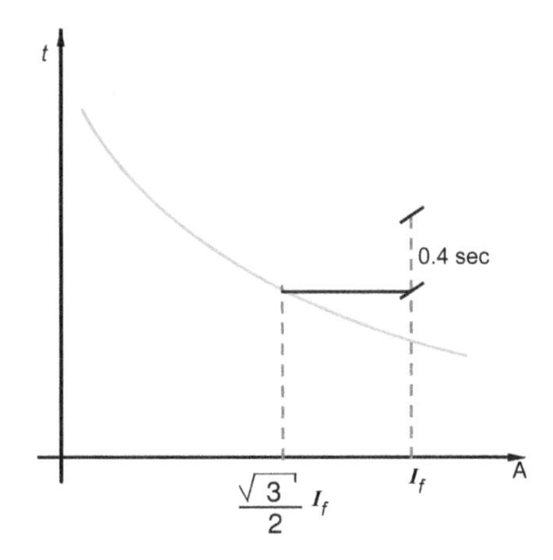

Figure 6.5 Coordination of overcurrent relays for a Dy transformer

time of the secondary relays at a current equal to $\sqrt{3}I_f/2$, and on the operating time for the primary relays for the full fault current value I_f, as shown in Figure 6.5.

Example 6.1

The power system of Figure 6.6 contains the element parameters as well as the short-circuit data for the main substation. The model also incorporates protection relays with the corresponding CTs.

Transformation ratios of the CTs (Ir1/Ir2) associated with breakers CB-1 to CB-10 are given in Figure 6.6. To determine the CTs saturation, use the data of Table 6.3 associated with each CT. The methodology employed is given for the Standard IEC61869-2 2012.

The following calculations will be done:

1. The three phase short-circuit levels on busbars 'Bus-1', 'Bus-2' and 'Bus-3'
2. CTs saturation using the Standard IEC61869-2 2012 methodology.
3. The settings of the instantaneous elements and the TAP and DIAL settings of the relays to guarantee a coordinated protection arrangement, allowing a discrimination margin of 0.2 s.

The summary results for the short-circuit calculation on busbar 'Bus-4' is shown in Table 6.4.

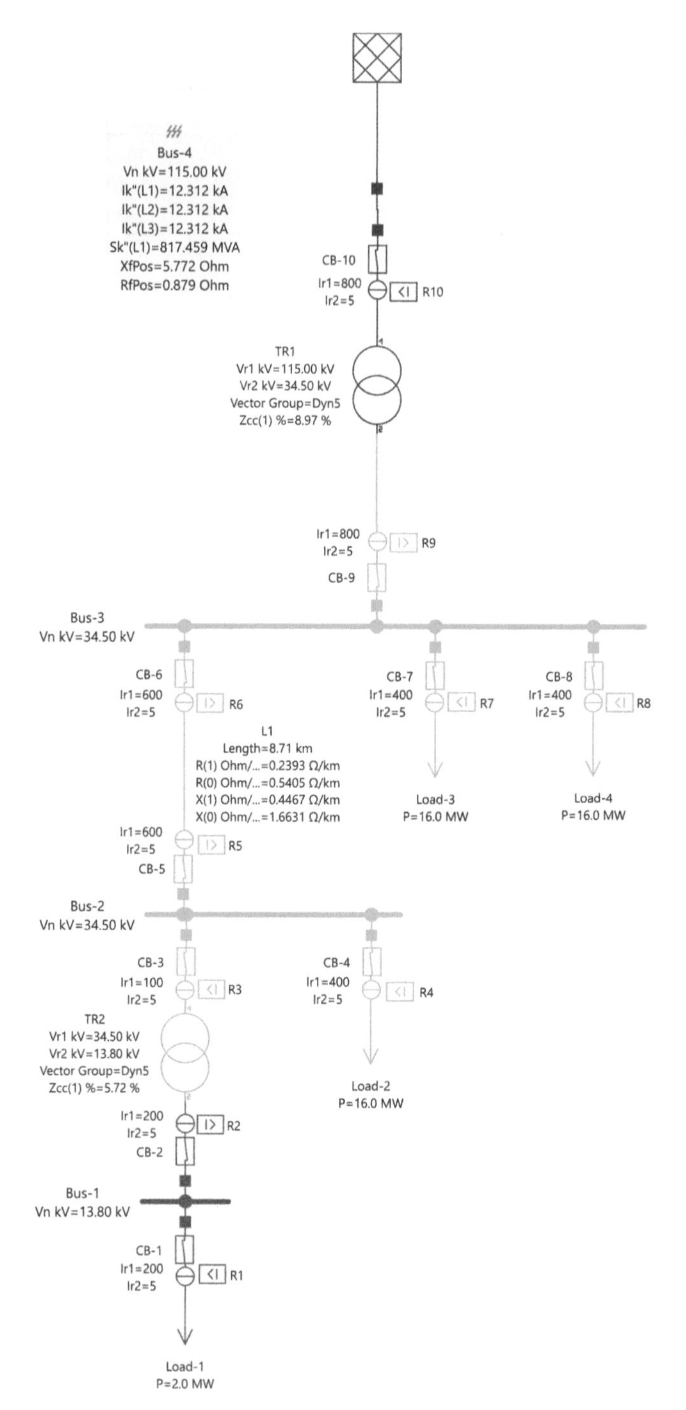

Figure 6.6 Power system of Example 6.1

Table 6.3 CTs associated to Example 6.1

Associated relay	Voltage (kV)	Transformation ratios of the CTs [IR1/IR2]	Type	Relay load (Ω)
R1	13.8	200/5	5P30 (30 VA)	1.6
R2	13.8	200/5	5P30 (30 VA)	1.6
R3	34.5	100/5	5P30 (30 VA)	0.6
R4	34.5	400/5	5P30 (30 VA)	1.6
R5	34.5	600/5	5P30 (30 VA)	1.6
R6	34.5	600/5	5P30 (30 VA)	1.0
R7	34.5	400/5	5P30 (30 VA)	1.0
R8	34.5	400/5	5P30 (30 VA)	1.0
R9	34.5	800/5	5P30 (30 VA)	1.6
R10	115.0	800/5	5P30 (30 VA)	1.0

Table 6.4 Short-circuit calculation summary for power system of Example 6.1

Name	Vn (kV)	Fault distance	$Ik''(R)$ (kA)	$\angle Ik''$ (R) (°)	Sk'' (R) (MVA)
Bus-4	115.0	0	12.312	−82.407	817.458
TR1	–	–	0.0	0.0	0.0
ExtGrid	–	–	12.312	97.59277	817.458

The p.u. impedances are calculated on the following bases:

$$V_b = 115\text{kV},\ S_b = 817.46\ \text{MVA}$$

The base impedance is as follows: $Z_b = \frac{V_b{}^2}{S_b} = 16.178\ \Omega$

The base current is as follows: $I_b = \frac{S_b}{V_b} = \frac{817.46MVA}{115kV} = 7108.35\ \ A$

Network equivalent

The system impedance is as follows: $Z_s = \frac{(0.87+j5.79)\Omega}{16.178\Omega} = 0.3619 \measuredangle 81.45\ p.u.$

Transformer TR1

The transformer TR1 impedance at system bases is

$$Z_{TR1} = Z_{CC} \times \frac{817.46MVA}{40MVA} \times \left(\frac{115kV}{115kV}\right)^2 = j1.833\ p.u.$$

Line L1

The line L1 impedance at system bases is

$$Z_{L1} = \frac{(2.084 + j3.89)\Omega}{16.178\Omega} = 0.273 \measuredangle 61.82 \ p.u.$$

Transformer TR2

The transformer TR2 impedance at system bases is

$$Z_{TR1} = Z_{CC} \times \frac{817.46MVA}{3MVA} \times \left(\frac{34.5kV}{115kV}\right)^2 = j1.403 \ p.u.$$

- For the three-phase fault on the bus 'Bus-3', Figure 6.7 presents the positive sequence network:

where:

I_b, is the base current.

Z_{th}, is the Thevenin impedance equivalent. In this case, Z_{th} is equal to:

$$Z_{TR1} = j1.833 \ p.u.$$

It is observed that the fault current on the bus 'Bus-3' corresponds to the same base current determined in the previous point. Thus,

$$Ifault \ three - phase_{Bus-3} = 7108.35 \ A \ prim$$

- The three-phase fault on the bus 'Bus-2' is

$$Ifault \ three - phase_{Bus-2} = I_b \times \frac{1}{Z_{th}} = 7108.35 \times \frac{1}{2.438 \measuredangle 85.705}$$

$$= 2915.35 \measuredangle - 85.706 \ A \ prim$$

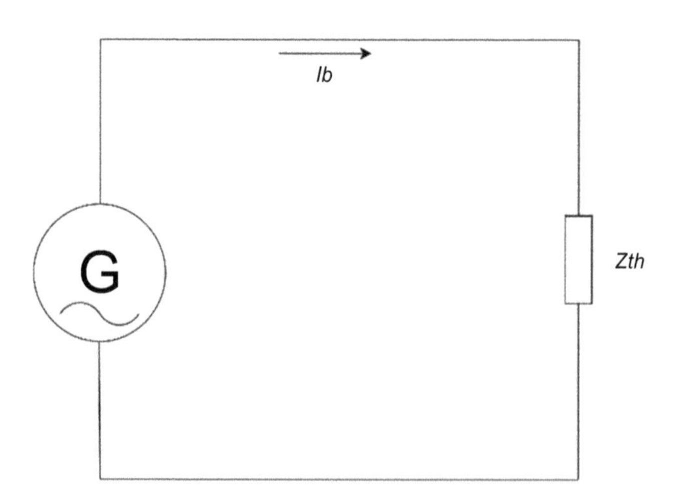

Figure 6.7 Positive sequence network – Example 6.1

where:

I_b is the base current.

Z_{th} is the Thevenin impedance equivalent. In this case, Z_{th} is equal to

$$Z_{th} = Z_s + Z_{TR1} + Z_{L1} = 2.438 \measuredangle 85.705$$

- The three-phase fault on the bus 'Bus-1' is

$$Ifault\ three - phase_{Bus-1} = I_b \times \frac{1}{Z_{th}} = 7108.35 \times \frac{1}{3.838 \measuredangle 87.27}$$
$$= 1851.85 \measuredangle - 87.27\ A\ prim$$

where:

I_b is the base current.

Z_{th} is the Thevenin impedance equivalent. In this case, Z_{th} is equal to:

$$Z_{th} = Z_s + Z_{TR1} + Z_{L1} + Z_{TR2} = 3.838 \measuredangle 87.27$$

A 10VA-5P20 TC, when subjected to a current 20 times greater than its nominal current and feeding its nominal load (10VA with I_n), has a guaranteed error of less than 5%.

To develop the saturation evaluation of a CT that has been included with Standard IEC 61869-2 of 2012, the following criteria are used:

a. It must be verified that the saturation voltage is greater than the required voltage.

$$V_X > I_S \times Z_s$$

where:

V_X is the saturation voltage.

I_s is the current seen in the secondary.

Z_s is the total secondary burden relay load in Ω.

In this way, for the TC associated with relay R1, we have the following:

$$I_S = Ifault\ three - phase_{Bus-1} = 1851.85 \measuredangle - 87.27\ A\ prim$$

$$I_S = 1851.85\ A\ prim \times \frac{5}{200} = 46.3\ A\ \sec$$

$$V_X = Precision\ limiting\ factor \times I_{r2} \times Relay\ Load$$

$$V_X = 30 \times 5\ A\ \sec \times 1.6\ \Omega = 240\ \mathrm{V}$$

The required voltage is: $V_S = I_S \times Z_s = 46.3\ A\ \sec \times 1.6\ \Omega = 74.08\ \mathrm{V}$

Therefore, it is observed that the first evaluation criterion is met.

b. It must be ensured that the required voltage is below the knee voltage.

$$V_{knee} > V_S$$

where:

V_{knee} is the knee voltage. The knee voltage is calculated as a percentage of the CT voltage rating. This percentage corresponds to the knee voltage in percentage, the saturation knee is between 50% and 75% of the nominal voltage of the CT. In this case, 50% is selected.

V_S is the required voltage.

$$V_{knee} = 50\% \times \frac{VA_{CT}}{I_{r2}} * Precision\ limiting\ factor$$

In this way, for the TC associated with relay R1, we have the following:

$$V_{knee} = 50\% \times \frac{30\ VA}{5\ \text{A s}} * 30 = 90\ \text{V}$$

Therefore, it is observed that the first evaluation criterion is met

$V_{knee} > V_S$.

c. It must be verified that the current seen in the secondary of the CT does not exceed the product of the precision limiting factor and the secondary current of the CT.

Precision limiting factor $\times I_{r2} > I_S$

In this way, for the CT associated with relay R1, we have the following:

$30 \times 5\ A\ \text{sec} > 46.3\ A\ \text{sec} \rightarrow 150 > 46.3$

Therefore, it is observed that the first evaluation criterion is met.

Table 6.5 presents the summary of the saturation verification for the CTs associated with Example 6.1.

Consider all the relays to be set are numerical type Beckwith M-7651 with the characteristics as shown in Figure 6.8. The relays have a standard inverse time current characteristic with the following constants of Table 6.6.

The time–current characteristic (TCC) is defined by $t = \frac{Time\ Dial \times 0.14}{(MULT)^{0.02} - 1}$ where MULT is fault current (in secondary amps)/tap. The following considerations have to be taken into account:

- For setting of the instantaneous element a value of ten times the maximum load current is used.
- The margin time for this relay can be 0.2 s since it is of numerical type.

Defining the pick-up for relays according to Tables 6.7–6.9.

Table 6.5 Summary results of saturation verification for the CTs

Associated relay	**Fault type**	**Transformation ratios of the CTs [IR1/IR2]**	$V_X > I_S \times Z_S$	$V_{knee} > V_S$	***Precision limiting f.*** $\times I_{r2} > I_S$
R1	Three phase	200/5	Meets criteria	Meets criteria	Meets criteria
R2	Three phase	200/5	Meets criteria	Meets criteria	Meets criteria
R3	Three phase	100/5	Meets criteria	Meets criteria	Meets criteria
R4	Three phase	400/5	Meets criteria	Meets criteria	Meets criteria
R5	Three phase	600/5	Meets criteria	Meets criteria	Meets criteria
R6	Three phase	600/5	Meets criteria	Meets criteria	Meets criteria
R7	Three phase	400/5	Meets criteria	Meets criteria	Meets criteria
R8	Three phase	400/5	Meets criteria	Meets criteria	Meets criteria
R9	Three phase	800/5	Meets criteria	Meets criteria	Meets criteria
R10	Three phase	800/5	Meets criteria	Meets criteria	Meets criteria

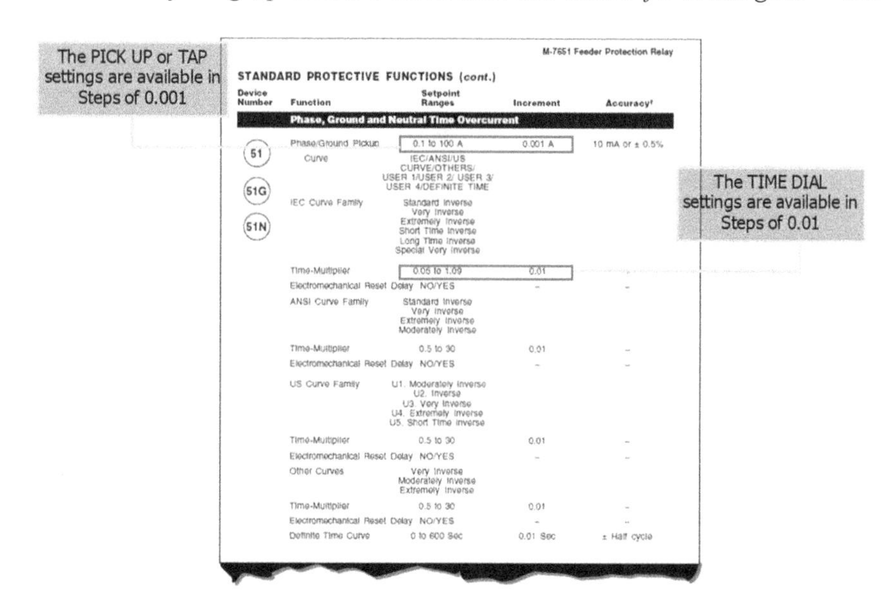

M-7651 Feeder Protection Relay

STANDARD PROTECTIVE FUNCTIONS (*cont.*)

Device Number	Function	Setpoint Ranges	Increment	Accuracy†
	Phase, Ground and Neutral Time Overcurrent			
51 51G 51N	Phase/Ground Pickup	0.1 to 100 A	0.001 A	10 mA or ± 0.5%
	Curve	IEC/ANSI/US CURVE/OTHERS/ USER 1/USER 2/ USER 3/ USER 4/DEFINITE TIME		
	IEC Curve Family	Standard Inverse Very Inverse Extremely Inverse Short Time Inverse Long Time Inverse Special Very Inverse		
	Time-Multiplier	0.05 to 1.09	0.01	
	Electromechanical Reset Delay	NO/YES	–	–
	ANSI Curve Family	Standard Inverse Very Inverse Extremely Inverse Moderately Inverse		
	Time-Multiplier	0.5 to 30	0.01	–
	Electromechanical Reset Delay	NO/YES	–	–
	US Curve Family	U1. Moderately Inverse U2. Inverse U3. Very Inverse U4. Extremely Inverse U5. Short Time Inverse		
	Time-Multiplier	0.5 to 30	0.01	–
	Electromechanical Reset Delay	NO/YES	–	–
	Other Curves	Very Inverse Moderately Inverse Extremely Inverse		
	Time-Multiplier	0.5 to 30	0.01	–
	Electromechanical Reset Delay	NO/YES	–	–
	Definite Time Curve	0 to 600 Sec	0.01 Sec	± Half cycle

Figure 6.8 *Specifications of relay Beckwith M-7651 (reproduced by permission of Beckwith Electric [3])*

Table 6.6 *IEC constants for standard inverse time–current characteristic curve*

IDMT curve description	Standard	α	β	L
Standard inverse	IEC	0.02	0.14	

Table 6.7 *Pick-up calculation for relay associated to loads*

Load	Relay	Nominal voltage (kV)	Nominal current (A)	Pick-up (A s)
Load-1	R1	13.8	93.0	1.25 × 93 × (5/200) = 2.91
Load-2	R4	34.5	298.0	1.25 × 298 × (5/400) = 4.66
Load-3	R7	34.5	298.0	1.25 × 298 × (5/400) = 4.66
Load-4	R8	34.5	298.0	1.25 × 298 × (5/400) = 4.66

Table 6.8 *Pick-up calculation for relay associated to transformers*

Transformer	Relay	Nominal voltage (kV)	Nominal current (A)	Pick-up (A s)
TR1	R10	115	200.82	1.25 × 208.82 × (5/400) = 3.26
TR1	R9	34.5	669.40	1.25 × 669.40 × (5/800) = 5.23
TR2	R3	34.5	50.20	1.25 × 50.20 × (5/100) = 3.14
TR2	R2	13.8	125.51	1.25 × 125.51 × (5/200) = 3.92

Table 6.9 Pick-up calculation for relay associated to line L1

Line	Relay	Nominal voltage (kV)	Nominal current (A)	Pick-up (A s)
L1	R5	34.5	348.2*	1.25 × 348.2 × (5/600) = 3.63
L1	R6	34.5	348.2*	1.25 × 348.2 × (5/600) = 3.63

Note: * *It corresponds to the sum between the nominal current of the transformer at 34.5 kV and the nominal current of Load-4.*

For relay R1:

$$I_{\text{ins. trip}} = 10 \times I_{\text{nom}} \times (1/\text{CTR})$$

$$I_{\text{ins. trip}} = 10 \times 93.0 = 930.0\ A_{\text{prim}}$$

$$I_{\text{ins. trip}} = 10 \times 93.0 \times (5/200) = 23.25\ A_{\text{sec}}$$

$$\text{MULT} = \frac{23.25}{2.91} = 7.99 \text{ times}$$

With dial = 0.05,

$$t = \frac{0.05 \times 0.14}{(7.99)^{0.02} - 1} = 0.165 \text{ s}$$

To coordinate with relay R1 at 930 A_{prim},
$t_{R2} = 0.165 + 0.2 = 0.365$ s.
For relay R2:

$$\text{MULT}_{R2} = 930 \times (5/200) \times (1/3.92) = 5.93 \text{ times}$$

At 5.93 times and $t_{R2} = 0.365$ s.

$$\text{dial} = \left(\frac{0.365}{0.14}\right) \times \left((5.93)^{0.02} - 1\right) = 0.094$$

However, the dial selected is 0.1.

$$t = \frac{0.1 \times 0.14}{(5.93)^{0.02} - 1} = 0.386 \text{ s}$$

This relay has no setting for the instantaneous.

Taking these settings to simulation software, the following results of Figure 6.9 are obtained.

By verifying selectivity through the simulation software, the dial for relay R3 is determined, as shown in Figure 6.10.

The procedure is the same for all the relays of the power. Table 6.10 shows the corresponding settings.

By verifying selectivity through the simulation software, the dial for relays R5, R6, R7, R8, R9 and R10 are determined, as shown in Figures 6.11 and 6.12.

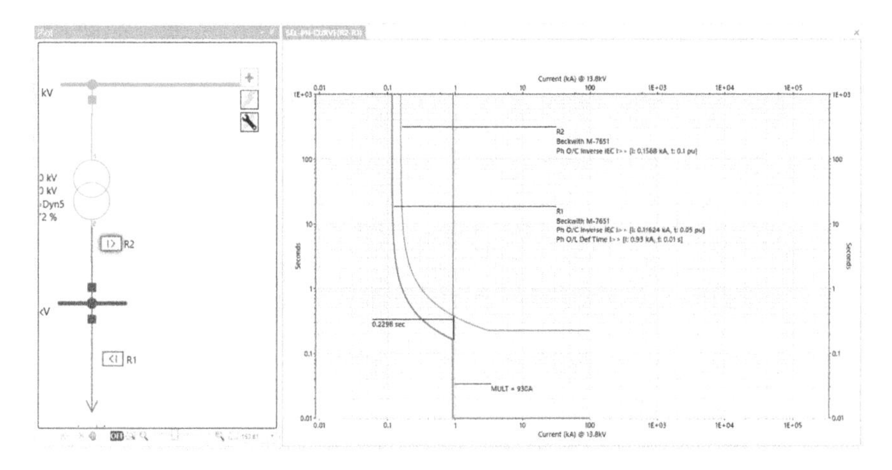

Figure 6.9 Overcurrent inverse-time relay curves associated relays R1 and R2

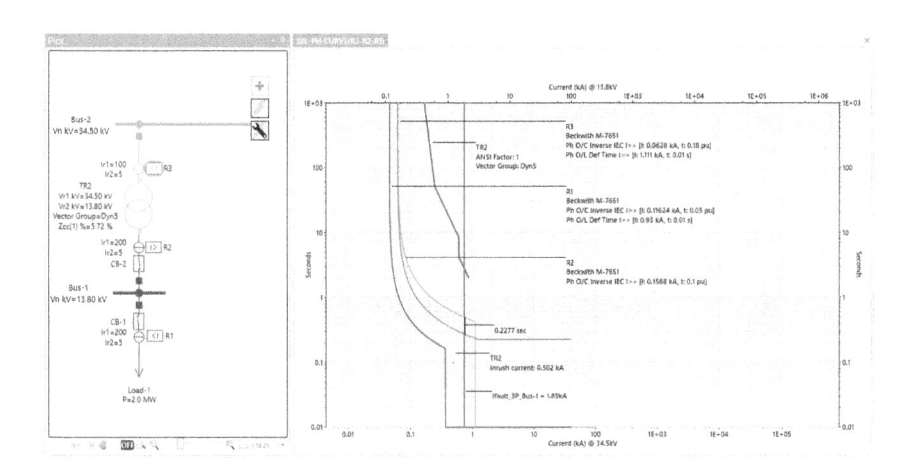

Figure 6.10 Overcurrent inverse-time relay curves associated relays R1, R2 and R3

Table 6.10 Summary of relay settings for Example 6.1

Relay	CT ratio	Pickup (A s)	Dial	Instantaneous (A s)
R1	200/5	2.91	0.05	23.25
R2	200/5	3.92	0.10	–
R3	100/5	3.14	0.18	55.56
R4	400/5	4.66	0.05	37.25
R5	600/5	3.63	0.08	–
R6	600/5	3.63	0.11	36.44
R7	400/5	4.66	0.05	37.25
R8	400/5	4.66	0.05	37.25
R9	800/5	5.23	0.13	–
R10	800/5	1.63	0.18	19.99

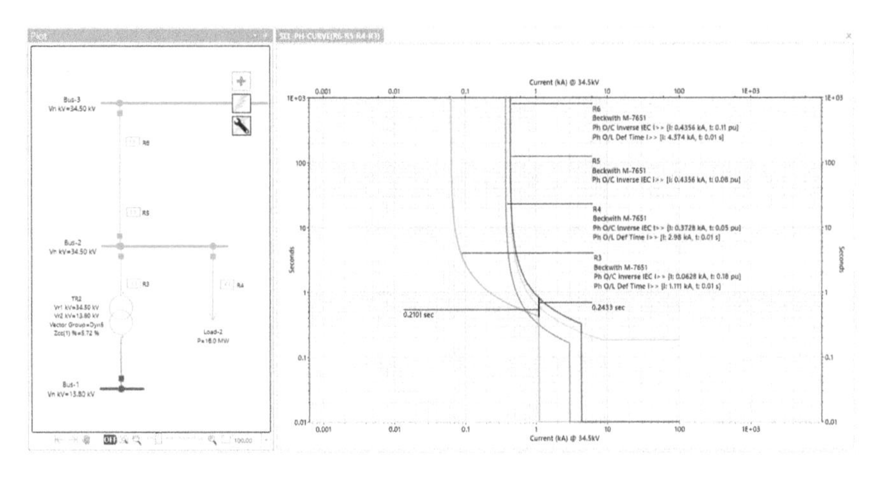

Figure 6.11 Overcurrent inverse-time relay curves associated relays R3, R4, R5 and R6

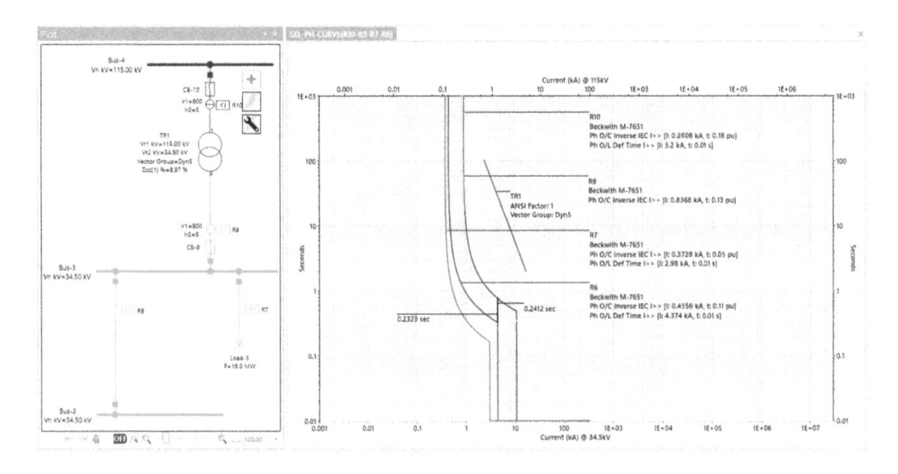

Figure 6.12 Overcurrent inverse-time relay curves associated relays R6, R7, R9 and R10

6.3 Protection equipment installed along the feeders

The devices most used for distribution system protection are overcurrent relays, reclosers, sectionalisers and fuses.

The coordination of overcurrent relays was dealt with in detail in the previous section, and this part will cover the other three devices referred to above. The three last devices can be mounted as stand-alone or as part of other switchgear like those known as vacuum fault interrupter (VFI). This type of switchgear can be arranged in different configurations where the type of equipment can be different (breakers or fuses) and also the number of inputs and outputs can vary. For example, Figure 6.13 illustrates a VFI switchgear components from EATON and Figure 6.14 is a picture of a typical Pad-Mounted Gear also from S&C.

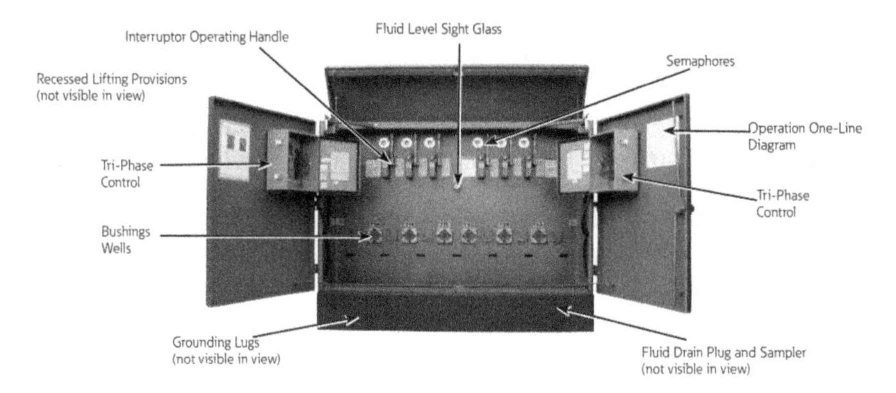

Figure 6.13 VFI switchgear components (reproduced from EATON)

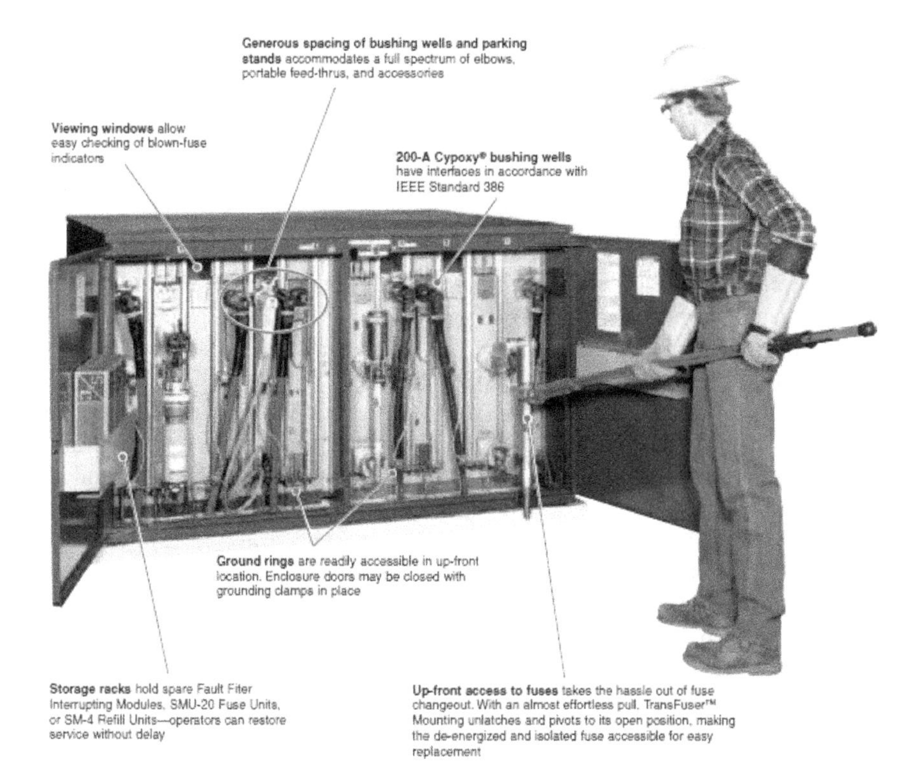

Figure 6.14 PME Pad-Mounted Gear (reproduced by permission of S&C)

6.3.1 Reclosers

6.3.1.1 General

Reclosers are playing an important role for the new trends in urban power grids since they offer the availability to reconfigure the network both normal and fault conditions. Some techniques for Volt/VAR control, FLISR and reliability are supported by different reclosers along the networks. A recloser is a device with the ability to detect phase and phase-to-earth overcurrent conditions, to interrupt the circuit if the overcurrent persists after a predetermined time, and then to automatically reclose to re-energise the line. A recloser can be seen as a set of elements that include an overcurrent relay and the associated breaker with the corresponding reclosing control. If the fault which originated the operation still exists, then the recloser will stay open after a preset number of operations, thus isolating the faulted section from the rest of the system. Protection coordination studies are usually supported by the corresponding units with the corresponding time characteristics as the overcurrent relays.

In an overhead distribution system between 80% and 95% of the faults are of a temporary nature and last, at the most, for a few cycles or seconds. Thus, the recloser, with its opening/closing characteristic, prevents a distribution circuit from being left out of service for temporary faults. Typically, reclosers are designed to have up to three open-close operations and, after these, a final open operation to lock out the sequence. One further closing operation by manual means is usually allowed. The counting mechanisms register operations of the phase or earth-fault units which can also be initiated by externally controlled devices when appropriate communication means are available.

The operating time–current characteristic curves of reclosers normally incorporate three curves, one fast and two delayed, designated as A, B and C, respectively. Figure 6.15 shows a typical set of time–current curves for reclosers. However, new reclosers with microprocessor-based controls may have keyboard-selectable time–current curves that enable an engineer to produce any curve to suit the coordination requirements for both phase and earth faults. This allows reprogramming of the characteristics to tailor an arrangement to a customer's specific needs without the need to change components.

Coordination with other protection devices is important in order to ensure that, when a fault occurs, the smallest section of the circuit is disconnected to minimise disruption of supplies to customers. Generally, the time characteristic and the sequence of operation of the recloser are selected to coordinate with mechanisms upstream towards the source. After selecting the size and sequence of operation of the recloser, the devices downstream are adjusted in order to achieve correct coordination. A typical sequence of a recloser operation for a permanent fault is shown in Figure 6.16. The first shot is carried out in instantaneous mode to clear temporary faults before they cause damage to the lines. The three later ones operate in a timed manner with predetermined time settings. If the fault is permanent, the time-delay operation allows other protection devices nearer to the fault to open, limiting the amount of the network being disconnected.

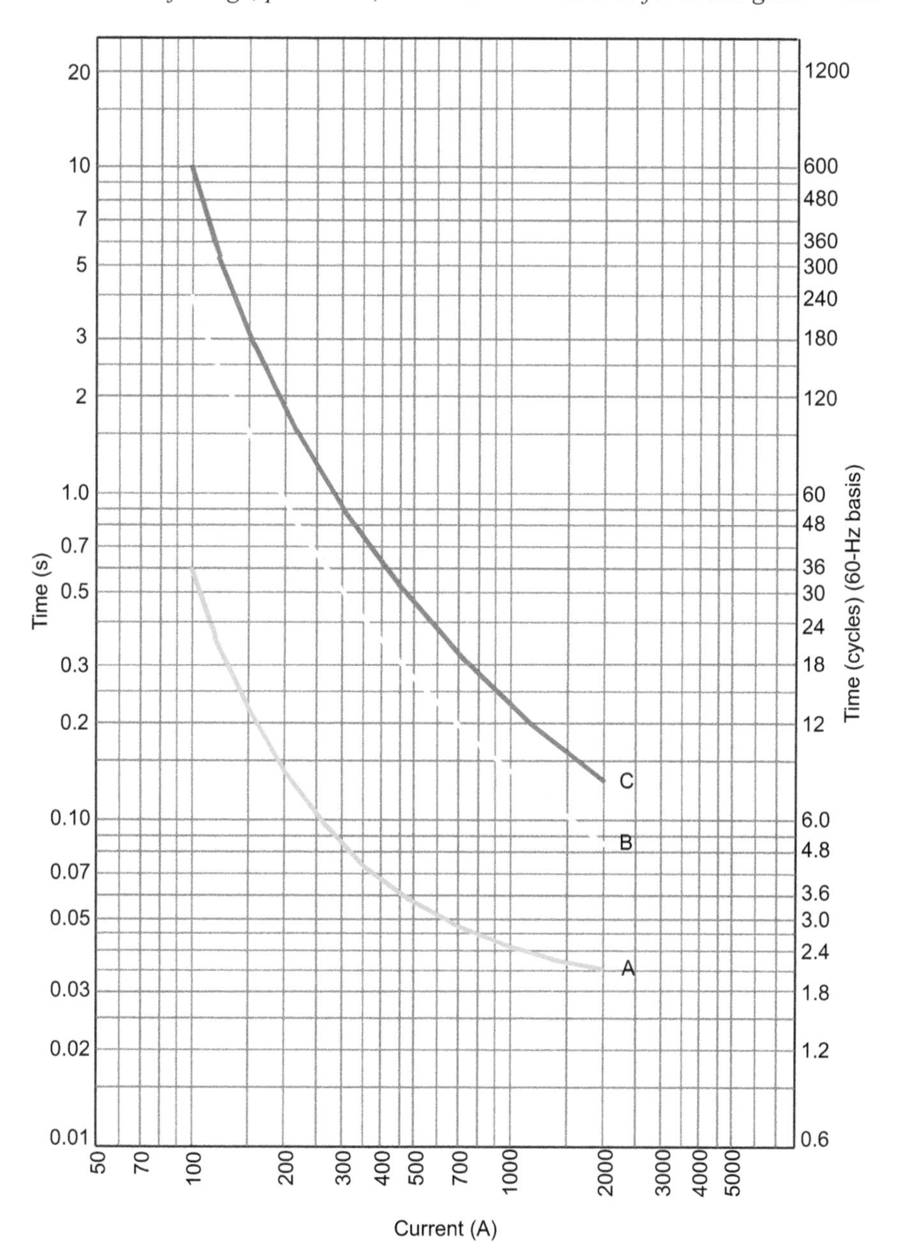

Figure 6.15 Time–current curves for reclosers

Earth faults are less severe than phase faults, and therefore, it is important that the recloser has an appropriate sensitivity to detect them. One method is to use CTs connected residually so that the resultant residual current under normal conditions is approximately zero. The recloser should operate when the residual current exceeds the setting value, as would occur during earth faults.

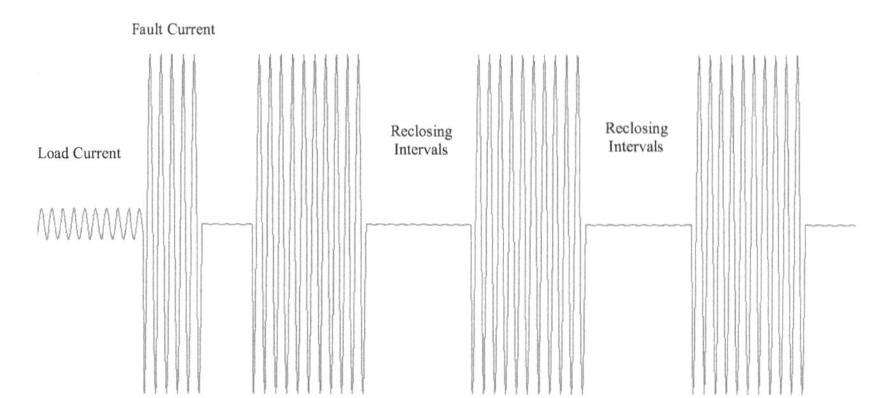

Figure 6.16 Typical sequence for recloser operation

Typically, reclosers are designed to have up to three open-close operations and, after these, a final open operation to lock out the sequence. One further closing operation by manual means is usually allowed. The counting mechanisms register operations of the phase or ground-fault units which can also be initiated by externally controlled devices when appropriate communication means are available.

6.3.1.2 Classification

Reclosers can be classified as follows:

- Number of phases: single-phase and three-phase;
- Arc-interrupting medium: oil or vacuum;
- Type of insulation: oil, epoxy or SF_6;
- Type of control: hydraulic, electronic or magnetic actuator.

Single-phase reclosers are used when the load is predominantly single phase. In such a case, when a single-phase fault occurs, the recloser should permanently disconnect the faulted phase so that supplies are maintained on the other phases. Three-phase reclosers are used when it is necessary to disconnect all three phases in order to prevent unbalanced loading on the system.

Reclosers with hydraulic operating mechanisms have a disconnecting coil in series with the line. When the current exceeds the setting value, the coil attracts a piston which opens the recloser main contacts and interrupts the circuit. The time characteristic and operating sequence of the recloser are dependent on the flow of oil in different chambers. The electronic type of control mechanism is normally located outside the recloser and receives current signals from a CT-type bushing. When the current exceeds the predetermined setting, a delayed shot is initiated which finally results in a tripping signal being transmitted to the recloser control mechanism. The control circuit determines the subsequent opening and closing of the mechanism, depending on its setting. Reclosers with electronic operating mechanisms use a coil or motor mechanism to close the contacts. Oil reclosers use the oil to extinguish the arc and also to act as the basic insulation. The same oil can be used in the control mechanism. Vacuum and SF_6 reclosers have the advantage of requiring less maintenance.

The following figures illustrate several features of reclosers. Figure 6.17 shows different types of single phase reclosers: (a) NOJA OSM – vacuum interrupting, epoxy insulated and electronically controlled; (b) COOPER NOVA – oil interrupting, epoxy insulated and electronically controlled; (c) COOPER D – oil interrupting, oil insulated and hydraulically controlled.

Figure 6.18 shows different types of three phase reclosers: (a) G&W Viper-LT – Vacuum interrupting, epoxy insulated and electronically controlled; (b) SCHNEIDER U – Vacuum interrupting, epoxy insulated and electronically controlled; (c) ABB OVR – Vacuum interrupting, epoxy insulated and electronically controlled; (d) Hawker Siddeley Switchgear Ltd's GVR – vacuum interrupting, SF_6 insulated and magnetic actuator.

Figure 6.19 presents the internal components of a three phase recloser, air-insulated, vacuum interrupting, electronically controlled recloser, manufactured by Cooper Systems.

Figure 6.20 presents a single phase recloser with magnetic-vacuum interruption technology by HUBBELL.

The coordination margins with hydraulic reclosers depend upon the type of equipment used. In small reclosers, where the current coil and its piston produce the opening of the contacts, a separation greater than 12 cycles ensures non-simultaneous operation.

With large capacity reclosers, the piston associated with the current coil only actuates the opening mechanism. In such cases, a separation of more than eight cycles guarantees non-simultaneous operation.

The principle of coordination between two large units in series is based on the time of separation between the operating characteristics, in the same way as for small units.

Electronically controlled reclosers can be coordinated more closely since there are no inherent errors such as those which exist with electromechanical

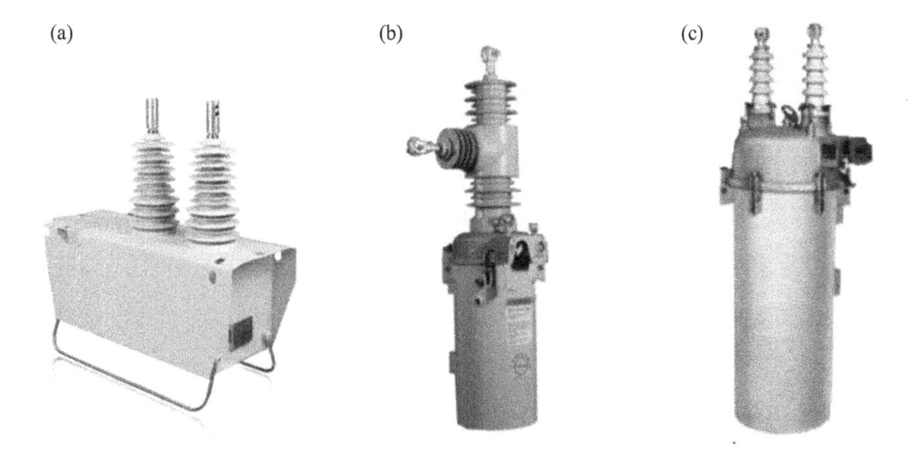

Figure 6.17 Single-phase reclosers

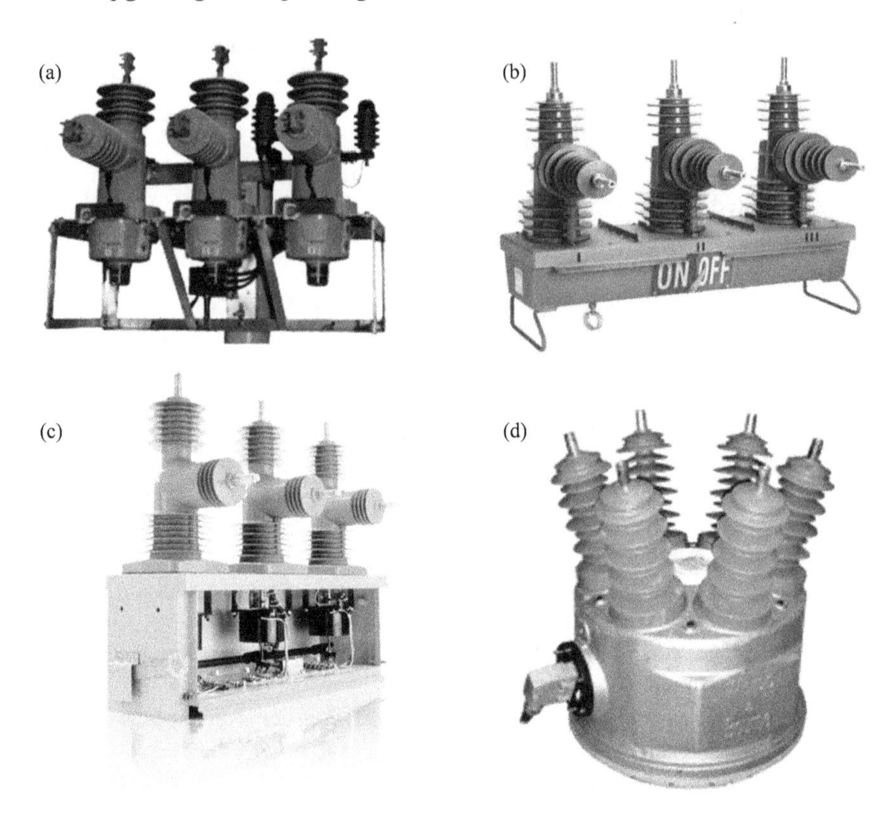

Figure 6.18 Three-phase reclosers

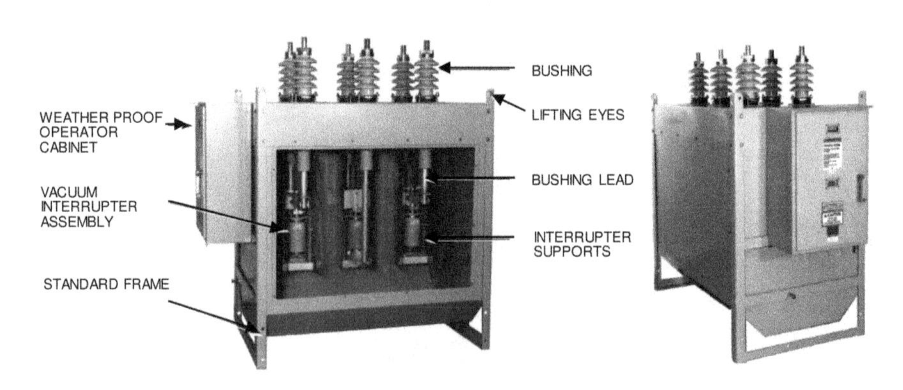

Figure 6.19 COOPER Kyle type VSA20A

mechanisms (due to overspeed, inertia, etc.). The downstream recloser must be faster than the upstream recloser, and the clearance time of the downstream recloser plus its tolerance should be lower than the upstream recloser clearance time less its tolerance.

Figure 6.20 Versa-Tech II HUBBELL

Normally, the setting of the recloser at the substation is used to achieve at least one fast reclosure in order to clear temporary faults on the line between the substation and the load recloser. The latter should be set with the same, or a larger, number of rapid operations as the recloser at the substation.

It should be noted that the criteria of spacing between the time–current characteristics of electronically controlled reclosers are different to those used for hydraulically controlled reclosers.

6.3.1.3 Applications

Reclosers are used at the following points on a distribution network:

- in substations, to provide primary protection for a circuit;
- in main feeder circuits, in order to permit the sectioning of long lines and thus prevent the loss of a complete circuit due to a fault towards the end of the circuit;
- in branches or spurs, to prevent the tripping of the main circuit due to faults on the spurs.

Figure 6.21 illustrates several options to locate reclosers in a distribution system.

6.3.1.4 Specifications

The voltage rating and the short-circuit capacity of the recloser should be equal to, or greater than, the values which exist at the point of installation. The same criteria

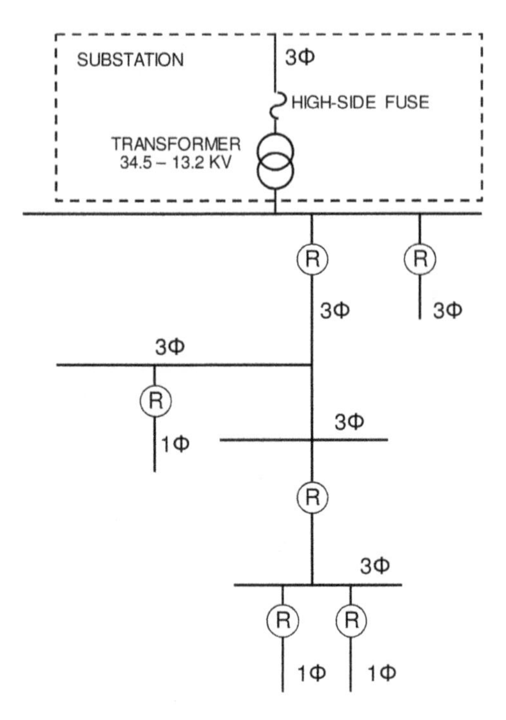

Figure 6.21 Option to locate reclosers

should be applied to the current capability of the recloser with respect to the maximum load current to be carried by the circuit. It is also necessary to ensure that the fault current at the end of the line being protected is high enough to cause operation of the recloser.

To properly apply automatic circuit reclosers, five major factors must be considered:

1. System voltage: system voltage will be known. The recloser must have a voltage rating equal to (or greater than) system voltage.
2. Maximum fault current available at the recloser location: maximum fault current will be known or can be calculated. The recloser interrupting must be equal to (or greater than) the maximum available fault current at the recloser location.
3. Maximum load current: the recloser continuous current rating must be equal to (or greater than) anticipated circuit load. For series-coil-type reclosers, the coil size can be selected to match the present load current, the anticipated future load current or the substation transformer capacity. Minimum-trip current is nominally twice the coil continuous-current rating. For electronically controlled reclosers, minimum-trip current must be greater than any anticipated peak load. Generally, a trip-current value of at least twice the expected load current is used.

4. Minimum-fault current within the zone to be protected. Minimum fault current that might occur at the end of the line section must be checked to confirm that the recloser will sense and interrupt this current.
5. Coordination with other protective devices on both the source and the load sides of the recloser.

After the first four application factors have been satisfied, coordination of the recloser with both the source and the load-side devices must be determined. Proper selection of time delays and sequences is vital to assure that any momentary interruption or longer-term outage due to faults is restricted to the smallest possible section of the system.

Generally, recloser timing and sequences are selected to coordinate with the source-side devices. After the size and sequence of the required recloser have been determined, the protective equipment farther down the line is selected to coordinate with it.

6.3.2 Fuses

6.3.2.1 General

A fuse is an overcurrent protection device; it possesses an element which is directly heated by the passage of current and which is destroyed when the current exceeds a predetermined value. A suitably selected fuse should open the circuit by the destruction of the fuse element, eliminate the arc established during the destruction of the element and then maintain circuit conditions open with nominal voltage applied to its terminals (i.e., no arcing across the fuse element). Figure 6.22 illustrates a reference of fuses for its installation on distribution networks.

Most fuses used in distribution systems operate on the expulsion principle, i.e. they have a tube to confine the arc, with the interior covered with deionising fibre, and a fusible element. In the presence of a fault, the interior fibre is heated up when the fusible element melts and produces de-ionising gases which accumulate in the

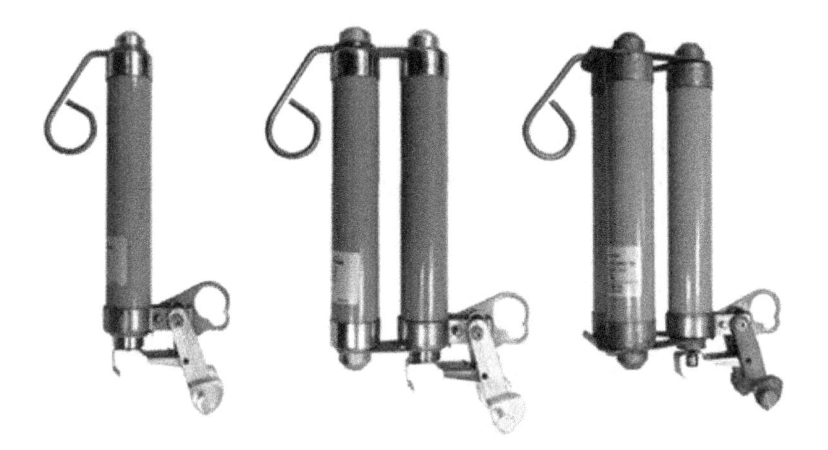

Figure 6.22 Example of a fuse for distribution (from EATON)

tube. The arc is compressed and expelled out of the tube; in addition, the escape of gas from the ends of the tube causes the particles which sustain the arc to be expelled. In this way, the arc is extinguished when current zero is reached. The presence of de-ionising gases, and the turbulence within the tube, ensures that the fault current is not re-established after the current passes through zero point. The zone of operation is limited by two factors: the lower limit based on the minimum time required for the fusing of the element (minimum melting time) and the upper limit determined by the maximum total time which the fuse takes to clear the fault.

6.3.2.2 Applications

Fuses are the most popular protective device used in electrical systems. In distribution systems, in particular, they are applied in most elements. Specific applications have been developed for the following elements:

- distribution transformers,
- capacitors,
- feeders.

6.3.2.3 Type

Power fuses: power fuses provide reliable and economical protection for transformers and capacitor banks. They are normally in outdoor substations served, and they incorporate silver or nickel-chrome fusible elements.

Fuse cutouts: a fuse cutout is a combination of a fuse and a switch, used in primary overhead feeder lines and distribution transformers to protect from current surges and overloads. The main components of fuse cutouts are the cutout body, the fuse holder and the fuse element.

The cutout body, an open ‘C’-shaped frame that supports the ‘fuse holder’ and a porcelain insulator that electrically isolates the conductive portions of the assembly from the support to which the insulator is fastened.

The fuse holder, often called the ‘fuse tube’, contains the interchangeable fuse element and acts as a simple knife switch. When the contained fuse operates or blows, the fuse holder drops open, disengages the knife switch and hangs from a hinge assembly.

The fuse element, or ‘fuse link’, is the replaceable portion of the assembly that operates due to high electrical currents.

Current-limiting fuse: a current-limiting fuse is a fuse that abruptly introduces a high resistance to reduce current magnitude and duration, resulting in subsequent current interruption. This type of fuse significantly reduces the current amplitude and the energy released in the event of a short circuit.

These fuses develop a positive internal gap of high dielectric strength after circuit interruption, thus precluding destructive re-ignitions when exposed to full system voltage such as are experienced with current-limiting fuses after clearing under low recovery-voltage conditions.

Type SM Power Fuses have helically coiled silver fusible elements that are of solderless construction and are surrounded by air. Because of this construction, the fusible element is free from mechanical and thermal stress and confining support, and therefore is not subject to damage – even by inrush currents that approach but do not exceed the fuse's minimum melting time–current characteristic curve. Current-limiting fuses, in contrast, have fusible elements which consist of a number of very fine diameter wires or one or more perforated or notched ribbons, surrounded by, and in contact with, a filler material such as silica sand. Because of this construction, current-limiting fuses are susceptible to element damage caused by current surges that approach the fuse's minimum melting time–current characteristic curve. This damage may occur in one or more of the following ways: the fusible element may melt, but not completely separate because the molten metal is constrained by the filler material – resulting, possibly, in resolidification of the element with a different cross-sectional area.

One or more, but not all, of the parallel wires or ribbons of the fusible element may melt and separate.

The fusible element may break as a result of fatigue brought about by current cycling that can cause localised buckling from thermal expansion and contraction.

Damage to fusible elements of current-limiting fuses, as described above, may shift or alter their time–current characteristics, resulting in a loss of complete coordination between the fuse and other downstream overcurrent protective devices. Moreover, a damaged current-limiting fuse element may melt due to an otherwise harmless inrush current, but the fuse may fail to clear the circuit due to insufficient power flow – with the fuse continuing to arc and burn internally due to load-current flow. Because of the potential for damage to the fusible element from inrush currents, and because of the effects of loading and manufacturing tolerances, current-limiting fuse manufacturers typically require that when applying such fuses, adjustments be made to the minimum melting time–current characteristic curves. These adjustments are referred to as 'safety zones' or 'setback allowances', and range from 25% in terms of time to 25% in terms of current. The latter can result in an adjustment of 250% or more in terms of time, depending on the slope of the time–current characteristic curve at the point where the safety zone or setback allowance is measured. Furthermore, most current-limiting fuses inherently have steep, relatively straight time–current characteristic curves which, together with the required large safety-zone or setback-allowance adjustments, force the selection of a current-limiting fuse ampere rating substantially greater than the transformer full-load current to withstand combined transformer-magnetising and load inrush currents, and also to coordinate with secondary-side protective devices. The selection of such large fuse ampere ratings results in reduced protection for the transformer and possible impairment of coordination with upstream protective devices. Also, since high-ampere-rated current-limiting fuses typically require the use of two or three lower ampere-rated fuses connected in parallel, increased cost and space requirements may be countered.

The value of the limited cutoff current is determined as a function of the prospective current for current values in the first half cycle where the short-circuit current is limited as shown in Figure 6.23.

The fuse elements or fuse links used in most distribution cutouts are tin or silver alloy fuse links that melt (or operate) when exposed to high current conditions. Ampere ratings of fuse elements vary from 1 A to 200 A.

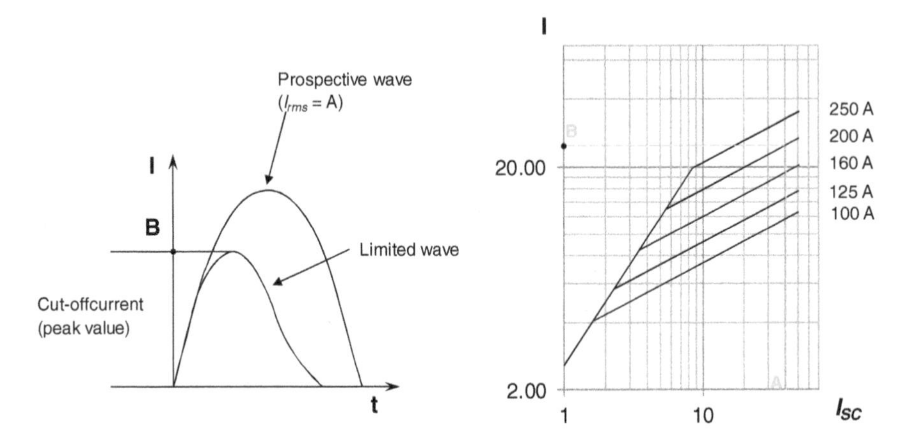

Figure 6.23 Fuse application where one branch system is protected

In distribution systems, it is common to designate fuse links as K and T for fast and slow types respectively, depending on the speed ratio.

The speed ratio is the ratio of minimum melt current which causes fuse operation at 0.1 s to the minimum-melt current for 300 s operation.

For the K link, a speed ratio (SR) of 6–8 is defined and, for a T link, 10–13.

6.3.2.4 Classification

There are a number of standards to classify fuses according to the rated voltages, rated currents, time–current characteristics, manufacturing features and other considerations. For example, there are several sections of UL 198-2023 Standards which cover low-voltage fuses of 600 V or less [4]. For medium- and high-voltage fuses within the range 2.3–138 kV, standards such as IEEE C37.40, 41, 42, 46, 47 and 48 apply [5–10]. Other organisations and countries have their own standards; in addition, fuse manufacturers have their own classifications and designations.

In distribution systems, the use of fuse links designated K and T for fast and slow types, respectively, depending on the speed ratio, is very popular. The speed ratio is the ratio of minimum melt current which causes fuse operation at 0.1 s to the minimum-melt current for 300 s operation. For the K link, a speed ratio (SR) of 6–8 is defined and for a T link, 10–13. Figure 6.24 shows the comparative operating characteristics of type 200K and 200T fuse links. For the 200K fuse, a 4400 A current is required for 0.1 s clearance time and 560 A for 300 s, giving an SR of 7.86. For the 200T fuse, 6500 A is required for 0.1 s clearance and 520 A for 300 s; for this case, the SR is 12.5.

6.3.2.5 Specifications

The following information is required in order to select a suitable fuse for use on the distribution system:

1. Voltage and insulation level,
2. Type of system,

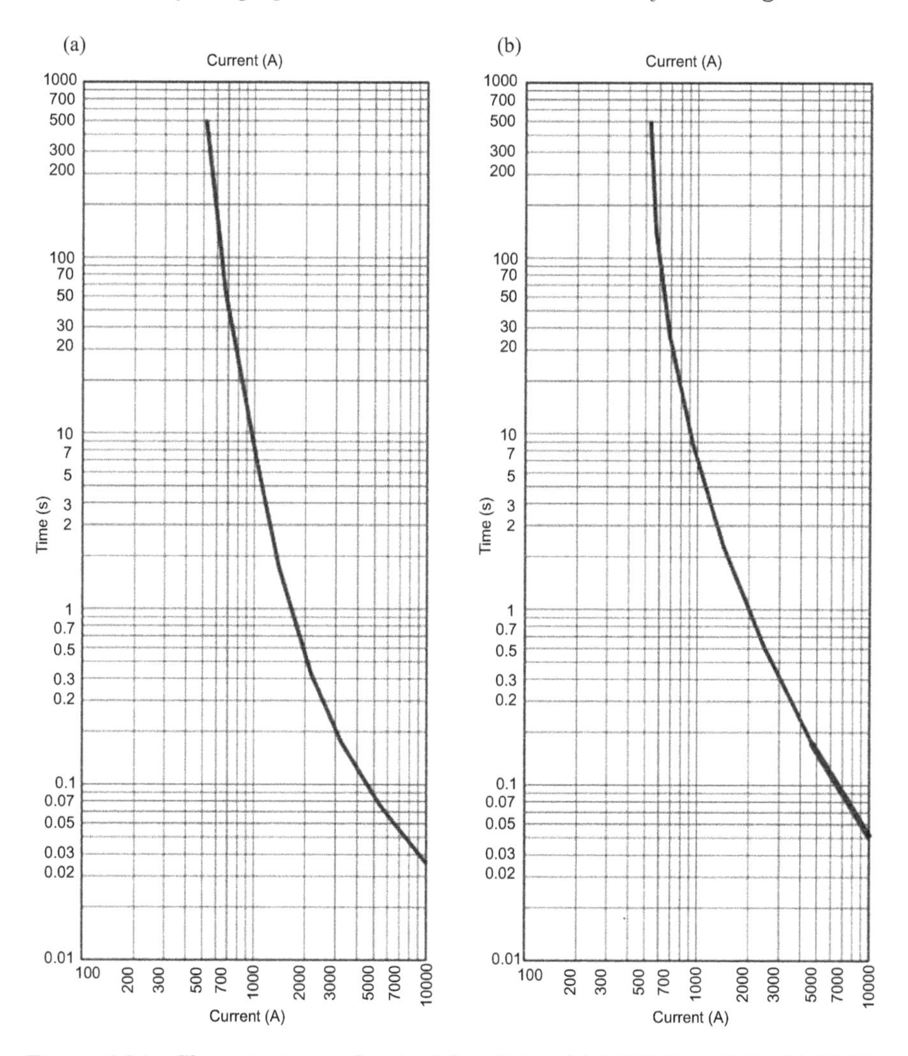

Figure 6.24 Characteristics of typical fuse links: (a) 200K Fuse link; (b) 200T Fuse link

3. Maximum short-circuit level,
4. Load current.

The above four factors determine the fuse nominal current, voltage and short-circuit capability characteristics.

Selection of nominal current

The nominal current of the fuse should be greater than the maximum continuous load current at which the fuse will operate. An overload percentage should be allowed according to the protected-equipment conditions. In the case of power transformers, fuses should be selected such that the time–current characteristic is above the inrush curve of the transformer and below its thermal limit. Some manufacturers have produced tables to assist in the proper fuse selection for different ratings and connection arrangements.

Selection of nominal voltage

The system characteristics determine the voltage seen by the fuse at the moment when the fault current is interrupted. Such a voltage should be equal to, or less than, the nominal voltage of the fuse. Therefore, the following criteria should be used:

- In unearthed systems, the nominal voltage should be equal to, or greater than, the maximum phase-to-phase voltage.
- In three-phase earthed systems, for single phase loads the nominal voltage should be equal to, or greater than, the maximum line-to-earth voltage and for three-phase loads the nominal voltage is selected on the basis of the line-to-line voltage.

Selection of short-circuit capacity

The symmetrical short-circuit capacity of the fuse should be equal to, or greater than, the symmetrical fault current calculated for the point of installation of the fuse.

Fuse notation

When two or more fuses are used on a system, the device nearest to the load is called the main protection, and that upstream, towards the source, is called the back up. The criteria for coordinating them will be discussed later.

6.4 Setting criteria

The following basic criteria should be employed when coordinating time–current devices in distribution systems:

1. The main protection should clear a permanent or temporary fault before the backup protection operates or continues to operate until the circuit is disconnected. However, if the main protection is a fuse and the back-up protection is a recloser, it is normally acceptable to coordinate the fast-operating curve or curves of the recloser to operate first, followed by the fuse, if the fault is not cleared (see Section 6.4.2).
2. Loss of supply caused by permanent faults should be restricted to the smallest part of the system for the shortest time possible.

Further in the section, criteria and recommendations are given for the coordination of different devices used on distribution systems.

6.4.1 Fuse–fuse coordination

The essential criterion when using fuses is that the maximum clearance time for a main fuse should not exceed 75% of the minimum melting time of the backup fuse, as indicated in Figure 6.25. This ensures that the main fuse interrupts and clears the fault before the back-up fuse is affected in any way. The factor of 75% compensates for effects such as load current and ambient temperature, or fatigue in the fuse element caused by the heating effect of fault currents which have passed through the fuse to a fault downstream, but which were not sufficiently large enough to melt the fuse. Keeping the coordination between fuses is the idea as shown in Figure 6.26.

The series fuse–fuse coordination is given by manufacturers based on information which is presented normally in tables that list maximum fault current values

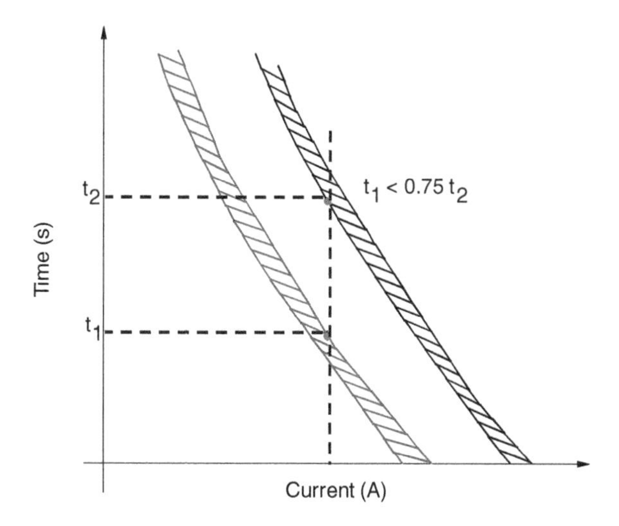

Figure 6.25 Criteria for fuse–fuse coordination

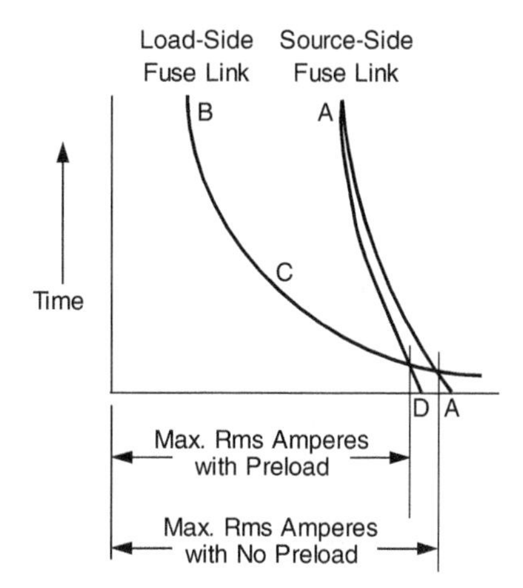

Figure 6.26 Time–current characteristics for fuse–fuse coordination

Table 6.11 Maximum fault current in amperes, rms. S&C Standard Speed Fuse Links

Load – Side Fuse Link Ampere Rating	Source – SideFuse Link Ampere Rating																	
	5	7	10	15	20	25	30	40	50	65	80	100	101*	102*	103*	125	150	200
1	120	220	370	590	750	890	1100	1500	1850	2250	2800	3700	5200	8900	15000	4300	5500	7100
2	95	205	360	580	750	890	1100	1500	1850	2250	2800	3700	5200	8900	15000	4300	5500	7100
3		175	335	570	740	880	1100	1500	1850	2250	2800	3700	5200	8900	15000	4300	5500	7100
5		60	280	530	700	850	1050	1450	1800	2250	2800	3700	5200	8900	15000	4300	5500	7100
7			170	490	680	830	1050	1450	1800	2250	2750	3700	5200	8900	15000	4300	5500	7100
10				330	560	740	970	1400	1750	2200	2750	3700	5200	8900	15000	4250	5500	7100
15					120	480	780	1250	1650	2100	2700	3650	5100	8900	15000	4250	5500	7100
20							520	1100	1600	2050	2600	3600	5100	8900	15000	4200	5500	7000
25								920	1450	1950	2550	3500	5100	8900	15000	4150	5400	7000
30								560	1200	1750	2450	3450	5000	8800	15000	4050	5400	7000
40									285	1300	2100	3200	4800	8800	14500	3850	5200	6800
50										290	1550	2850	4650	8600	14500	3650	5100	6700
65											365	2400	4300	8500	14500	3250	4800	6500
80												1350	3750	8200	14000	2550	4350	6200
100													2300	7600	14000		3100	5200
101*															13000			1650
102*															6200			
125															13000			3150
150															11500			

[1]The maximum fault – current values are based on published total clearing time – current characteristics curves and on minimum melting fault – current characteristic for preloading of the source – side fuse link to this ampere rating.
*S&C Coordinating Speed Fuse Links, TCC No.172-6.

that represent the intersection of the total clearing time–current characteristic curve of the load-sided fuse link with the minimum melting time–current characteristic curve of the source-side fuse link.

Typically, the tables present the maximum fault current with source-side fuse links arranged horizontally and the load-side fuse links arranged vertically. Coordination between two fuses is guaranteed when the short-circuit values are equal or lower to that given in the table for the intersection of the values corresponding to the two fuses.

Some manufacturers give the tables considering both preload and no preload. In the first case, the curve is steeper as shown in Figure 6.26. Table 6.11 presents the coordination of S&C Standard Speed Fuse Links, based on preloading of source-side fuse links.

6.4.2 *Recloser–fuse coordination*

The criteria for determining recloser–fuse coordination depend on the relative locations of these devices, i.e. whether the fuse is at the source side and then backs up the operation of the recloser which is at the load side or vice versa. These possibilities are treated in the following paragraphs.

6.4.2.1 Fuse at the source side

When the fuse is at the source side, all the recloser operations should be faster than the minimum melting time of the fuse. This can be achieved through the use of multiplying factors on the recloser time–current curve to compensate for the fatigue of the fuse link produced by the cumulative heating effect generated by successive recloser operations. The recloser opening curve modified by the appropriate factor then becomes slower but, even so, should be faster than the fuse curve. This is illustrated in Figure 6.27.

The multiplying factors referred to above depend on the reclosing time in cycles and on the number of the reclosing attempts. Some values proposed by Cooper Power Systems are reproduced in Table 6.12.

It is convenient to mention that if the fuse is at the high-voltage side of a power transformer and the recloser at the low-voltage side, either the fuse or the recloser curve should be shifted horizontally on the current axis to allow for the transformer turns ratio. Normally, it is easier to shift the fuse curve based on the transformer tap which produces the highest current on the high-voltage side. However, if the transformer connection group is delta-star, the considerations given in Section 6.2 should be taken into account.

6.4.2.2 Fuses at the load side

The procedure to coordinate a recloser and a fuse, when the latter is at the load side, is carried out with the following rules:

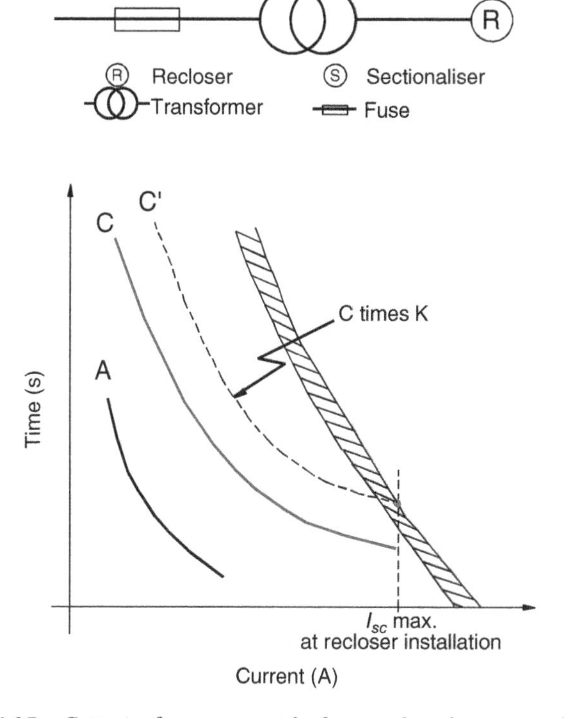

Figure 6.27 Criteria for source-side fuse and recloser coordination

Table 6.12 K factor for the source side fuse link

Reclosing time in cycles	Multipliers for Two-fast, two-delayed sequence	One-fast, three-delayed sequence	Four-delayed sequence
25	2.70	3.20	3.70
30	2.60	3.10	3.50
50	2.10	2.50	2.70
90	1.85	2.10	2.20
120	1.70	1.80	1.90
240	1.40	1.40	1.45
600	1.35	1.35	1.35

Table 6.13 K factor for the load side fuse link

Reclosing time in cycles	Multipliers for One fast operations	Two fast operations
25–30	1.25	1.80
60	1.25	1.35
90	1.25	1.35
120	1.25	1.35

- The minimum melting time of the fuse must be greater than the fast curve of the recloser times the multiplying factor, given in Table 6.13 and taken from the same reference as above.
- The maximum clearing time of the fuse must be smaller than the delayed curve of the recloser without any multiplying factor; the recloser should have at least two or more delayed operations to prevent loss of service in case the recloser trips when the fuse operates.

The application of the two rules is illustrated in Figure 6.28.

Better coordination between a recloser and fuses is obtained by setting the recloser to give two instantaneous operations followed by two timed operations. In general, the first opening of a recloser will clear 80% of the temporary faults, while the second will clear a further 10%. The load fuses are set to operate before the third opening, clearing permanent faults. A less effective coordination is obtained using one instantaneous operation followed by three timed operations.

6.4.3 Recloser–sectionaliser coordination

Since the sectionalisers have no time–current operating characteristic, their coordination does not require an analysis of these curves.

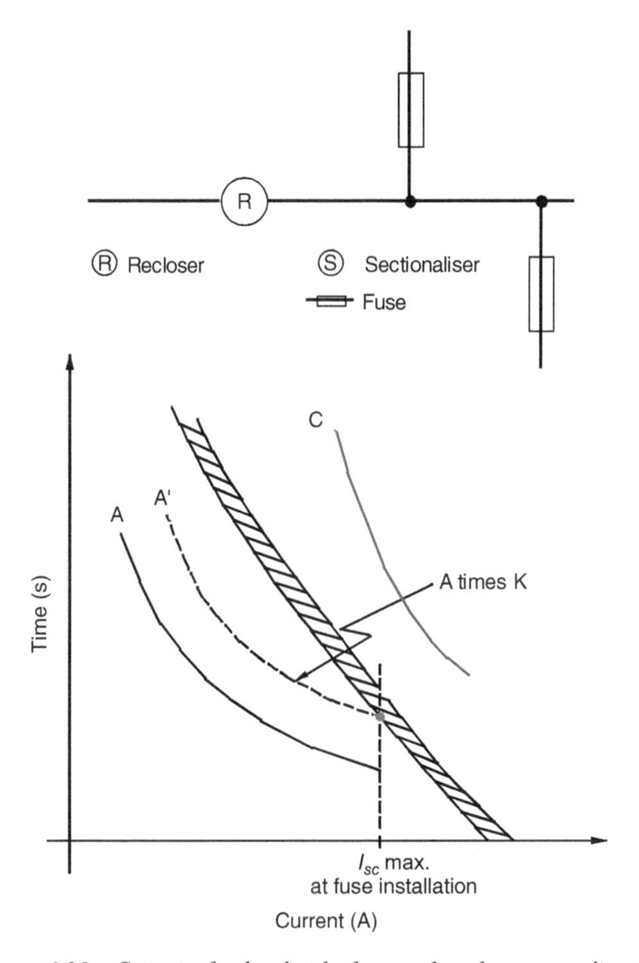

Figure 6.28 Criteria for load-side fuse and recloser coordination

The coordination criteria in this case are based upon the number of operations of the backup recloser. These operations can be any combination of rapid or timed shots as mentioned previously, e.g. two fast and two delayed. The sectionaliser should be set for one shot less than those of the recloser, e.g. three disconnections in this case. If a permanent fault occurs beyond the sectionaliser, the sectionaliser will open and isolate the fault after the third opening of the recloser. The recloser will then re-energise the section to restore the circuit. If additional sectionalisers are installed in series, the furthest recloser should be adjusted for a smaller number of counts. A fault beyond the last sectionaliser results in the operation of the recloser and the start of the counters in all the sectionalisers. Figure 6.29 shows an example of coordination between three sectionalisers and their setting.

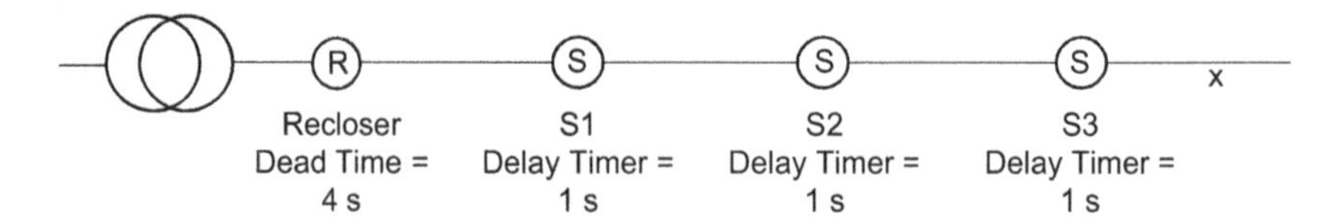

Figure 6.29 Coordination of one recloser with three sectionalisers

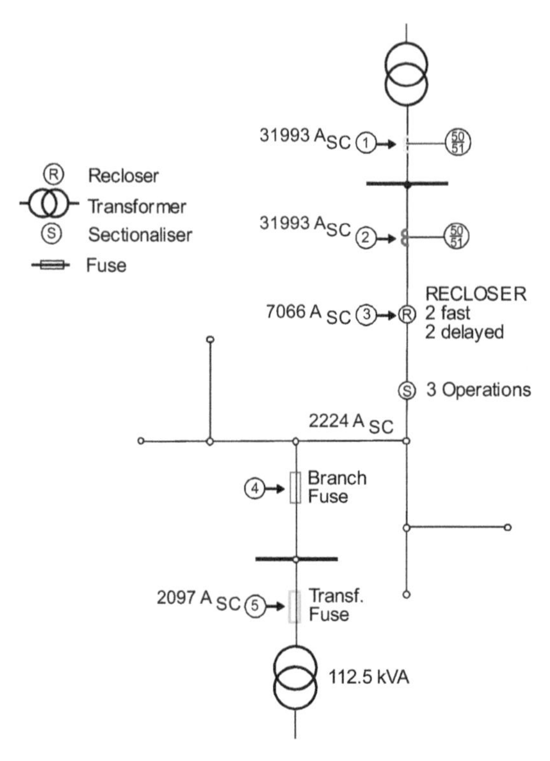

Figure 6.30 Portion of a distribution feeder

6.4.4 Recloser–sectionaliser–fuse coordination

Each one of the devices should be adjusted in order to coordinate with the recloser. In turn, the sequence of operation of the recloser should be adjusted in order to obtain the appropriate coordination for faults beyond the fuse by following the criteria already mentioned.

Figure 6.30 shows a portion of a 13.2 kV distribution feeder which is protected by a set of overcurrent relays at the substation location. A recloser and a sectionaliser have been installed downstream to improve the reliability of supply to customers.

The recloser chosen has two fast and two delayed operations with 90 cycles intervals.

The time–current curves for the transformer and branch fuses, the recloser, and the relays are shown in Figure 6.30. For a fault at the distribution transformer, its fuse should operate first, being backed up by the recloser fast operating shots. If the fault is still not cleared, then the branch fuse should operate next followed by the delayed opening shots of the recloser and finally by the operation of the feeder relay. The sectionaliser will isolate the faulted section of the network after the full number of counts has elapsed, leaving that part of the feeder upstream still in service.

As the nominal current of the 112.5 kVA distribution transformer at 13.2 kV is 4.9 A, a 6T fuse was selected on the basis of allowing a 20% overload. The fast curve of the recloser was chosen with the help of the following expression based on the criteria already given, which guarantees that it lies in between the curves of both fuses:

$$t_{recloser} \cdot k \leq t_{MMT\ of\ branch\ fuse} \cdot 0.75$$

where $t_{\text{MMT of branch fuse}}$ is the minimum melting time. The 0.75 factor is used in order to guarantee the coordination of the branch and transformer fuses, as indicated in Section 6.4.1.

At the branch fuse location, the short-circuit current is 2224 A, which results in operation of the branch fuse in 0.02 s. From Table 6.13, the k factor for two fast operations and a reclosing time of 90 cycles is 1.35. With these values, the maximum time for the recloser operation is (0.02 0.75/1.35) = 0.011 s.

This time, and the pickup current of the recloser, determines the fast curve of the recloser.

The feeder relay curve is selected so that it is above that of the delayed curve of the recloser, and so that the relay reset time is considered. The curves of Figure 6.31 show that adequate coordination has been achieved.

6.4.5 Recloser–recloser coordination

The coordination between reclosers is obtained by appropriately selecting the amperes setting of the trip coil in the hydraulic reclosers, or of the pick-ups in electronic reclosers.

6.4.6 Recloser–relay coordination

Two factors should be taken into account for the coordination of these devices; the interrupter opens the circuit some cycles after the associated relay trips, and the relay has to integrate the clearance time of the recloser. The reset time of the relay is normally long and, if the fault current is reapplied before the relay has completely reset, the relay will move towards its operating point from this partially reset position.

For example, consider a recloser with two fast and two delayed sequences with reclosing intervals of 2 s, which is required to coordinate with an inverse time delay overcurrent relay which takes 0.6 s to close its contacts at the fault level under question and 16 s to completely reset. The impulse margin time of the relay is neglected for the sake of this illustration. The rapid operating time of the recloser is 0.030 s, and the delayed operating time is 0.30 s. The percentage of the relay operation during which each of the two rapid recloser openings takes place is

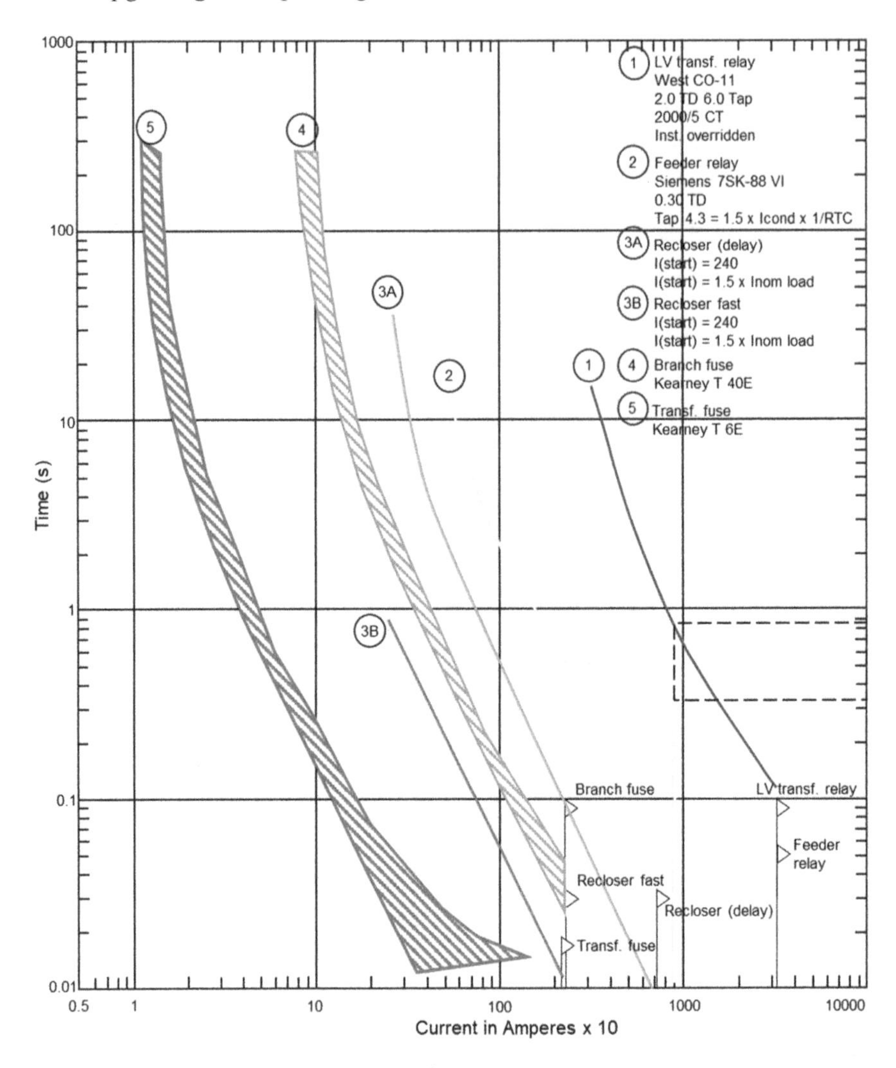

Figure 6.31 Phase–current curves

(0.03 s/0.6 s) × 100% = 5%. The percentage of relay reset which takes place during the recloser interval is (2 s/16 s) × 100% = 12.5%. Therefore, the relay completely resets after both of the two rapid openings of the recloser.

The percentage of the relay operation during the first time-delay opening of the recloser is (0.3 s/0.6 s) × 100% = 50%. The relay reset for the third opening of the recloser = 12.5%, as previously, so that the net percentage of relay operation after the third opening of the recloser = 50% − 12.5% = 37.5%. The percentage of the relay operation during the second time delay opening of the recloser takes place = (0.3 s/0.6 s) × 100% = 50%, and the total percentage of the relay operation after the fourth opening of the recloser = 37.5% + 50% = 87.5%.

From the above analysis, it can be concluded that the relay does not reach 100% operation by the time the final opening shot starts, and therefore coordination is guaranteed.

6.5 Protection considerations when distributed generation is available

As previously indicated, distributed generation (DG) brings about important benefits to the operation of distribution systems. The main benefit, of course, is the possibility of having generation at the user level that increases service reliability. If the source comes from green power, not only the price is lowered but also the pollution emission. However, important considerations have to be done to make sure that the generation is properly handled as follows.

6.5.1 Short-circuit levels

Short-circuit levels increase along the feeders with the DG that increase the withstand capabilities of breakers, sectionalisers, reclosers, capacitors, etc. In particular, extra care has to be exercised with breakers and reclosers to make sure that their duties are above the maximum total short-circuit currents.

6.5.2 Synchronisation

The reclosing functionality at the substation that feeds distribution lines should be disabled if there is not a proper way to open the generators connected along the feeders upon the occurrence and clarification of a fault. This prevents the possibility of energising a line with generation at the other end without following an appropriate synchronisation procedure. This of course reduces the flexibility of the operation but is required to avoid accidents that could be fatal.

6.5.3 Overcurrent protection

The use of overcurrent relays should be examined as the short-circuit currents can flow in and out of the substation if distributed generators are present. In this case, it could be convenient to replace them with directional overcurrent protection to achieve better coordination.

6.5.4 Adaptive protection

Due to the inherent nature of DG, machines can be on and off during normal operation. This is even more possible if the generation is from solar or wind power. In these cases, the relays should accommodate different topologies that make their operation risky or too slow unless adaptive protection is used. Most numerical relays have four or more setting groups that could be selected according to an equal number of operating scenarios.

Microgrids or grids that contain secondary local generation are a great case where adaptive protection should be applied. From the diagram below, you can see that two sources are present. During normal operation, the utility grid that operates the relays requires one set of protection settings. When the utility grid is interrupted the diesel generator starts. In order to protect the system, the relays must change settings in order to be coordinated correctly. Having adaptive protection will allow for continuous protection of the distribution system regardless of the source (Figure 6.32).

Example 6.2

Figure 6.33 will allow to illustrate one adaptive protection case with a local generator and the external grid contribution.

The system is modelled in a NEPLAN to coordinate with the overcurrent protections, as illustrated in Figure 6.34.

One fault location called F1 is also illustrated in Figure 6.34, according to this, the system satisfies the following equations:

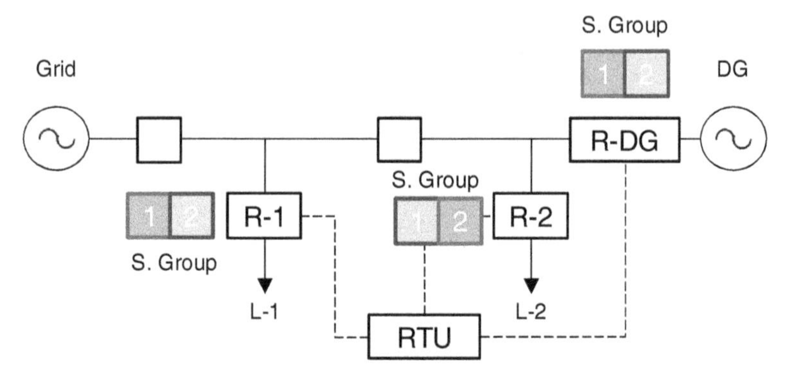

Figure 6.32 Example of adaptive protection setting with a RTU device

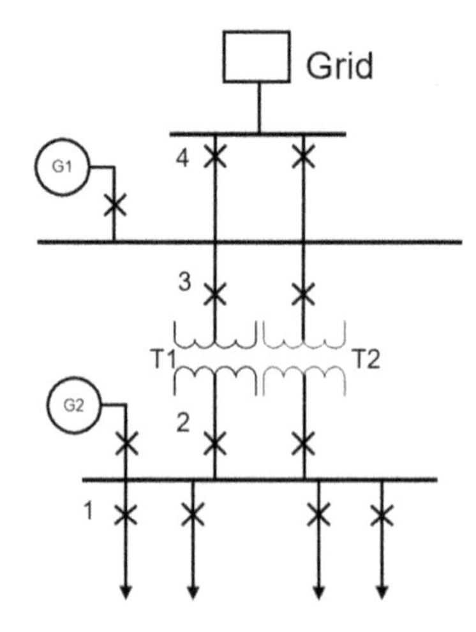

Figure 6.33 Single line diagram to illustrate de adaptive protection

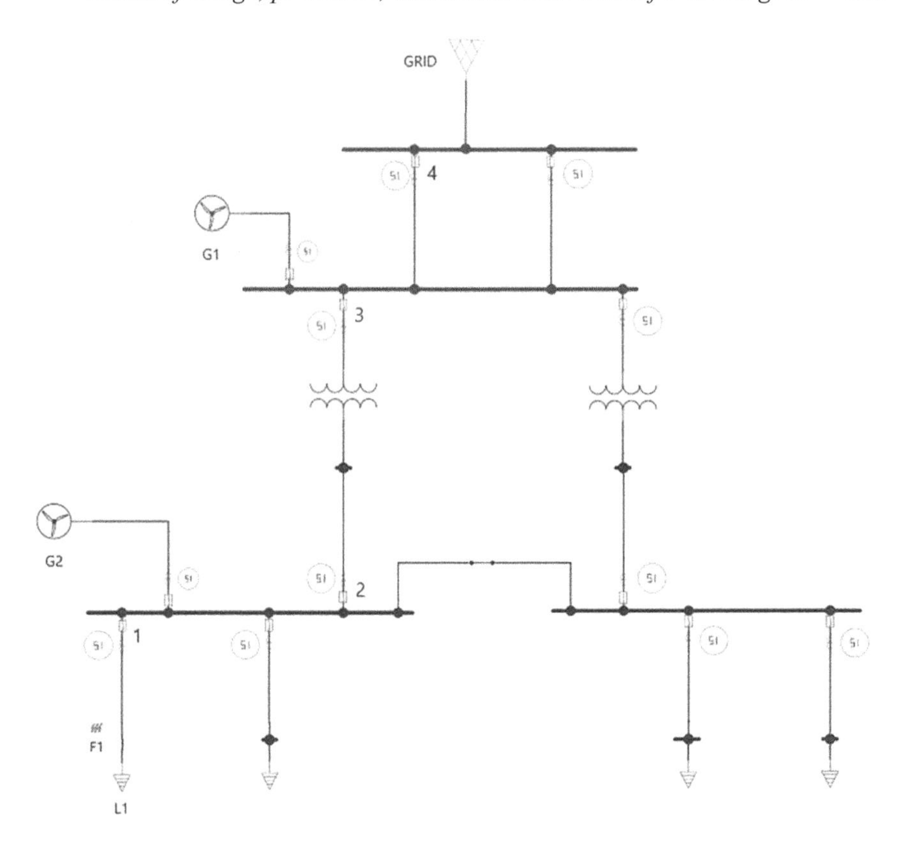

Figure 6.34 Single line diagram in NEPLAN

Normal condition:

$$I_{r1} = I_{T1} + I_{T2} + I_{G2}$$

$$I_{r1} = I_{T2}$$

where I_{r1} and I_{r2} correspond to the magnitude of the current measured by the relays in 1 and 2.

However, these conditions are affected by changes in the topology. For example, if there is an outage of G2, the system will satisfy these new conditions:

Outage of G2:

$$I_{r1}' = I_{T1} + I_{T2}$$

$$I_{r2} = I_{T1}$$

According to this, the main affectations are due to the current through relay 1 because it will be higher when G2 is operating; this event is more significative when the fault contribution of the generator is higher than the contribution from the grid. Figure 6.35 illustrates the coordination of the tripping times for both topologies (normal, outage of G2).

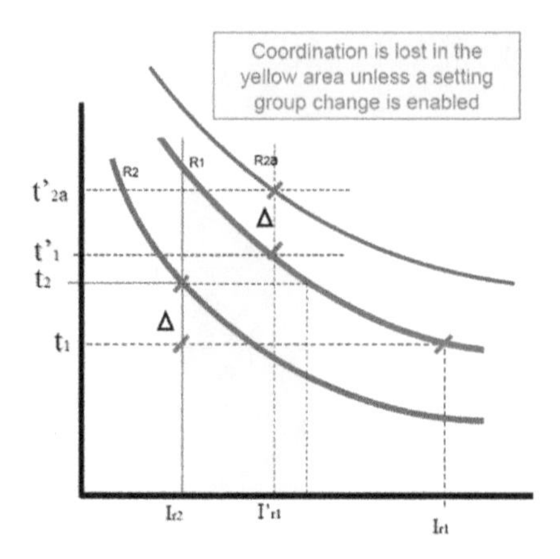

Figure 6.35 Coordination affectation due to an outage of G2

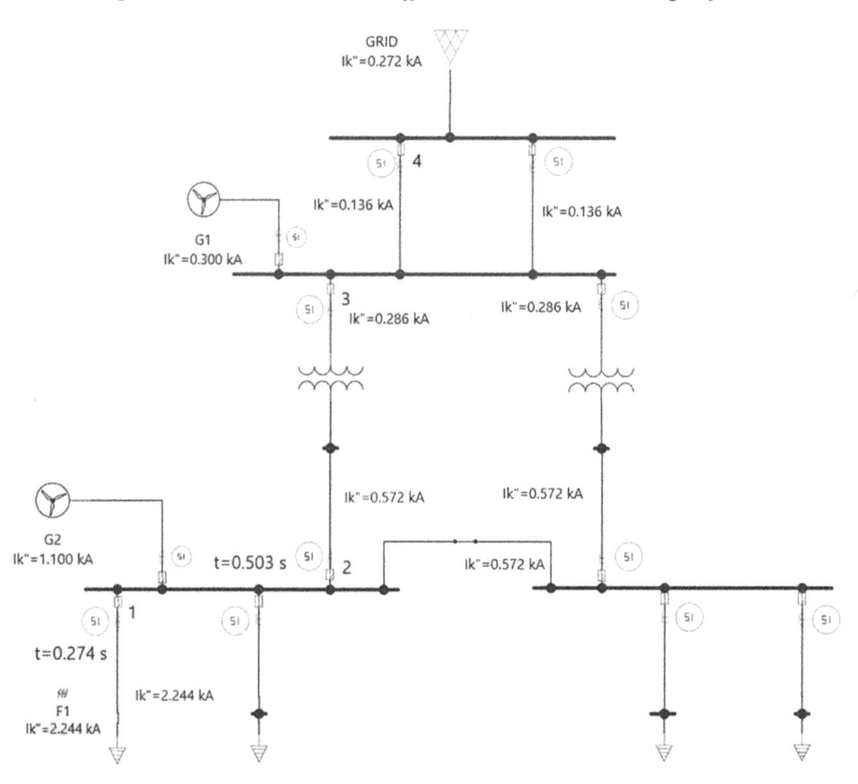

Figure 6.36 Short circuit and times under normal condition

As illustrated in Figure 6.35, t_1 and t_2 correspond to the tripping times of relays 1 and 2 under normal condition, respectively. However, if the settings are the same for the outage of G2 condition, the tripping time of relay 2 (t_2) will be faster than relay 1 (t_1'). The proper coordination comes again once the setting group of relay 2 changes to a new position. For example, R_{2a} where the new tripping time is t_2'.

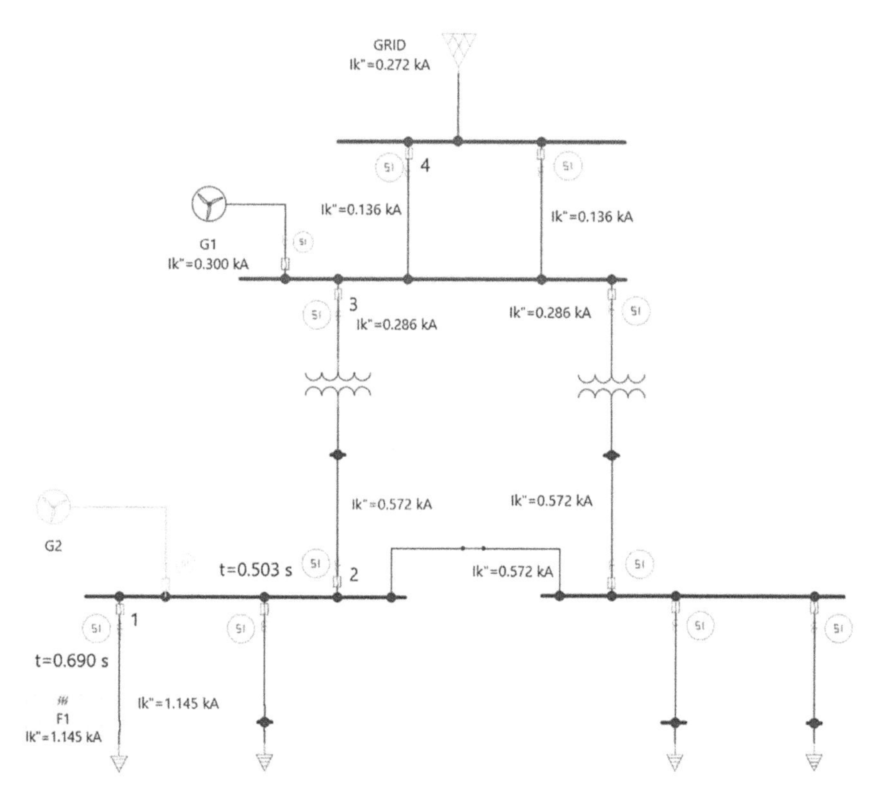

Figure 6.37 Short circuit and times in the outage of G2

To illustrate this phenomenon, protections are modelled with inverse time characteristics and the short-circuit contributions have been modelled in the software where the short-circuit level of the generator G2 is close to the sum of the other contributions. Figure 6.36 presents the results in relays 1 and 2 when the topology is under normal condition. Figure 6.37 presents the results when the topology is in outage of G2. As it is noted, selectivity is obtained in Figure 6.36, but the coordination is lost in Figure 6.37 if the settings don't change.

To illustrate the proper coordination in both scenarios, Figure 6.38 presents the selectivity diagrams with a new setting group defined in relay 1 when the topology changes.

6.6 Networks automation with communications

For years, analogue communication networks have facilitated long-distance information exchange channels. However, the advent of digital communications has enabled the use of physical mediums with the inherent capacity to transfer large data packets reliably and efficiently.

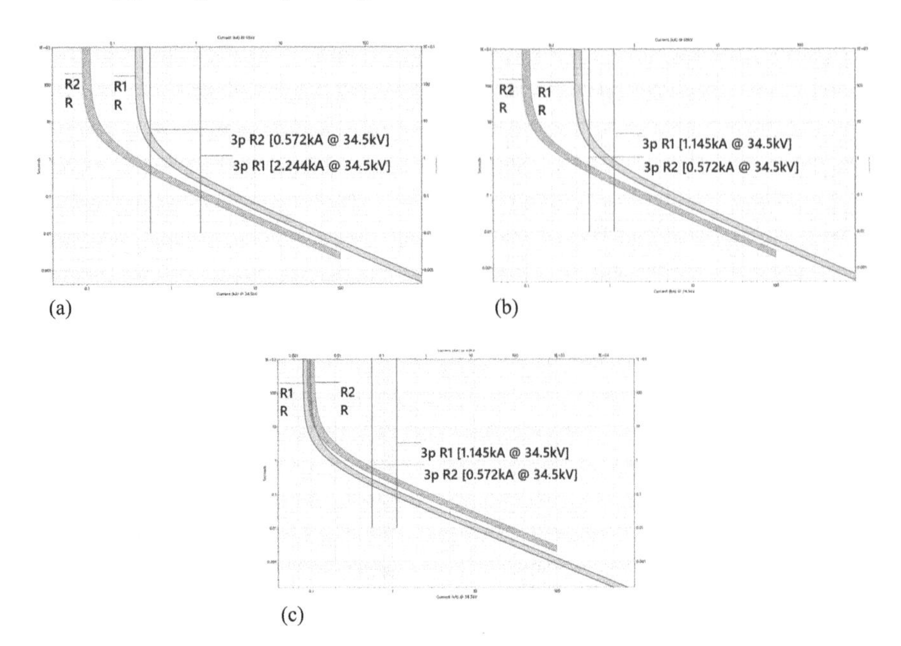

Figure 6.38 Selectivity diagrams for different scenarios. (a) Coordination in normal operation; (b) Loss of coordination in outage of G2 and (c) New setting group for coordination in outage of G2.

Digital communication networks and data services have offered solutions that optimise numerous processes globally. In smart network environments, these solutions have proven to be successful. A prime example is their application in protective relaying. While the theoretical foundations of protection functionalities have remained constant since their inception, the operation of protective relaying has significantly improved in response time over time. This improvement is attributed to the availability of much faster relays based on numerical technology and the remarkable development of communication capabilities in recent years.

The standardisation of protocols has provided a convergent platform for implementing solutions based on openness and flexibility. However, given the high requirements for reliability, availability and security in electrical system networks, protocols needed adaptation to offer lower latency and broader broadband.

The spectrum of smart grid solutions continues to expand, with each new solution introducing additional communication solutions tailored to specific cases. Traditional communications, such as wired communication between field equipment, auxiliary services, and controllers, serial communication between intelligent electronic devices (IEDs), communication between equipment in different locations and communication through a wide area network (WAN) for integrating substations in a control centre station, are considered in the power system environment. Communications have evolved towards the convergence of multiple

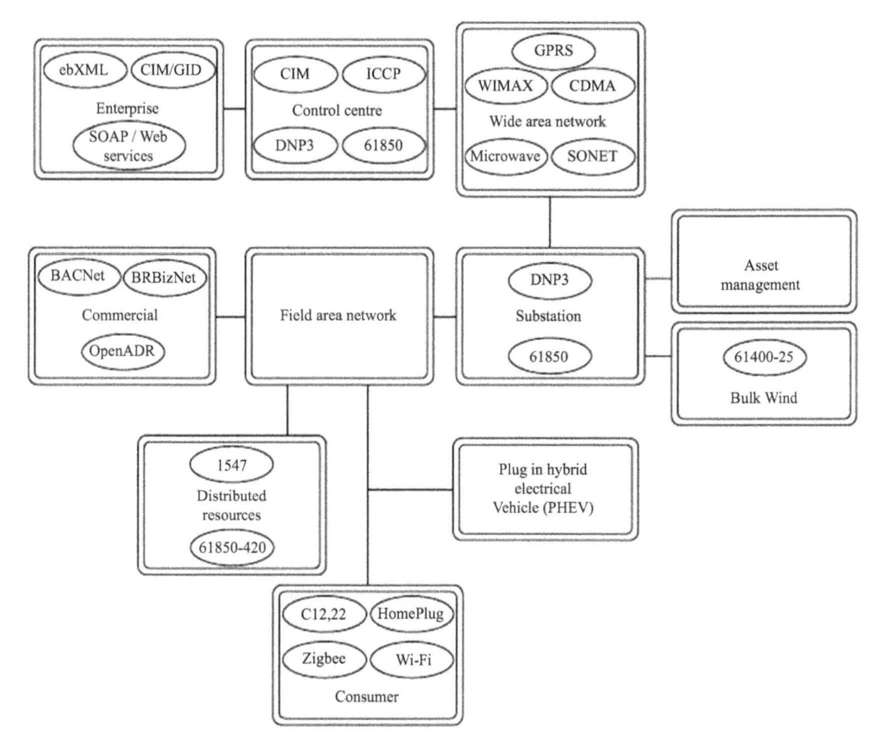

Figure 6.39 Communication options in the smart grids

designs suitable for each domain within an information model, offering interoperability between systems and a distributed computing system to achieve optimum levels of usage. Consequently, standards such as the Common Information Model (CIM), specified in IEC 61968 and IEC 61970, have been developed [11,12].

Figure 6.39 illustrates sample communication solutions applied to smart grid domains. This figure depicts a communication infrastructure for smart grid that enables utilities to interact with their devices and grid systems, as well as with customers and facilities for DG and energy storage. To fully realise the vision of a smart grid, companies need to support multiple communication networks: a home area network (HAN) for energy efficiency on the client side, a neighbourhood area network (NAN) for advanced measurement applications and a WAN for distributed automation and the Smart Grid backbone.

Communication applications in smart grid have evolved rapidly, particularly in two areas: intelligent metering infrastructure and distribution network, and substation communications.

6.6.1 Communication solutions in AMI

The advanced metering infrastructure (AMI) offers diverse solutions such as phone lines, GPRS/3G/4G, PLC, fibre optic and ethernet, which may initially utilise traditional proprietary protocols. However, in response to the demands of the smart grid, these solutions are transformed into Device Language Message Specification (DLMS)/COSEM open standards for data interchange with metering equipment.

Additionally, noteworthy emerging technologies fall under the Internet of Things (IoT) philosophy, combining smart meters and cloud computing. These technologies are supported by low-speed wireless communications, utilising either the existing cellular network or satellite communications. This creates a comprehensive platform that integrates the real-time collection and analysis of large amounts of data through the power of big data and cloud computing. Some examples of these technologies include LoRaWan, SigFox and NB-IoT. Unlike SigFox or LoRaWan, NB-IoT is not IP-based, allowing it to handle more data.

DLMS is an application-level protocol integrated into IEC 62056 for device message specification [13]. Similarly, COSEM is defined as the Companion Specification for energy metering. These standards play a crucial role in ensuring interoperability and data interchangeability in the context of smart grids.

6.6.2 Distribution network communications

6.6.2.1 IEC 61850

One of the smart network objectives is the interoperability among systems, as is defined by the National Institute of Standards and Technology (NIST) Framework and Roadmap for Smart Grid Interoperability Standards, Release 3.0 [14]. Substation automation has been restricted to the solution offered by the various protection and control equipment manufacturers that needed to interface and integrate their units without restrictions. IEC started to develop a common standard for substation communications in 1994. Similarly, IEEE developed a common communication standard known as UCA (Utility Communication Architecture). In 1997, IEEE and IEC agreed to work in conjunction to create a common substation communication standard IEC 61850 [15].

The IEC 61850 Standard in Section 8-1 discusses the critical and non-critical data interchange method in time across local area networks (LANs) employing Abstract Communication Service Interface (ACSI) mapping in multimedia messaging (multimedia messaging service, MMS) over frames ISO/IEC 8802-3 [16]. Similarly, in section IEC 61850-9-2, the specific communication service mapping (SCSM) is defined for the transmission of sampled values according to the IEC 61850-7-2, also using frames ISO/IEC 8802-3.

Figure 6.40 illustrates how critical data like generic object-oriented substation events (GOOSE) messages and sampled values are directly mapped to lower ethernet layers, and how metered values or time synchronisations pass through TCP/IP protocols.

IEC 61850 provides the multicast-based generic substation events (GSE) as a way to quickly transfer event data over an entire substation network. Furthermore, parts 9-1 and 9-2 specify a process bus for use with IEDs.

6.6.2.2 DNP3-IEEE Standard 1815

The IEEE 1815-2012 standard [17] defines the DNP3 protocol used for communications media employed in utility automation systems where it describes the

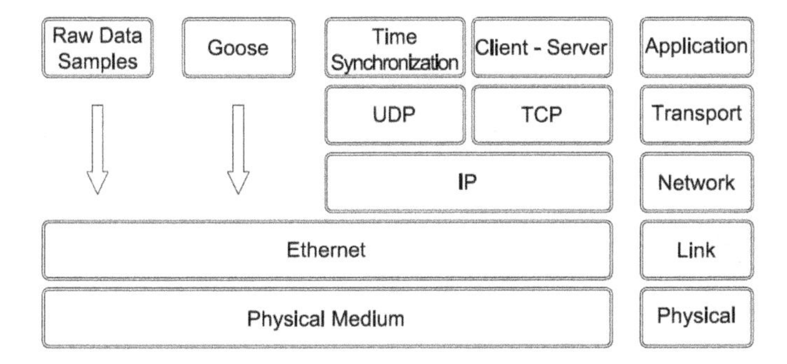

Figure 6.40 Message communication OSI-7 stack

structures, functions or applications. The revision of the IEEE 1815-2010 standard includes improved protocols that aid in preventing cybersecurity concerns particular to the communication media employed in utility automation systems.

6.6.2.3 IEC 60870-5 as the standard for remote control

The trend is to adopt IEC 61850 architecture for substation communications and remote control. IEC 60870 has experienced a massive expansion and wide application [18].

IEC 60870-5 is the standard covering telecontrol equipment and systems. Part 5 covers transmission protocols. The standard is well established worldwide.

The IEC 60870-5-104 or IEC 104 protocol is a standard based on IEC 60870-5-101 or IEC 101. It employs the TCP/IP network interface to provide LAN and WAN connectivity. The original IEC 101 itself is also retained for some particular data and services utilised.

The IEC 60870-5-101 and 104 parts are mainly used for exchanging information between substations and control centres. Application areas range from primary substations (high and medium levels of voltage) to the secondary substations (medium or low voltage).

6.7 Information security as the crucial element in smart networks

It is understandable that in an interconnected world, reliability risks increase and security varies according to protocols and smart grid domains where they are implemented. Some security standards are applicable to certain protocols while others are applied to particular profiles.

When extremely sensible and critical data is transferred, security levels should be elevated to meet basic information security like reliability, confidentiality and integrity.

In the electric field, associations like North American Electric Reliability Corporation (NERC) have developed the NERC 1300 Standard for power system security. These standards are divided into eight specific areas from CIP-002-1 to CIP-009-2 focused on establishing requirements for power system owners, operators and users to guard critical assets with the best security practices. Also, IEC 62351 [19] or FIPS180-4, 186-3 must be mentioned. For example, IEC 62351 is developed for handling the security of protocols including IEC 60870-5 series, IEC 60870-6 series, IEC 61850 series, IEC 61970 series and IEC 61968 series.

It must be noted that cybersecurity standards in the application levels are part of applicable models in the computer sector such as ISO 27002 or when employing the common critical model known as Common Criteria for Information Technology Security Evaluation, ISO 15408 [20].

6.8 Interoperability

A complete integration of all the components of the power system from generation to the end-user is required to achieve the objectives of a smart grid. For instance, the integration of renewable energy sources will affect existing electricity grid infrastructure, operations and the functioning of the electricity market itself. This can be tackled by electricity grids operating smartly and cost-efficiently. They should relate to each other in a transparent way and independently of the applied technologies.

Utilities business need to integrate control centre applications (SCADA, Alarm Processing, History, etc.) and the new and legacy systems in Transmission and Distribution (EMS, DMS, Asset Management, Outage Managing, GIS, Planning Systems, etc.). Also, it is expected to include the control centre into an enterprise-wide integration strategy (business integration) and the exchange model and status information with other utilities/sites for security and open access. In order to achieve this, it is required to implement a common power system model, a common information exchange format and an integration framework based on standards.

There are many limitations on communication among the components of a power system, because of incompatibilities between protocols and technologies [32–36]. Additionally, it is required to combine a large number of autonomous IT systems into a homogeneous IT landscape. However, conventional network control systems could only be integrated with considerable effort. Interfaces solve this problem but they have the disadvantage of limiting the information system's growth. Therefore, it is required to standardise the representation of the power system's information, in such a way that each component accesses the data independently of the source. A seamless and efficient information exchange is necessary at various stages, between an increasing number of companies (TSOs, DSOs, generators, etc.). A solution for this is achieved when power systems fulfil the conditions of interoperability.

IEEE Std. 2030-2011 [26] defines interoperability as the capability of two or more networks, systems, devices, applications or components to externally exchange and readily use information securely and effectively. The same standard defines a methodology to guarantee the interoperability in power systems from the perspectives of three systems: the power system itself, the information system and the communication system. The process has to be done in such a way as to guarantee a long-life cycle permitting upgrading.

This concept applied to smart grid networks ensures efficient communication whether the information systems are used on different types of infrastructure or even at a distance and protects the investment costs. Such information exchanges have become indispensable in network planning, power system operation (real-time information on the generation output, balancing control, etc.), market (generation schedules, trades, balancing resource management, etc.). Interoperability must be vertical and horizontal of the involved products and systems and a common definition of data model must be done.

The various functions in IT have the responsibility of distributing safely and reliably information to all points of the network where this is required for monitoring or decision-making. Thanks to these information technologies, smart grid networks will become much more dynamic in their configuration and will approach operating conditions that give greater opportunities for real-time analysis and optimisation technologies.

The utilities must process large amounts of operational and non-operational information, both in real-time and stored, in either static or dynamic form. They also receive large amounts of information remotely, directly from the analysis done at fault sites and events, and AMI deployments along the system, thanks to the installation of IED and other smart grid devices.

A model of well-defined data elements allows not only the simple direct exchange of information, but it also assists in fulfilling requests from development and security. The existence of a data model for the architecture is almost always a good indicator of a well-governed process of facilitating the use of the data by multiple applications because of the establishment of a common semantics.

Two global standards exist for the integration of businesses in the electricity sector. One of them is MultiSpeak®, sponsored by National Rural Electric Cooperative Association (NRECA) [27], and the other is the CIM, an international standard maintained by Technical Committee 57 (TC 57) of the International Electrotechnical Commission (IEC) [28].

The CIM data model describes the electrical network, the connected electrical components, the additional elements and the data needed for network operation as well as the relations between these elements. Working Group 14 (WG14) [29] of TC 57 which deals with work processes to distribution utilities has recently defined the NetworkDataSet message that specifies how the CIM should be used to exchange detailed models for distribution engineering analysis [29]. CIM-compatible NetworkDataSet model exchanges are in operation at several utilities.

Good data models will enhance the return on investment of the smart grid by enabling more applications to use the data, and improving its value for newly developed analytics that may examine the information in novel ways.

Ontology-based strategies are commonly used with success in creating and manipulating data models since they provide easy export or translation to Extensible Markup Language (XML) or Unified Modelling Language (UML). Data without a clear and intuitive meaning cannot be used. Within the context of a single user interface application, developers strive to make the meaning clear, but when data is transferred to another system, the meaning may be lost.

The objects of the real world are represented by means of classes, which conform to the basic unit that encompasses all their information. Thanks to these classes, the elements of power systems (transformers, sectionalisers, circuit breakers generators, etc.) can be modelled. UML language is used to represent the different relations among the classes that the model uses. These relations among classes are known as association. For each class, the attributes can be observed, i.e. the information or data that represents the object.

The UML, a standardised, object-oriented method that is supported by various software tools, is used as the descriptive language. CIM is used primarily to define a common language for exchanging information via direct interfaces or an integration bus and for accessing data from various sources.

The CIM model is subdivided into packages such as basic elements, topology, generation, load model, measurement values and protection. The sole purpose of these packages is to make the model more transparent. Relations between specific types of objects being modelled may extend beyond the boundaries of packages.

The electrically conductive connections between the elements are defined via terminals and nodes (connectivity nodes). Every conductive element has one or more terminals. A terminal connects the element, such as a generator, or one side of, for example, a circuit-breaker, to a node. A node can hold any number of terminals and provides an impedance-free connection linking all elements connected to it. A topology processor can determine the current network topology via these relations and with the current states of the circuit-breakers. This topology model can also be used to describe gas, water, district heating and other networks for tasks such as modelling interconnected control centres.

When the elements to be modelled have been already defined, the XML language is utilised to create instances, which constitute the physical model of the elements of the real-world power network. Figure 6.41 shows the XML representation of some elements of a power system modelled with NEPLAN. In this case, it is clear that the modelled elements have been already defined, and it is possible to create the instances in the physical model of the power system. The illustration here represents a 13.8 kV – Bus 9 in XML language. The rectangle in red of part (a) of the figure indicates the node that will be represented in XML. The rectangle in red of part (b) is a portion of the code generated in XML to illustrate how the information is represented with this format.

A modern network control system provides a service-oriented architecture with standardised process, interface and communication specifications based on standards IEC 61968 and IEC 61970 [31,32]. They form the basis for integrating the network control system in the enterprise service environment of the utility.

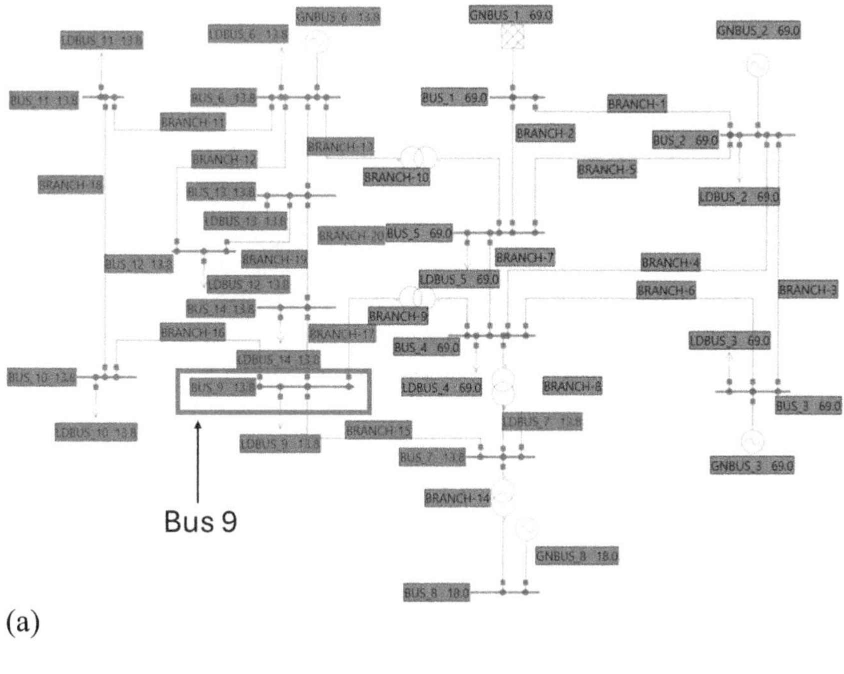

(a)

```
▼<rdf:RDF xmlns:rdf="http://www.w3.org/1999/02/22-rdf-syntax-ns#"
 xmlns:cim="http://iec.ch/TC57/2013/CIM-schema-cim16#"
 xmlns:entsoe="http://entsoe.eu/CIM/SchemaExtension/3/1#" xmlns:md="http://iec.ch/TC57/61970-
 552/ModelDescription/1#">
 ▼<md:FullModel rdf:about="urn:uuid:61a10afc-35e1-4fd1-887b-537db224aa24">
    <md:Model.created>2019-09-25T08:03:52</md:Model.created>
    <md:Model.scenarioTime>2019-09-25T07:58:37</md:Model.scenarioTime>
    <md:Model.version>1</md:Model.version>
    <md:Model.DependentOn rdf:resource="urn:uuid:e1089fef-a88f-46d0-8b95-240e556a4b3c"/>
    <md:Model.DependentOn rdf:resource="urn:uuid:4f193c8a-0eda-4480-ac31-acd50084ac08"/>
    <md:Model.DependentOn rdf:resource="urn:uuid:ed3cce61-77ac-4b71-874a-f4a4c3db3ae2"/>
    <md:Model.description/>
    <md:Model.modelingAuthoritySet>http://neplan.ch/CGMES/2.4.14</md:Model.modelingAuthoritySet>
    <md:Model.profile>http://entsoe.eu/CIM/StateVariables/4/1</md:Model.profile>
  </md:FullModel>
 ▼<cim:SvVoltage rdf:ID="_4af4a3d0-144b-418c-8e9e-31940f583596">
    <cim:SvVoltage.v>14.076400</cim:SvVoltage.v>
    <cim:SvVoltage.angle>-16.055200</cim:SvVoltage.angle>
    <cim:SvVoltage.TopologicalNode rdf:resource="#_c9a75968-fa19-4363-b6ff-3a28d36caf43"/>
  </cim:SvVoltage>
 ▼<cim:SvVoltage rdf:ID="_8c9c722b-0c11-4cec-9d42-3135d58f9f71">
    <cim:SvVoltage.v>14.237560</cim:SvVoltage.v>
    <cim:SvVoltage.angle>-14.808600</cim:SvVoltage.angle>
    <cim:SvVoltage.TopologicalNode rdf:resource="#_bba01e38-aefd-492c-af04-7e74bd10410c"/>
  </cim:SvVoltage>
 ▼<cim:SvVoltage rdf:ID="_0d575c75-87cb-46b6-b503-401b5ae00323">
    <cim:SvVoltage.v>14.226410</cim:SvVoltage.v>
    <cim:SvVoltage.angle>-15.023000</cim:SvVoltage.angle>
    <cim:SvVoltage.TopologicalNode rdf:resource="#_9e44b610-ce35-493b-a1b4-f5b89754db98"/>
  </cim:SvVoltage>
```

Bus Voltage

(b)

Figure 6.41 Example of (a) power system representation and (b) CIM/XML representation [30]

6.9 Cybersecurity

One important aspect that must be analysed after information is flowing around the system is the role of the actors and parties that are involved. Cybersecurity in smart grids is required in order to guarantee that data around the system and its components is shared with security, integrity and reliability. Also, aspects like unauthorised access, physical integrity of equipment, user's privacy and protection of information against cyber-attacks must be considered. The standard NIST IR 7628-2014 shows relevant aspects to be considered in cybersecurity [33]. Table 6.14 summarises the requirements considered in cybersecurity.

Some standards related to cybersecurity are: Department of Homeland Security – DHS-2009, 'DHS Cyber Security Procurement Language for Control Systems' [34]; IEC 62351-2018, 'Power Systems Management and Associated Information Exchange – Data and Communications Security' [19]; IEEE C37.240-2014, 'Cyber Security Requirements for Substation Automation, Protection and Control Systems' [35]; IEEE 1686-2013 'IEEE Standard for Intelligent Electronic Devices Cyber Security Capabilities' [36]; ISO/IEC 27019:2017 'Information Technology – Security Techniques – Information Security Controls for the Energy Utility Industry' [37]; NIST 1108-2014, 'NIST Framework and Roadmap for Smart Grid Interoperability Standards, Release 3.0' [14]; NISTIR 7628-2014, 'Guidelines for Smart Grid Cyber Security' [33].

Table 6.14 Requirements for cybersecurity [33,38]

Requirement	Description	Requirement	Description
Access control	The focus of access control is ensuring that resources are accessed only by the appropriate personnel, and that personnel are correctly identified.	Security program management	The security program lays the groundwork for securing the organisation's enterprise and smart grid information system assets. Security procedures define how an organisation implements the security program.
Audit and accountability	Periodic audits and logging of the smart grid information system need to be implemented to validate that the security mechanisms present validation testing are still installed and operating correctly.	Personnel security	The organisation may consider implementing a confidentiality or nondisclosure agreement that employees and third-party users of facilities must sign before being granted access to the smart grid information system.
Security assessment and authorisation	Security assessments include monitoring and reviewing the performance of smart grid information systems. Internal checking methods, such as compliance audits and incident investigations, allow the organisation to determine the effectiveness of the security program.	Risk management and assessment	Risk management planning is a key aspect of ensuring that the processes and technical means of securing smart grid information systems have fully addressed the risks and vulnerabilities in the smart grid information system.

(Continues)

Table 6.14 (Continued)

Requirement	Description	Requirement	Description
Continuity of operations	The ability for the smart grid information system to function after an event is dependent on implementing continuity of operations policies, procedures, training and resources.	Smart grid information system and services acquisition	Smart grid information systems and services acquisition covers the contracting and acquiring of system components, software, firmware and services from employees, contactors and third parties.
Identification and authentication	Identification and authentication is the process of verifying the identity of a user, process or device, as a prerequisite for granting access to resources in a smart grid information system.	Smart grid information system and communication protection	Smart grid information system and communication protection consists of steps taken to protect the smart grid information system and the communication links between smart grid information system components from cyber intrusions.
Smart grid information system development and maintenance	Maintenance activities encompass appropriate policies and procedures for performing routine and preventive maintenance on the components of a smart grid information system.	Smart grid information system and information integrity	Maintaining a smart grid information system, including information integrity, increases assurance that sensitive data have neither been modified nor deleted in an unauthorised or undetected manner.
Media protection	The security requirements under the media protection family provide policy and procedures for limiting access to media to authorised users.	Testing and certification of smart grid cybersecurity	The testing and certification of the smart grid cybersecurity requirements provide assurance that systems and system components are conformant to the requirements selected by the organisation.
Physical and environmental security	Physical and environmental security encompasses protection of physical assets from damage, misuse or theft.		

6.10 IEC 61850 overview

Substations designed in the past made use of protection and control schemes implemented with single-function, electromechanical or static devices, and hard-wired relay logic. SCADA functions were centralised and limited to monitoring of circuit loadings, bus voltages, aggregated alarms, control of circuit breakers and tap changers. Disturbance recording and sequence-of-event data if available were centralised and local to the substation [39–46].

With the advent of microprocessor-based multi-function IEDs, more functionality into fewer devices was possible, resulting in simpler designs with reduced wiring. In addition, owing to communication capabilities of the IEDs, more information could be accessed remotely, translating into fewer visits to the substation.

Microprocessor-based protection solutions have been successful because they offered substantial cost savings while fitting very well into pre-existing frameworks of relay application. A modern microprocessor-based IED replaces an entire panel of electromechanical relays with external wiring intact, and internal DC wiring replaced by integrated relay logic. Users retained total control over the degree of integration of various functions, while interoperability with the existing environment (instrument transformers, other relays, control switches, etc.) has been maintained using traditional hard-wired connections.

In terms of SCADA integration, the first generation of such systems achieved moderate success especially in cases where the end-user could lock into a solution from a single vendor. Integrating systems made up of IEDs from multiple vendors invariably led to interoperability issues on the SCADA side. Integration solutions tended to be customised.

Owners of such systems were faced with long-term support and maintenance issues. During this period, two leading protocols emerged: DNP 3.0 and IEC 60870.

Beginning in the early 1990s, initiatives were undertaken to develop a communication architecture that would facilitate the design of systems for protection, control, monitoring and diagnostics in the substation. The primary goals were to simplify development of these multi-vendor substation automation systems and to achieve higher levels of integration reducing even further the amount of engineering and wiring required.

In 1994, EPRI/IEEE started a work UCA2 with focus on the station bus. In 1996, IEC TC57 (technical committee 57) began work on IEC 61850 to define a station bus. This led to a combined effort in 1997 to define an international standard that would merge the work of both groups that were focused on the development of a standard in which devices from all vendors could be connected together to share data, services and functions. The result was the international IEC 61850 Standard Edition 1 ‘Communication Networks and Systems in Substation Automation’.

The International IEC 61850 Standard was issued in 2005 and it was developed to control and protect power systems by standardising the exchange of information between all IEDs within an automated substation and a remote control link. IEC 61850 provides a standardised framework for substation integration that specifies the communication requirements, the functional characteristics, the structure of data in devices, the naming conventions for the data, how applications interact and control the devices and how conformity to the standard should be tested.

The development of the IEC 61850 Standard is continuing. This work resulted in Edition 2 of the standard, which was published during 2010. IEC 61850 was originally defined exclusively for substation automation systems, but has since been extended to other application areas – as is reflected in its changed title 'IEC 61850 Ed. 2. Communication Networks and Systems for Power Utility Automation'.

The IEC 61850 is increasingly being used for the integration of electrical equipment into distributed control systems in process industries. The fact that new application areas such as hydro and wind power are being added is yet another indication of its success.

Some of the benefits of the IEC 61850 Standard are:

- reduce dependence on multiple protocols
- higher degree of integration
- reduce construction cost by eliminating most copper wiring
- flexible programmable protection schemes
- communication networks replacing hard-wired connections
- advanced management capability
- high-speed peer-to-peer communications
- improved security/integrity
- reduced construction and commissioning time

Eliminating copper wiring with ethernet/fibre cables means no more binary inputs and outputs for control and protection functions. The traditional method of tripping a breaker via a contact could be replaced by GOOSE messages sent via ethernet or fibre optic cables.

6.10.1 Standard documents and features of IEC 61850

In IEC 61850 Ed. 2, Part 9-1 is deprecated as per conformance table. Ethernet-based sampled value transmission (IEC 61850 9-2) will take over process bus communication. Now all IEC 61850 specific communication service mappings are using ethernet technology. Some of the features of IEC 61850 include:

1. data modelling,
2. reporting schemes,
3. fast transfer of events, GSE – GOOSE and GSSE,
4. commands,
5. sampled data transfer,
6. setting groups,
7. data storage – SCL (Substation Configuration Language).

1. Data modelling: In IEC 61850, series functionalities that the real devices comprise are decomposed into the smallest entities, which are used to exchange information between different devices. These entities are called logical nodes. Complete functionality of the substation is modelled into different standard logical nodes. The logical nodes are the virtual representation of the real functionalities. The intent is that all data that could originate in the substation can be assigned to one of these logical nodes. Several logical nodes from different real devices build up a logical device.

Logical devices, logical nodes and data objects are all virtual terms. They represent real data, which is used for communication. One device (e.g., a control unit) only communicates with the logical nodes or its data objects of another device (e.g., an IED). The real data, which the logical nodes represent, is hidden and it is not accessed directly. This approach has the advantage that communication and information modelling is not dependent on operating systems, storage systems and programming languages. The concept of virtualisation is shown in Figure 6.42, where the real device on the right-hand side is modelled as a virtual model in the middle of the figure. The logical nodes (e.g. XCBR, circuit breaker) defined in the logical device (Bay) correspond to well-known functions in the real devices. In this example, the logical node XCBR represents a specific circuit breaker of the bay to the right.

Based on their functionality, a logical node contains a list of data (e.g., position) with dedicated data attributes. The data has a structure and a predefined semantic. The information represented by the data and their attributes are exchanged by the communication services according to the well-defined rules and the requested performances.

To illustrate more clearly how the logical devices, logical nodes, classes and data concepts map to the real world, imagine an IED that is a container, as shown in Figure 6.43.

The container is the physical device, which is containing one or more logical devices. Each logical device contains one or more logical nodes, each of which contains a predefined set of data classes. Every data class contains many data attributes (status value, quality, etc.).

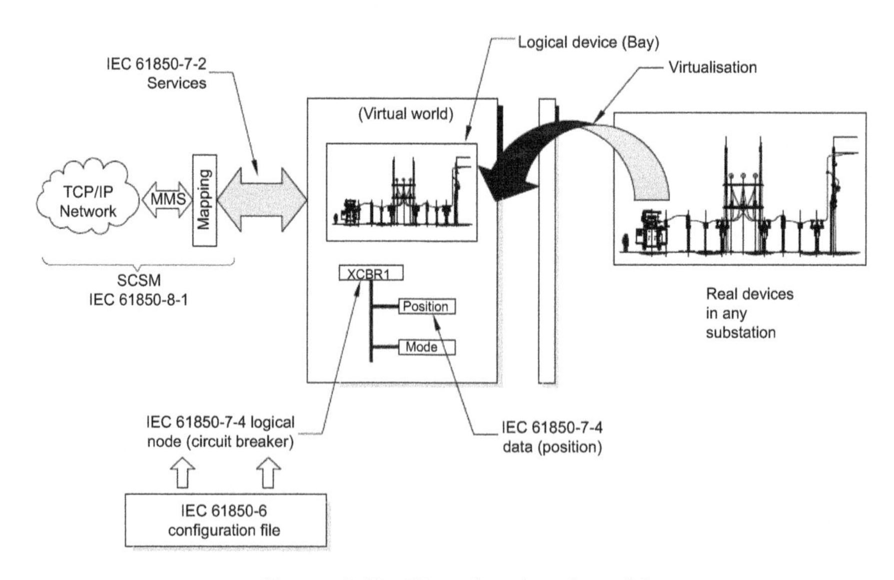

Figure 6.42 Virtual and real world

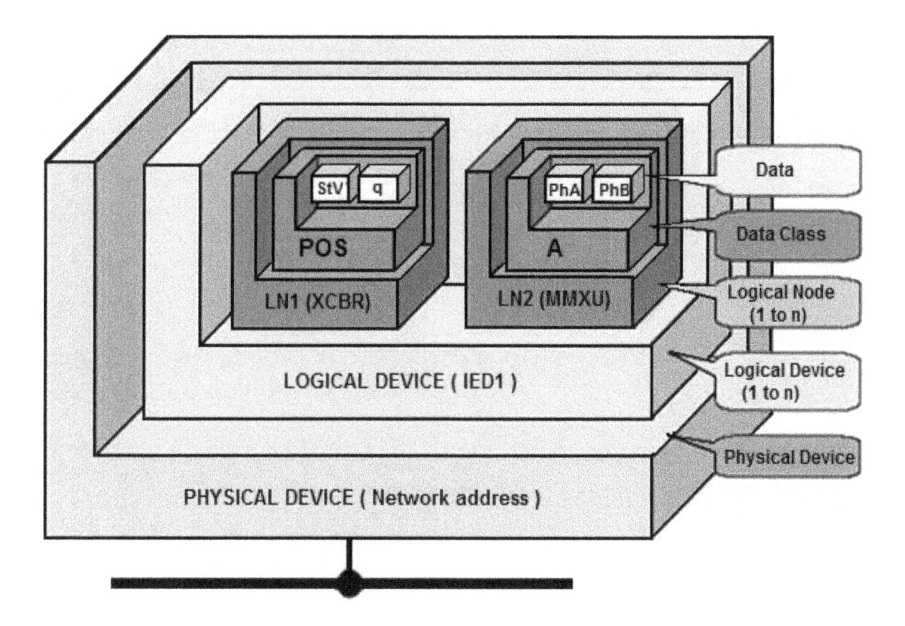

Figure 6.43 Physical and logical device

IEC 61850-7-4 Ed. 2 defines a list of logical node groups as shown in Table 6.15.

In Ed. 2, logical nodes (LN) list has been extended from 92 to 208 based on new requirements. New system logical nodes are introduced to indicate the status of GOOSE and sampled value subscription. Grouping of logical nodes is also extended to include relevant areas.

IEC 61850-5 Ed. 2 defines the protection logical nodes that are shown in Table 6.16.

2. Reporting schemes: There are various reporting schemes (BRCB and URCB) for reporting data from server through a server–client relationship which can be triggered based on predefined trigger conditions.

3. Fast transfer of events: GSE is defined for fast transfer of event data for a peer-to-peer communication mode. This is subdivided into GOOSE & GSSE (Generic Substation Status Event) which provide backward compatibility with the UCA GOOSE.

In IEC 61850 Ed. 2, GSSE is deprecated and moved to Annex of 7 2. This will cause older less flexible GSSE-based systems to move out from IEC 61850 scope. Hence, GOOSE will have added importance in InterBay Communication.

4. Setting groups: the setting group control blocks (SGCB) are defined to handle the setting groups so that users can switch to any active group according to the requirement.

5. Sampled data transfer: schemes are also defined to handle transfer of sampled values using sampled value control blocks (SVCB).

Table 6.15 Protection logical nodes defined by IEC 61850-7-4 Edition 2

Group indicator	**Logical node groups**
A	Automatic control
B	Reserved
C	Supervisory control
D	Distributed energy resources
E	Reserved
F	Functional blocks
G	Generic function references
H	Hydro power
	Interfacing and archiving
J	Reserved
K	Mechanical and non-electrical primary equipment
L	System logical nodes
M	Metering and measurement
N	Reserved
O	Reserved
P	Protection functions
Q	Power quality events detection related
R	Protection related functions
S[a]	Supervision and monitoring
T[a]	Instrument transformer and sensors
U	Reserved
V	Reserved
W	Wind power
X[a]	Switchgear
Y[a]	Power transformer and related functions
Z[a]	Further (power system) equipment

[a]LNs of this group exist in dedicated IEDs if a process bus is used. Without a process bus, LNs of this group are the I/Os in the hardwired IED one level higher (e.g., in a bay unit), representing the external device by its inputs and outputs (process image).

Table 6.16 Protection logical nodes defined by IEC 61850-5 Edition 2

Functionality allocated to LN	**IEC**	**IEEE**	**LN function**	**LN class**	**LN class naming**
Transient earth fault protection			PTEF	PTEF	Transient earth fault
Sensitive directional earth fault		(37)(67N)	PSDE	PSDE	Sensitive directional earth fault
Thyristor protection			PTHF	PTHF	Thyristor protection
Protection trip conditioning			PTRC	PTRC	Protection trip conditioning
Checking or interlocking relay		3	CILO	CILO	Interlocking
Over speed protection	$w >$	12	POVS		
Zero speed and under speed protection	$w <$	14	PZSU	PZSU	Zero speed or underspeed
Distance protection	$Z <$	21	PDIS	PDIS	Distance protection
				PSCH	Protection scheme
Volt per Hz protection		24	PVPH	PVPH	Volts per Hz

(Continues)

Table 6.16 (*Continued*)

Functionality allocated to LN	IEC	IEEE	LN function	LN class	LN class naming
Synchronism check		25	RSYN	RSYN	Synchronism-check
Over temperature protection	>	26	PTTR	PTTR	Thermal overload
(Time) Undervoltage protection	U <	27	PTUV	PTUV	Undervoltage
Directional power/reverse power protection	$\vec{P} >$	32	PDPR	PDOP	Directional over Power
				PDUP	Directional underpower
Undercurrent/underpower protection	$P <$	37	PUCP	PTUC	Undercurrent
				PDUP	Directional underpower
Loss of field/underexcitation protection		40	PUEX	PDUP	Directional under Power
				PDIS	(distance)
Reverse phase or phase balance current protection, Negative sequence current relay	$I_2 >$	46	PPBR	PTOC	Time overcurrent
Phase sequence or phase balance voltage protection, Negative sequence voltage relay	$U_2 >$	47	PPBV	PTOV	Overvoltage protection
Motor start-up protection		48, 49, 51LR66	PMSU	PMRI	Motor restart inhibition
				PMSS	Motor starting time supervision
Thermal overload protection	$\Theta >$	49	PTTR	PTTR	Thermal overload
		49R	PROL	PTTR	Thermal overload
Rotor thermal overload protection					
Rotor protection		49R	PROT	PTTR	Thermal overload
		64R			Time overcurrent
		(40)		PTOC	Ground detector
		50		PHIZ	Directional under
		51		PDUP	Power
					Distance
				PDIS	(impedance)
Stator thermal overload protection		49S	PSOL	PTTR	Thermal overload
Instantaneous overcurrent or rate of rise protection	$I \gg$	50	PIOC	PIOC	Instantaneous overcurrent
AC time overcurrent protection	$I >, t$	50TD51	PTOC	PTOC	Time overcurrent
Voltage controlled/dependent time overcurrent protection		51V	PVOC	PVOC	Voltage controlled time overcurrent
Power factor protection	$\cos\varphi >$	55	PPFR	POPF	Over power factor
	$\cos\varphi <$			PUPF	Under power factor
(Time) Overvoltage protection	$U >$	59	PTOV	PTOV	Overvoltage
DC-overvoltage protection		59DC	PDOV	PTOV	Overvoltage
Voltage or current balance protection		60	PVCB	PTOV	Overvoltage
				PTOC	Time overcurrent
Earth fault protection, ground detection	$I_E >$	64	PHIZ	PTOC	Time overcurrent
				PHIZ	Ground detector
Rotor earth fault protection		64R	PREF	PTOC	Time overcurrent
				PHIZ	Ground detector
Stator earth fault protection		64S	PSEF	PTOC	Time overcurrent
				PHIZ	Ground detector

(Continues)

Table 6.16 (Continued)

Functionality allocated to LN	IEC	IEEE	LN function	LN class	LN class naming
Interturn fault protection		64W	PITF	PTOC	Time overcurrent
AC directional overcurrent protection	$\vec{I} >$	67	PDOC	PTOC	Time overcurrent
Directional protection		87B	PDIR	PDIR	Direction comparison
Directional earth fault protection	$\vec{I}_E >$	67N	PDEF	PTOC	Time overcurrent
DC time overcurrent protection		76	PDCO	PTOC	Time overcurrent
Phase angle or out-of-step protection	$\phi >$	78	PPAM	PPAM	Phase angle measuring
Frequency protection		817	PFRQ	PTOF	Overfrequency
				PTUF	Under-frequency
				PFRC	Rate of change of frequency
Differential protection		87	PDIF	PDIF	Differential (Impedance)
Busbar protection		87B	PBDF	PDIF	Differential
				PDIR	Direction comparison
Generator differential protection		87G	PGDF	PDIF	Differential
Differential line protection		87L	PLDF	PDIF	Differential
Motor differential protection		87M	PMDF	PDIF	Differential
Restricted earth fault protection		87N	PNDF	PDIF	Differential
Phase comparison protection		87P	PPDF	PDIF	Differential
Differential transformer protection		87T	PTDF	PDIF	Differential
Harmonic restraint			PHAR	PHAR	Harmonic restraint

6. Commands: various command types are also supported by IEC 61850 which include direct and select before operate (SBO) commands with normal and enhanced securities.

7. Data storage: SCL is defined for complete storage of configured data of the substation in a specific format. SCL originally was the abbreviation for substation configuration language. Now that the use of IEC 61850 beyond substations has become reality, the name changed to system configuration language.

6.10.2 System configuration language

The fact that IEC61850 is a worldwide standard makes it also possible that relays from different manufacturers are able to exchange information as long as the relays conform to the IEC61850 standard.

To guarantee interoperability and enhance the configuration phase, the IEC 61850-6 introduced a common language which can be used to exchange information between different manufacturers using System Configuration Language.

Each proprietary tool must have a function that allows the export of the IED's description into this common, XML-based language. The ICD file (IED capability description) contains all information about the IED, which allows the user now to configure a GOOSE message.

The development process of a project based on IEC 61850 depends on the availability of software tools that make use of the SCL language. The SCL

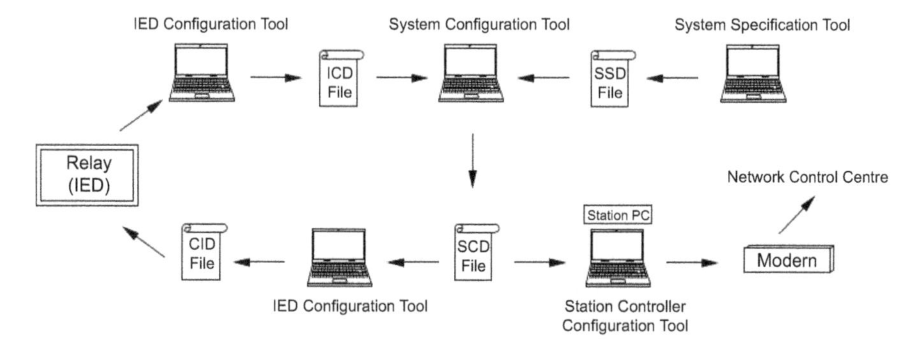

Figure 6.44 Substation engineering process using SCL language

specifies a common file format for describing IED capabilities, a system specific caption that can be viewed in terms of a single line diagram, and a substation automation system description. IEC 61850-6 introduces four types of common files. These files are the IED Capability Description (ICD), Configured IED Description (CID), Substation Configuration Description (SCD) and System Specification Description (SSD) files. IEC 61850-6 Ed. 2 introduced the new 'Instantiated IED Description' (IID) file. Figure 6.44 describes the complete engineering process using SCL language.

The configuration can be performed by a manufacturer independent tool, the so-called IEC61850 System Configurator. Some manufacturers improved their proprietary tools in a way that they can be used as IEC61850 System Configurator; however, there are also some third-party tools available. All ICD files get imported into the IEC61850 System Configurator and the GOOSE messages can be programmed by specifying the sender (publisher) and the receiver (subscriber) of a message. In the end, the whole description of the system, including the description of the GOOSE messages, gets stored in the SCD file (Substation Configuration Description). Each proprietary tool must be able to import this SCD file and extract the information needed for the IED. The information out of the SCD file is typically the list of the GOOSE messages.

SCL language does not cover all features of today's IEDs. In fact, it was not the intention of the editors of IEC 61850 standard to standardise all aspects of IEDs. This is because of the wide variety of functionality provided by each manufacturer.

It is important to know that in addition to the configuration information specified by part 6 of IEC 61850, the complete configuration of an IED should be completed with a proprietary IED configuration tool provided by the manufacturer in order to configure the device parameters and gain access to all of the internal functions supported.

The following list shows examples of features that can only be set up with a vendor's tool:

- logic and trips equations;
- graphical display on an IED's HMI;
- internal mappings;
- non-IEC 61850 and vendor-specific parameters.

Although the present practice of connecting a notebook computer directly to the front port of the IED will likely be eventually phased out in favour of remote access solutions via the LAN, it is assumed that some activities will continue to need to be performed at the relay, particularly during initial commissioning. It is also assumed that each individual vendor's IED configuration methodology will continue to be unique to each device or product family. Continuous innovation in IED design and competitive market forces would tend to preclude standardisation of individual IED configuration software, though some activity is underway in this regard.

The final configuration done by the proprietary tool is individualised per IED and can have either a proprietary format or a standard CID format. Nowadays, many vendors have decided to have a CID file using a proprietary format while a few vendors create the CID file in a similar way as the ICD file.

6.10.3 Configuration and verification of GOOSE messages

This section will describe how to configure and verify GOOSE messages for an automated IEC 61850 substation. The type of scheme being configured to illustrate the process is a communications-based breaker failure scheme. Figure 6.45 shows the one-line diagram of the system.

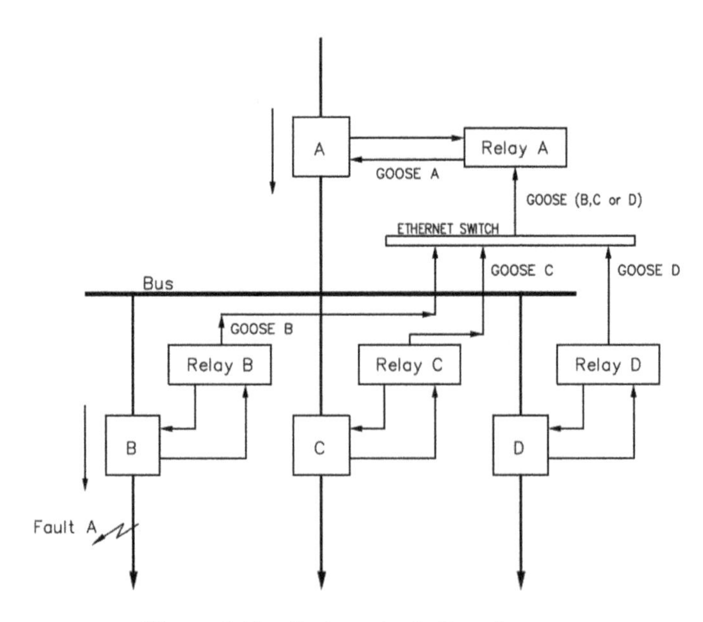

Figure 6.45 System single line diagram

The system has four IEDs from different vendors which are shown as relays A, B, C and D. Relays B, C and D are the main feeder protection and relay A is the backup protection.

Not shown are the IEDs in the breaker cabinet that take the GOOSE messages from each relay and convert them to a physical output to energise the breaker coil. In this example, a fault has been placed at location A, marked in the one line diagram. The instantaneous overcurrent element in relay B will transmit a trip GOOSE to breaker B. Simultaneously, relay A will receive the trip GOOSE from relay B as the breaker failure initiates signal. Relay A will receive this GOOSE and it will initiate the breaker failure timer, which will expire in 10 cycles. If for any reason the breaker failure initiate signal is not received and breaker B does not open, relay A will trip on a definite time overcurrent in 15 cycles. This section will only show the configuration between relays A and B. This same process is repeated for the other IEDs on the system. Figure 6.46 shows the logic of the breaker failure scheme.

The programming of this scheme can be achieved in two parts. The first part is the programming of the individual IEDs in order to obtain the SCL file required for the configuration of the system. These files will contain the published GOOSE messages for each IED. With this information, the full system configuration can be accomplished. Before any configuration can occur, it is good practice to create a virtual wiring map. This map will show what messages are going to be required in order to configure the scheme of interest. The map provides such information as to what GOOSE messages each IED will publish and subscribe.

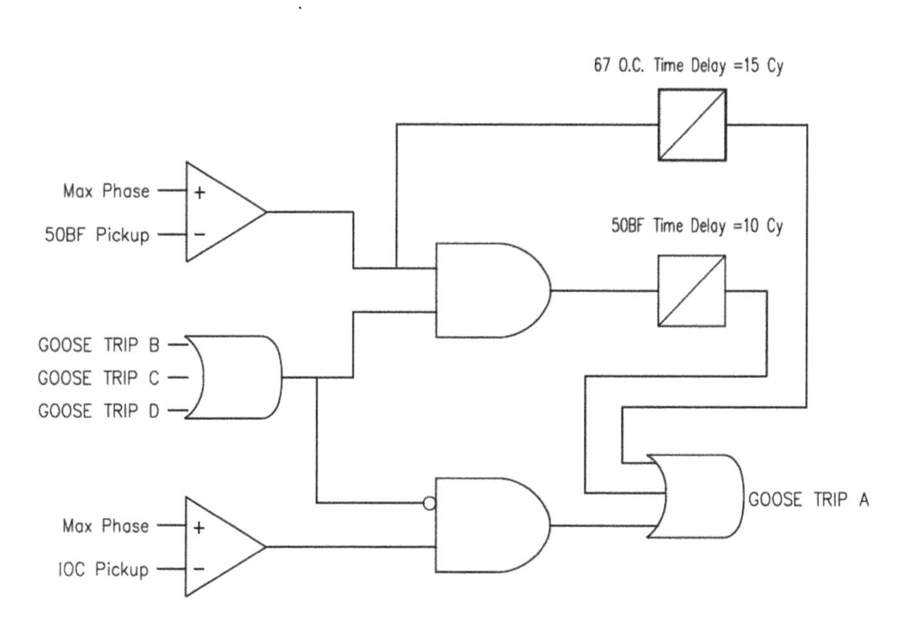

Figure 6.46 Logic of breaker failure scheme

The IED is configured via its own proprietary configuration tool. One of the biggest challenges in configuring an IEC 61850 based protection scheme is becoming familiar with the different IED configuration software packages. It is important to get familiar with how each configuration tool works. This will save time in the long run, especially when trying to configure the system scheme. The primary goal of configuring the IED is to obtain the SCL file types necessary for the system configuration. Figure 6.47 shows examples of the configuration tools used to configure the individual IEDs.

The SCL files provide the overall information as to how the relay is configured but most importantly, what GOOSE messages the particular IED will publish or subscribe to. The configuration process would begin by configuring the IEDs that will only be publishing and subscribing to the least number of GOOSE messages.

When the IED is configured, an SCL file is exported. This same process is repeated until all IEDs of interest have been configured and an SCL file is generated.

If desired, a simple test can be performed to verify that the published GOOSE messages are correct. This can be done by connecting some form of network analyser and capturing the GOOSE messages on the network. An illustration of proprietary configuration tools used for the configuration of IEDs is shown in Figure 6.47.

Some of the details of the GOOSE configuration were left out due to differences in the configuration of the IEDs. For example, in some IEDs, the GOOSE messages are going to be published through GGIO (generic GOOSE I/O), while others will be published through the protection node such as PDIS or PTOC. This difference can sometimes lead to the GOOSE message being delayed by some fixed

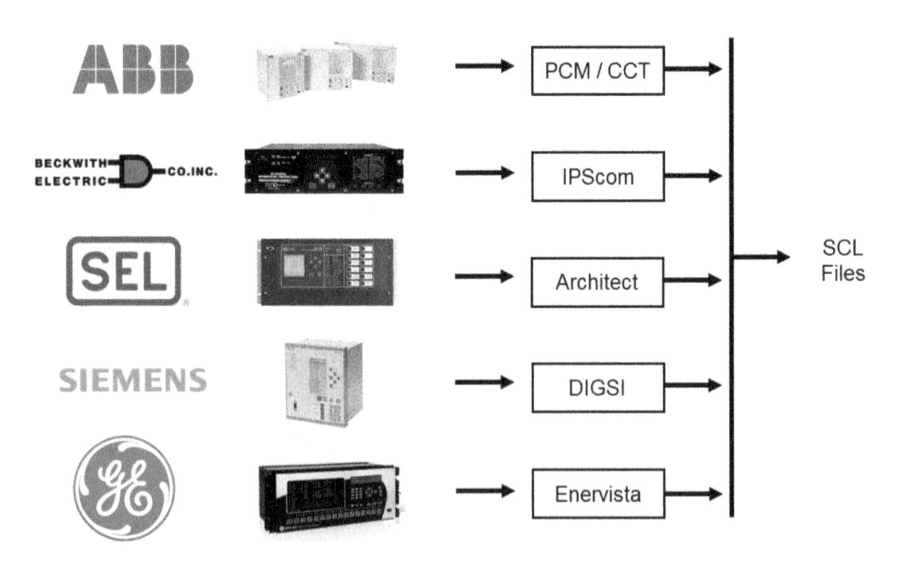

Figure 6.47 Proprietary configuration tools used for the configuration of IEDs

time set in the IED. To avoid these problems, it is recommended to read the IED's manual and see where the IED will publish its high-speed GOOSE.

6.10.4 Configuration of the system

In order to configure the system, it is necessary to have all ICD files available. These files are going to be used by a system configuration tool or the IED's individual configuration tool. The system configuration can be done by a substation configuration tool. This type of tool imports all the ICD files required for the system configuration. The substation configuration tool will generate an SCD file. Another method is by opening up the configuration tool of the IEDs that will subscribe to a GOOSE message.

For example, the configuration of relay A is done by importing all the ICD files into its configuration tool. The ICD file is intended to be a template for an IED type, which could be instantiated multiple times – e.g. per feeder – to build a system.

When the system is configured, a system verification test should be performed. This test will help identify any problems with the system configuration. The first time a system is configured, some configuration problems may occur. Many of these problems are due to the differences in the configuration of the IEDs. Some of the IEDs require a more manual approach, while others are a bit more automatic.

6.10.5 Publisher/subscriber concept

In IEC 61850, data is organised into logical entities called data objects. Data objects are further grouped into data sets that are collections of related data objects. Publishers and subscribers interact with these data objects and data sets. It is a key aspect of IEC 61850 and it plays a crucial role in the communication architecture of the standard.

Publisher: A publisher is an IED that generates data. This data can be measurements, status information, event notifications and others. The publisher is responsible for sending this data to the network.

Subscriber: A subscriber is an IED or a server that is interested in receiving specific types of data. Subscribers subscribe to particular data types or data sets. Once subscribed, a subscriber will receive updates from publishers whenever there is a change or an event related to the subscribed data.

6.10.6 System verification test

The system verification test will identify any configuration conflicts in the system. The tools required are going to be a network analyser (software) and a modern test set that is IEC 61850 compliant. The test set must be able to receive and send GOOSE messages via the substation LAN. This would require the test system to interrogate the network, acquire the right GOOSE message and stop a timer with minimal effect on time.

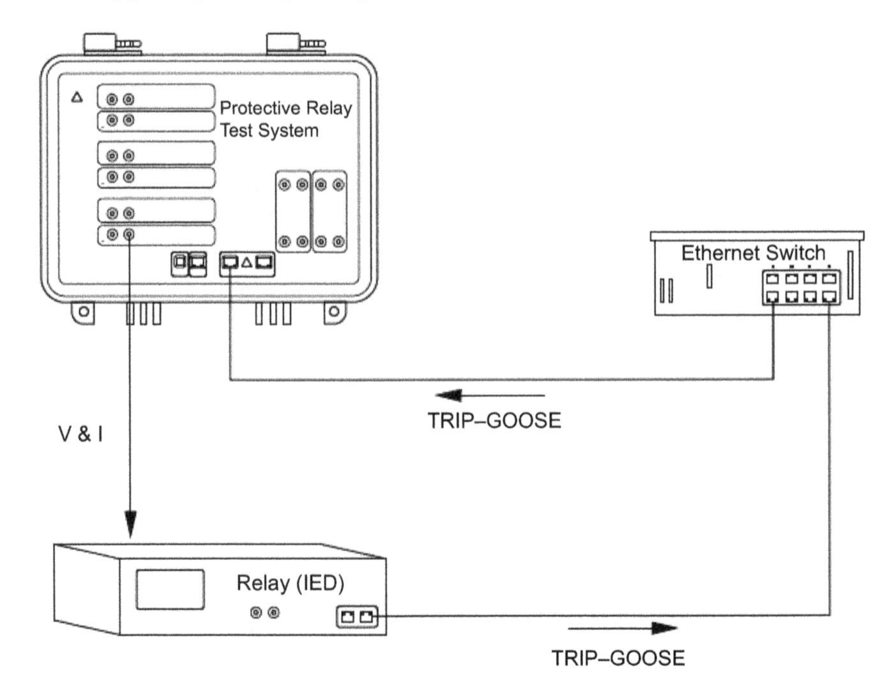

Figure 6.48 Test connections used for standalone IEC 61850 based IED

Figure 6.48 presents the test connections used for standalone IEC 61850 and illustrates how the trip-GOOSE messages are sent from a numerical relay to an ethernet switch and from this to a protective relay test system.

Modern test systems must be able to receive and send GOOSE messages via the substation LAN. This would require the test system to be able to interrogate the network, acquire the right GOOSE message and stop injections or timer in less than 2 ms. Also the test system would have to be able to read SCL files and map inputs to the various GOOSE messages available in the SCL file. If an SCL file is not available, then the test system would have to be able to interrogate the network, and display all available GOOSE messages on the network, to allow the user to be able to map these messages to binary inputs on the test system.

6.10.7 Substation IT network

As communications in the substation take on more critical roles in the protection and control task of the utility, it is important for the protection engineer to understand the basics of the IT network. The protection engineer must also understand the behaviour and characteristics of components like ethernet switches, ethernet ports and router, as well as being familiar with terminology such as LAN, VLAN, Mac address, network topology, latency, priority tag, firewalls, RSTP and HSR/PRP.

Many experienced protection engineers find discussion of IT network issues to be dense and perhaps intimidating, because until now they have not faced the need to understand the behaviour and performance characteristics of IT networks. In the modern substations, ethernet switches are as important to understand as protective relays in order to achieve availability, dependability, security and maintainability goals of the substation.

6.10.8 Process bus

Process bus is defined by IEC 61850 Ed. 2 Part 9-2. Process bus is the digitalisation of all analogue signals in the substation. This is achieved by connecting all current transformers, potential transformers and control cables to merging units. These units convert the analogue signals to binary signals and send the information via the process bus to all the devices that subscribe for that information. There are some pilot projects installed around the world with this technology reporting successful results. Figure 6.49 shows a full implementation of IEC 61850 – process bus and station bus.

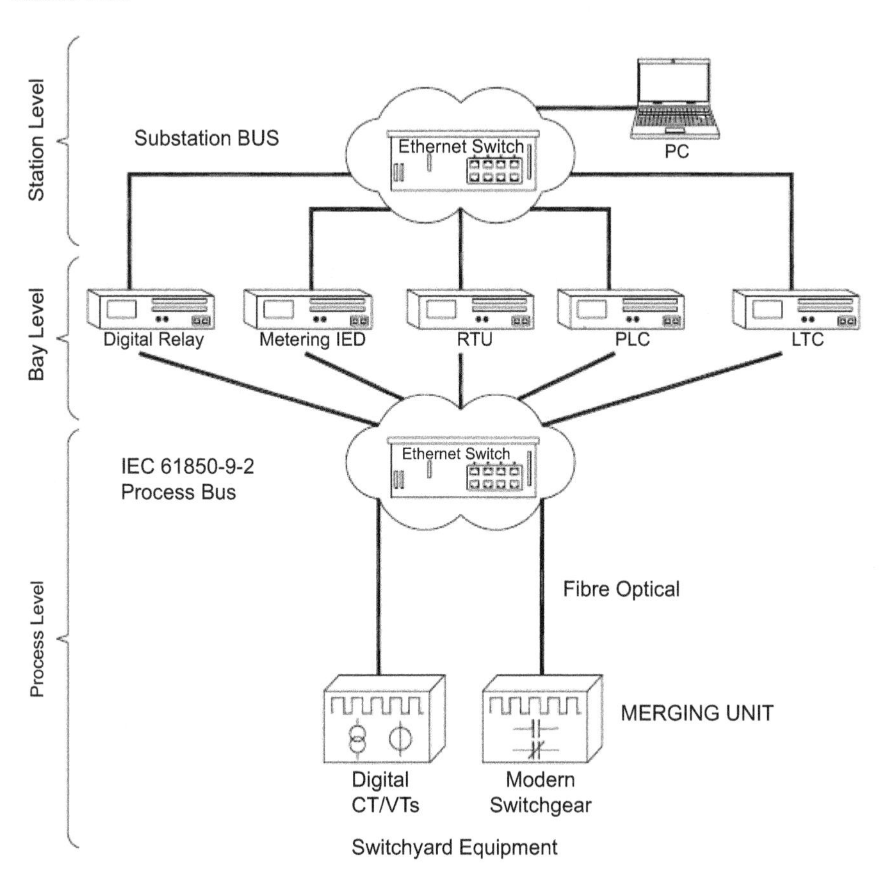

Figure 6.49 Substation network – process bus and station bus

6.10.9 Communications redundancy networks IEC 618590

It is shown how GOOSE messages replace the wired logic between IED protection relays by implementing an ethernet network designed for substation communications [47–56]. The relevance of communications engineering in substation networks is defined based on the criticality of packets circulating between the different automation elements from either an IED to the SCADA system or another IED within the same network. When an event is detected, IEDs use a multicast transmission to notify the event to any device in the network that uses IEC61850. This requires that sending of messages is done as fast as possible and the data is transmitted in a reliable way.

This gives the requirement for communication to be redundant, which allows ensuring the survival of data in case of a failure and providing routes of alternative data when a communication path between devices fails. In data networks, there are two basic redundancy techniques: active redundancy, in which both connection links are active at the same time; and passive redundancy, where one link is active and the other is in stand-by mode. Choosing one or the other will depend on the criticality of the network and the type of application. For instance, passive redundancy would not be suitable for high availability environments such as substations, metal refineries, chemical plants and other applications.

Critical GOOSE messages and control commands must be received by subscribing devices within a specific time window to ensure the reliability of protection and control performance. IEC 61850 has adopted the recommendations of IEC 62439-3 and specifies the active redundancy protocols in its implementation. This corresponds to the parallel redundancy protocol (PRP) and the seamless high availability redundancy protocol (HSR). Both protocols duplicate the data transmitted through the network and allow an immediate switching time when a fault occurs in a cable or switch.

In order to ensure a correct, secure and scalable implementation of the systems, compliance of reliability, cybersecurity and other communications requirements is required.

Example 6.3

The system of Figure 6.50 contains a protection relay (IED) that will trip the overcurrent unit due to the fault illustrated. The corresponding output is sent to the circuit breaker (CB) control to open it using GOOSE.

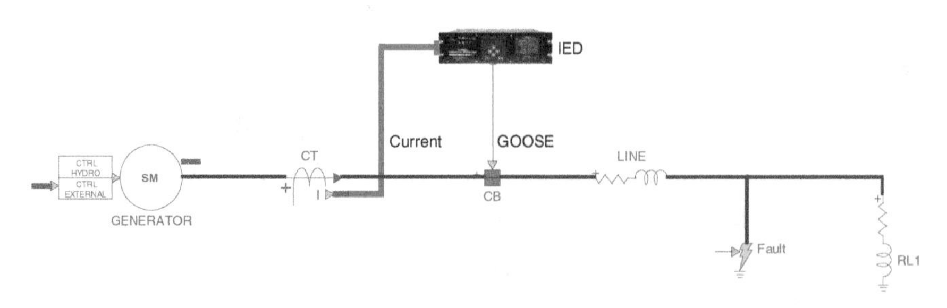

Figure 6.50 Overcurrent protection trip using GOOSE – Example 6.2

The relay has the phase inverse time overcurrent unit with following settings:

- Phase CT ratio: 60
- Curve: US very inverse
- Pickup: 1.66
- Multiplier Dial: 2.00

As shown in Table 6.16, the Logical Node for the Phase Inverse Time Overcurrent unit in IEC61850 is PTOC. In addition, considering that the IED will trip the CB using GOOSE, the CB control will be subscribed to the GOOSE published by the IED. Figure 6.51 illustrates how the IED is publishing GOOSE messages once it was set with the corresponding vendor software.

The operation of the PTOC logical node is a Boolean variable. According to this, the IED will be publishing the operation as False and it will change to True once the overcurrent unit trips. Figure 6.52 illustrates the GOOSE dataset in False (the relay has not tripped) and in True (when the relay trips).

The change from False to True is illustrated in Figure 6.53 where the chart has been obtained in real time, the fault occurs at 500 ms and the relay trips at 931 ms.

```
7278 214.794949  Beckwith_09:08:d1  Iec-Tc57_01:00:12  GOOSE
7279 214.842447  Beckwith_09:08:d1  Iec-Tc57_01:00:12  GOOSE
7280 214.876850  Beckwith_09:08:d1  Iec-Tc57_01:00:07  GOOSE
7281 214.878743  Beckwith_09:08:d1  Iec-Tc57_01:00:12  GOOSE
7282 214.918133  Beckwith_09:08:d1  Iec-Tc57_01:00:12  GOOSE
7283 214.962500  Beckwith_09:08:d1  Iec-Tc57_01:00:12  GOOSE
7284 215.003243  Beckwith_09:08:d1  Iec-Tc57_01:00:12  GOOSE
7285 215.035824  Beckwith_09:08:d1  Iec-Tc57_01:00:12  GOOSE
7286 215.084631  Beckwith_09:08:d1  Iec-Tc57_01:00:12  GOOSE
7288 215.119217  Beckwith_09:08:d1  Iec-Tc57_01:00:12  GOOSE
7289 215.161627  Beckwith_09:08:d1  Iec-Tc57_01:00:12  GOOSE
```

Figure 6.51 GOOSE messages in the network – Example 6.2

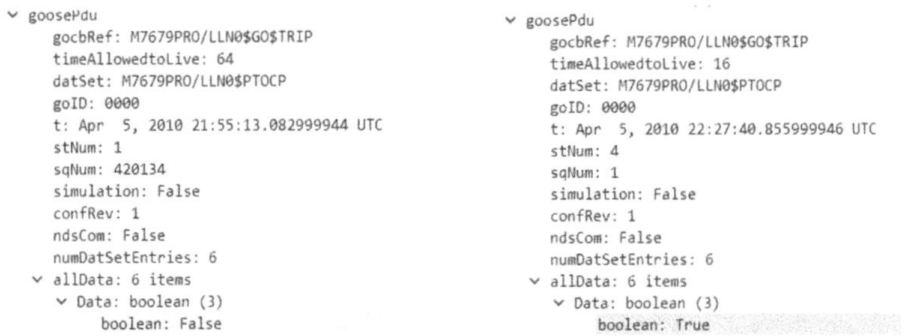

```
v goosePdu
    gocbRef: M7679PRO/LLN0$GO$TRIP
    timeAllowedtoLive: 64
    datSet: M7679PRO/LLN0$PTOCP
    goID: 0000
    t: Apr  5, 2010 21:55:13.082999944 UTC
    stNum: 1
    sqNum: 420134
    simulation: False
    confRev: 1
    ndsCom: False
    numDatSetEntries: 6
  v allData: 6 items
    v Data: boolean (3)
        boolean: False
```

```
v goosePdu
    gocbRef: M7679PRO/LLN0$GO$TRIP
    timeAllowedtoLive: 16
    datSet: M7679PRO/LLN0$PTOCP
    goID: 0000
    t: Apr  5, 2010 22:27:40.855999946 UTC
    stNum: 4
    sqNum: 1
    simulation: False
    confRev: 1
    ndsCom: False
    numDatSetEntries: 6
  v allData: 6 items
    v Data: boolean (3)
        boolean: True
```

Figure 6.52 Boolean PTOC operation in False and True – Example 6.2

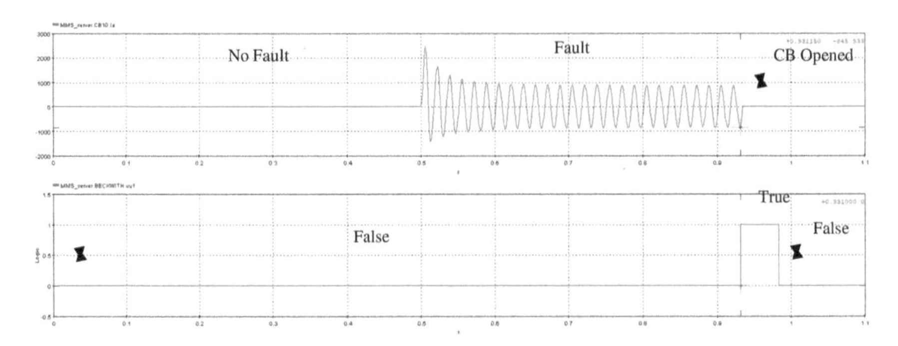

Figure 6.53 GOOSE message real time operation – Example 6.2

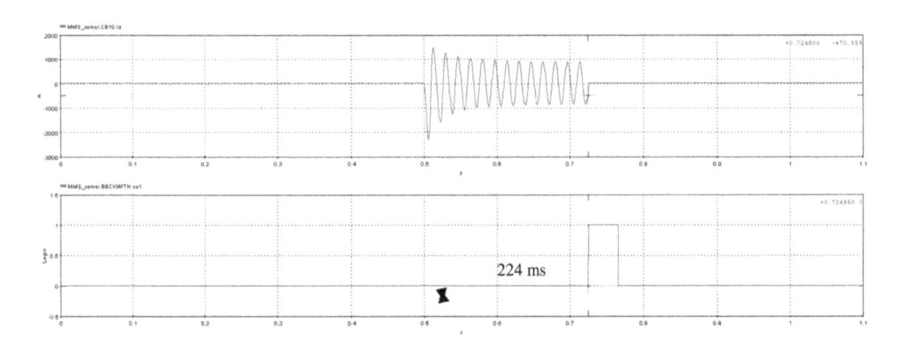

Figure 6.54 Operating time reduction – Example 6.2

As noticed, the relay operation time is around 431 ms (931 ms – 500 ms). Figure 6.54 illustrates how the operating time is reduced to 224 ms if the multiplier dial setting is changed from 2.00 to 1.00.

Example 6.4

The system illustrated in Figure 6.55 contains a substation with three feeders and a source. Two IEDs will be coordinated for the fault indicated in this figure.

The fault occurs at 0.5 s; it can be seen in Figure 6.56 that CB1 and CB2 will see the same fault current.

Using the LN PIOC of the overcurrent protection, the operation time of the IEDs for the fault can be seen. Figure 6.57 illustrates that both IEDs are tripping at the same time due to the same current and the same setting of the PIOC units (the CB controls have been temporally disabled, that's why the breaker keep closed).

In order to allow selectivity, different coordination criteria have been used for many years. For example, disabling the instantaneous unit for the IED 1 and setting a time delay.

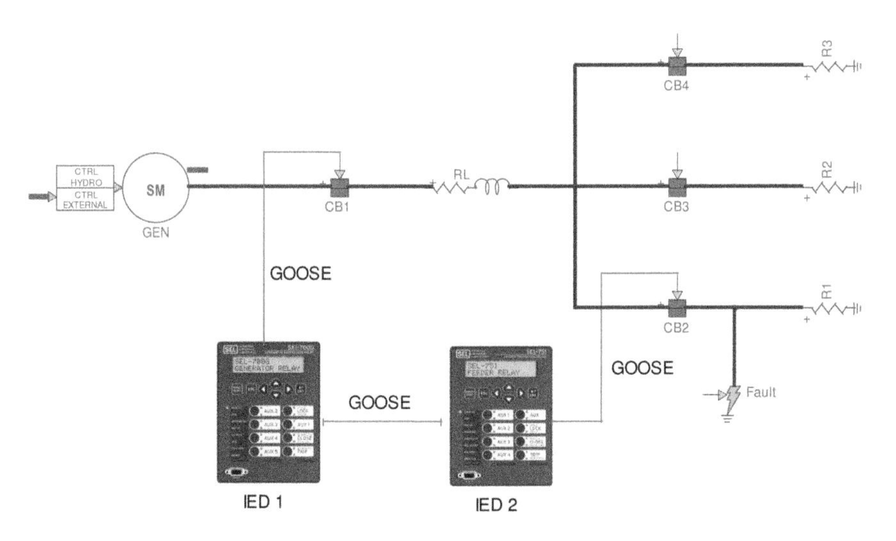

Figure 6.55 Single line diagram – Example 6.3

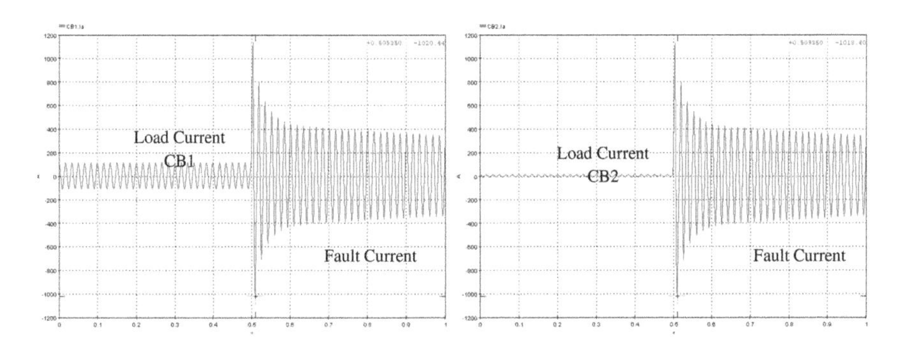

Figure 6.56 Fault currents through CB1 and CB2 – Example 6.3

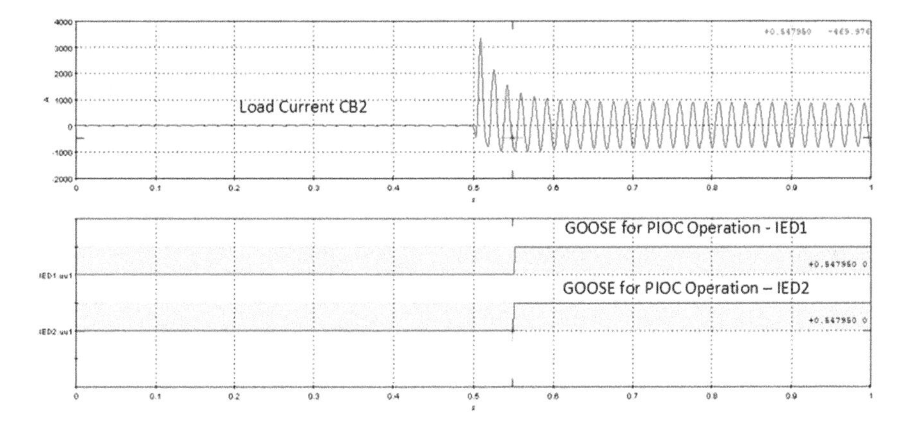

Figure 6.57 Tripping times for IEDs – Example 6.3

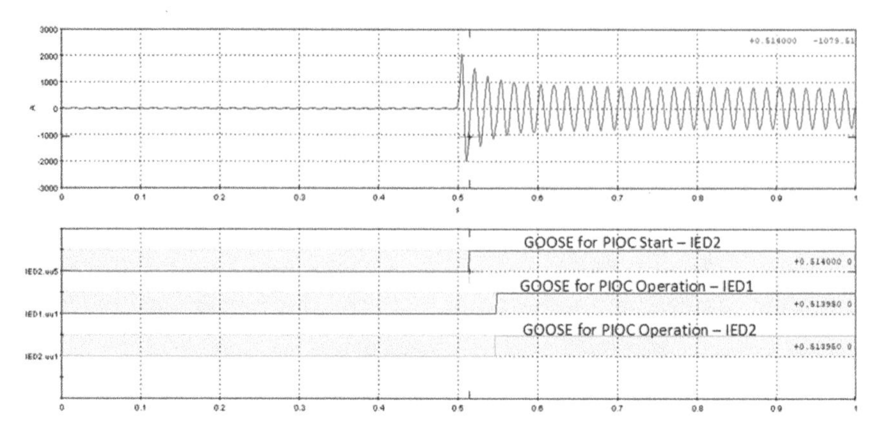

Figure 6.58 Starting time of IED 2 and tripping times of both IEDs – Example 6.3

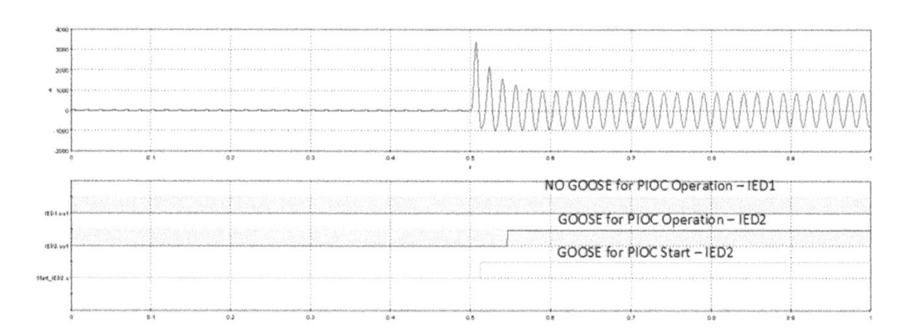

Figure 6.59 Blocking the GOOSE for IED 1 – Example 6.3

In this study case, IEC61850 will be used to block the operation of IED 1 once the IED 2 detects the fault. In Figure 6.58, the starting GOOSE for the LN PIOC of the IED 2 is anticipated to the operation of both protections. Therefore, IED 1 will be subscribed to the GOOSE of the IED2 in order to block its operation based on the start of the PIOC unit.

Figure 6.59 presents that IED 1 doesn't trip since the GOOSE has been blocked (in this graphic the PIOC start is in green).

Finally, if the breaker controls are enabled, the current flow on both CBs after the fault is indicated in Figure 6.60.

Proposed exercises

1. The radial circuit shown in Figure 6.61 presents two IDMT relays of 5 A. The settings of the relays are 115% and the time multiplier of relay A is 0.08 s. Find the time multiplier setting of relay B to coordinate two relays for a fault of 2300 A. Assume a grading margin of 0.35 s.

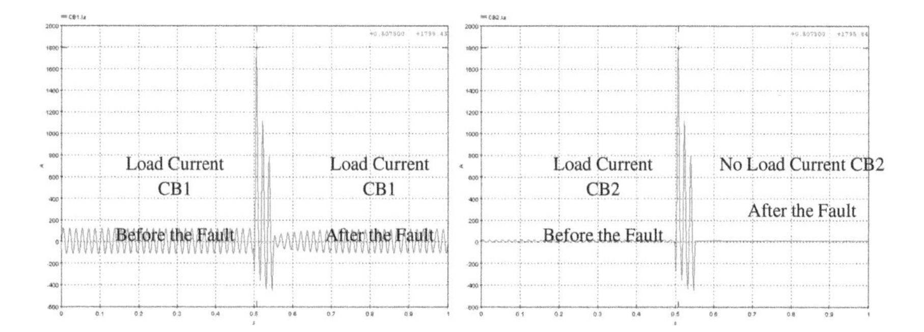

Figure 6.60 Currents through CB1 and CB2 with the protections

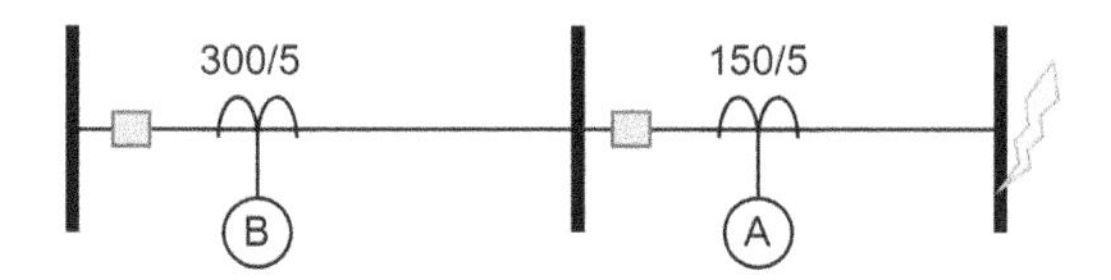

Figure 6.61 Single line diagram for exercise 1

2. A three-phase, 12500 kVA, 34.5/13.2 kV transformer is connected as delta-star. The CTs on the 13.2 kV side have a turns ratio of 600/5. What should be the CT ratio on the HV side?
3. Calculate the pick-up setting, time dial setting and the instantaneous setting of the phase relays installed in the high voltage and low voltage sides of the 115/13.2 kV transformers T1 and T3, in the substation illustrated in Figure 6.62. The short-circuit levels, CT ratios and other data are shown in the same diagram.
4. For the system shown in Figure 6.63, carry out the following calculations:
   (a) The maximum values of short-circuit current for three-phase faults at busbars A, B and C, taking into account that busbar D has a fault level of 12906.89 A rms symmetrical (2570.87 MVA).
      - The maximum peak values to which breakers 1, 5 and 8 can be subjected.
      - The rms asymmetrical values which breakers 1, 5 and 8 can withstand for 5 cycles for guarantee purposes.

        For these calculations, assume that the L/R ratio is 0.2.

   (b) The turns ratios of the CTs associated with breakers 1–8. The CT in breaker 6 is 100/5. Take into account that the secondaries are rated at 5 A and that the ratios available in the primaries are multiples of 50 up to 400, and from then on are in multiples of 100.
   (c) The instantaneous, pick-up and time dial settings for the phase relays in order to guarantee a co-ordinated protection system, allowing a time discrimination margin of 0.4 s.

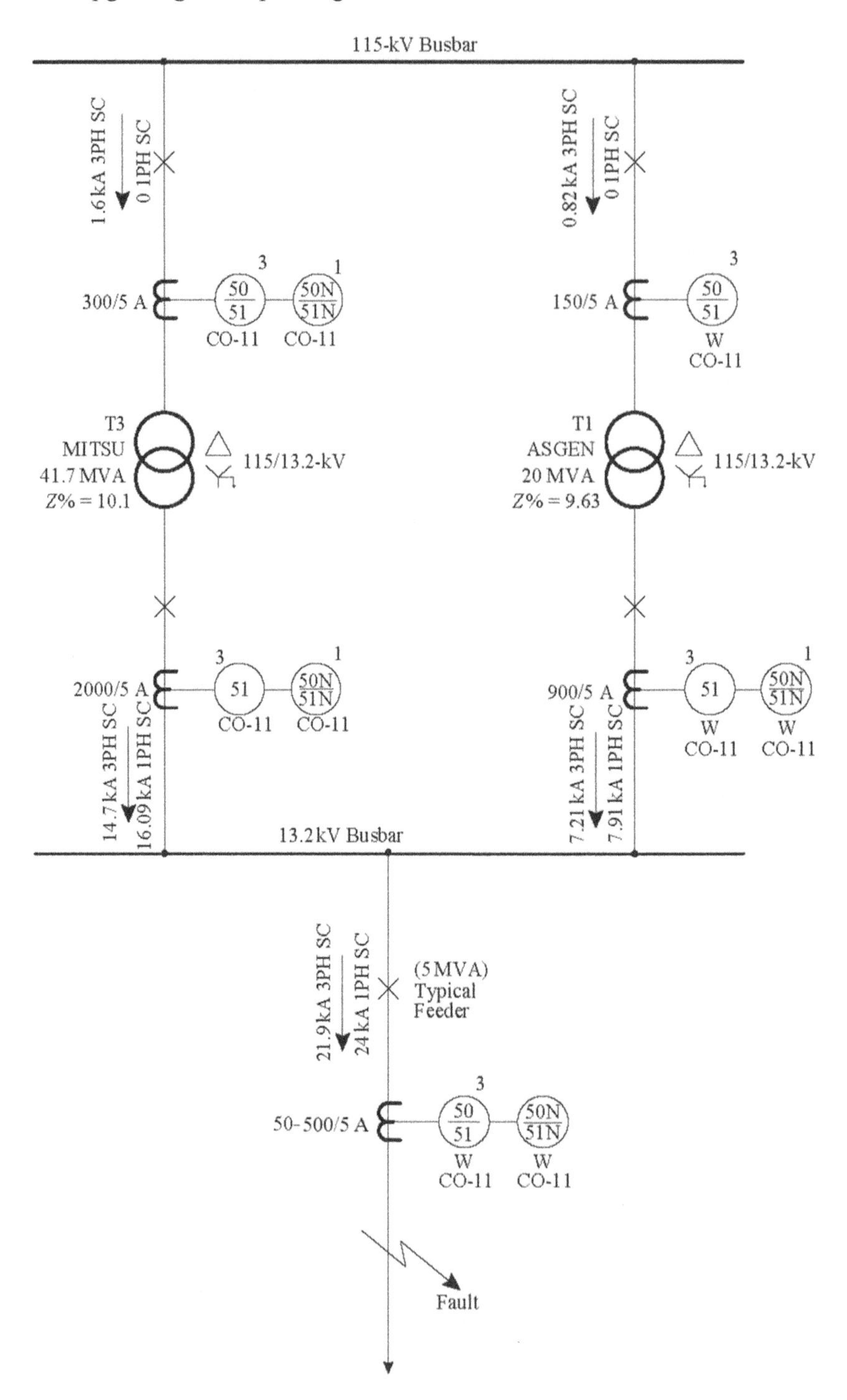

Figure 6.62 Single line diagram for Exercise 3

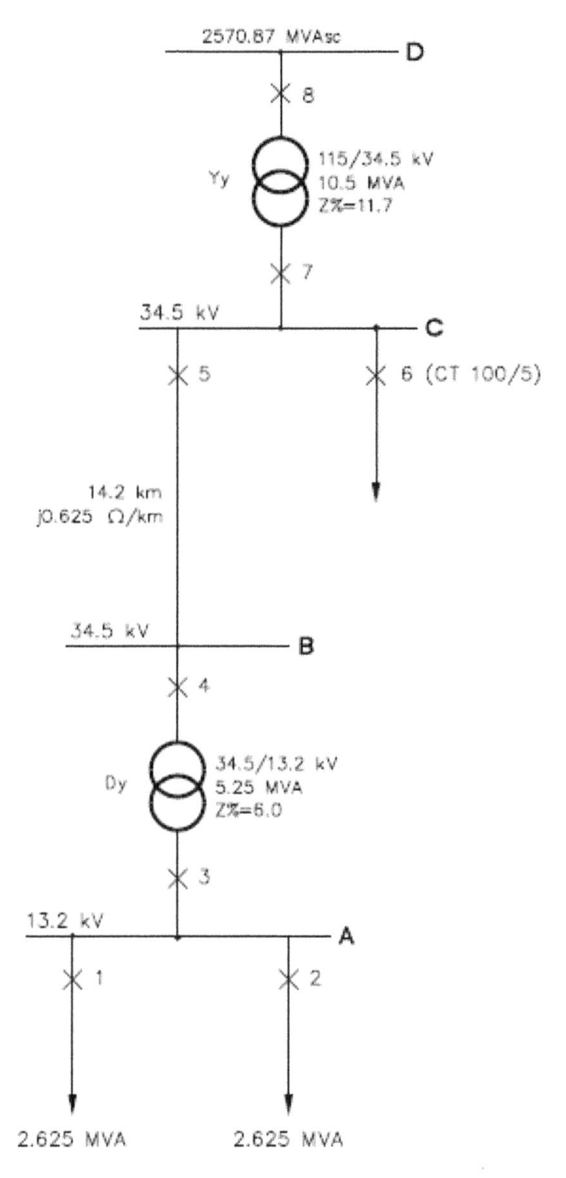

Figure 6.63 Single line diagram for Exercise 4

(d) The percentage of the 34.5 kV line which is protected by the instantaneous element of the overcurrent relay associated with breaker 5.
Bear in mind the following additional information:

- The settings of relay 6 are as follows: pick-up 7 A, time dial 5, instantaneous setting 1000 A primary current.

- All the relays are inverse time type, with the following characteristics:

  Pick-up: 1 to 12 in steps of 2 A
  Time dial: as in Figure 6.64
  Instantaneous element: 6 to 144 in steps of 1 A

  Calculate the setting of the instantaneous elements of the relays associated with the feeders assuming 0.5 I_{sc} on busbar A.

5. For the system shown in Figure 6.55 of Example 6.3, define a protection scheme for the breaker failure unit using IEC 61850.

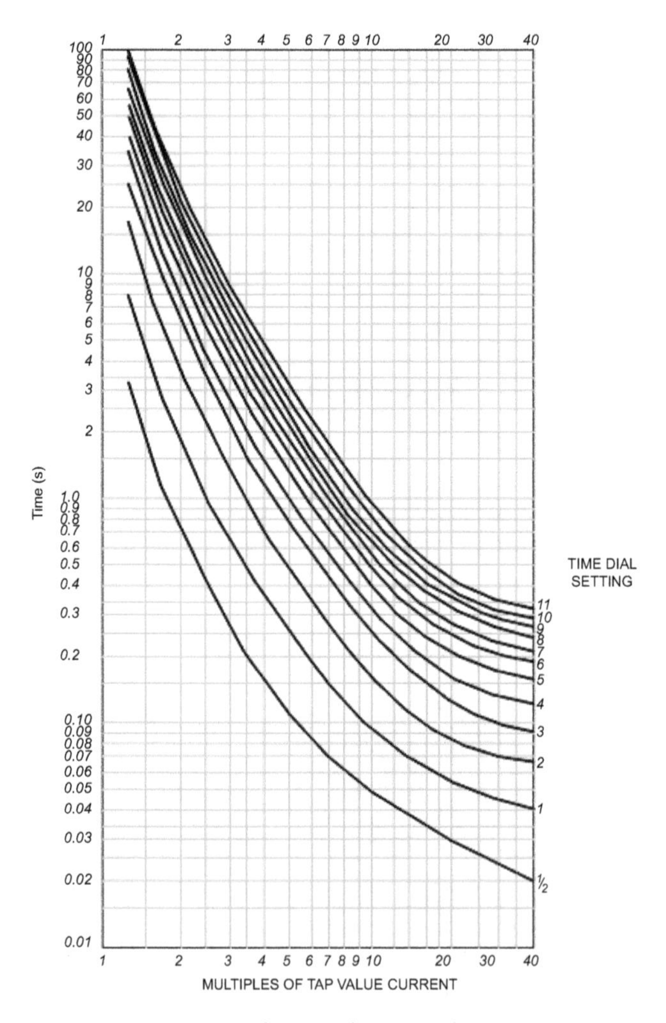

Figure 6.64 Relay time/current characteristic

References

[1] International Electrotechnical Commission, 'IEC 60255-1, 2022 – Measuring Relays and Protection Equipment – Part 1 Common Requirements', 2022.
[2] IEEE, 'IEEE Std C37.112-2018 – IEEE Standard for Inverse-Time Characteristics Equations for Overcurrent Relays', 2018.
[3] Beckwith Electric Co., 'M-7651A D-PAC Protection, Automation and Control System for Power Distribution Applications - Specification', Florida, USA, 2012, Available: https://beckwithelectric.com/products/m-7651a/ [accessed 24 May 2024].
[4] UL 198M, 'UL Standard for Safety Mine-Duty Fuses. 5th Ed', 2023.
[5] IEEE, 'IEEE Std C37.40-2003 – IEEE Standard Service Conditions and Definitions for High-Voltage Fuses, Distribution Enclosed Single-Pole Air Switches, Fuse Disconnecting Switches, and Accessories', 2003.
[6] IEEE, 'IEEE Std C37.41-2016 – IEEE Standard Design Tests for High-Voltage (>1000V) Fuses and Accessories', 2016.
[7] IEEE, 'IEEE Std C37.42-2016 – IEEE Standard Specifications for High-Voltage (>1000 V) Fuses and Accessories', 2016.
[8] IEEE, 'IEEE Std C37.47-2011 – IEEE Standard Specifications for High-Voltage (>1000 V) Distribution Class Current-Limiting Type Fuses and Fuse Disconnecting Switches', 2011.
[9] IEEE, 'IEEE Std C37.48-2005 – IEEE Guide for Application, Operation, and Maintenance of High-Voltage Fuses, Distribution Enclosed Single-Pole Air Switches, Fuse Disconnecting Switches, and Accessories', 2005.
[10] Institute of Electrical and Electronics Engineers. and IEEE-SA Standards Board, 'IEEE std C37.46-2010. Specifications for High-Voltage (>1000 V) Expulsion and Current-Limiting Power Class Fuses and Fuse Disconnecting Switches', p. 19, 2010.
[11] International Electrotechnical Commission, IEC 61968-11:2013. 'Application Integration at Electric Utilities – System Interfaces for Distribution Management – Part 11: Common Information Model (CIM) Extensions for Distribution'. CIM/Distribution Management, 2013.
[12] International Electrotechnical Commission, 'IEC 61970-301:2020. Energy Management System Application Program Interface (EMS-API) – Part 301: Common Information Model (CIM) Base', CIM/Energy Management', 2020.
[13] International Electrotechnical Commission, 'Electricity Metering Data Exchange – The DLMS/COSEM Suite,' IEC 62056, 2023.
[14] N. I. O. F. Standards and Technology, 'NIST Framework and Roadmap for Smart Grid Interoperability Standards, Release 3.0', USA, NIST Special Publication, vol. 1108r3, 2014.
[15] International Electrotechnical Commission, 'IEC 61850:2024 SER. Series Communication Networks and Systems for Power Utility Automation – ALL PARTS', 2024.

[16] ISO/IEC/IEEE, 'ISO/IEC/IEEE 8802-3:2021 Telecommunications and Exchange between Information Technology Systems Requirements for Local and Metropolitan Area Networks Part 3: Standard for Ethernet', 2021.
[17] IEEE, 'IEEE Standard for Electric Power Systems Communications-Distributed Network Protocol (DNP3)', IEEE Std 1815-2012 (Revision of IEEE Std 1815-2010), pp. 1–821, 2012, doi:10.1109/IEEESTD.2012.6327578.
[18] International Electrotechnical Commission, 'IEC 60870-5:2024 SER Series Telecontrol Equipment and Systems – Part 5: Transmission Protocols – All PARTS', 2024.
[19] International Electrotechnical Commission, 'IEC 62351-6:2020. Power Systems Management and Associated Information Exchange – Data and Communications Security – Part 6: Security for IEC 61850', 2020.
[20] ISO, ISO 15408, 'Information Technology – Security Techniques'. 2009.
[21] P. P. Parikh, M. G. Kanabar, and T. S. Sidhu, 'Opportunities and Challenges of Wireless Communication Technologies for Smart Grid Applications', in *IEEE PES General Meeting*, Piscataway, NJ: IEEE, pp. 1–7, 2010, doi:10.1109/PES.2010.5589988.
[22] V. Aravinthan, B. Karimi, V. Namboodiri, and W. Jewell, 'Wireless Communication for Smart Grid Applications at Distribution Level – Feasibility and Requirements', in *2011 IEEE Power and Energy Society General Meeting*, Piscataway, NJ: IEEE, pp. 1–8, 2011, doi:10.1109/PES.2011.6039716.
[23] Electric Power Research Institute (EPRI), 'Report to NIST on the Smart Grid Interoperability Standards Roadmap', 2009. Available:https://www.naesb.org/pdf4/interimsmartgridroadmapnistrestructure_061709.pdf.
[24] IEEE, 'Std 242-2001 – IEEE Recommended Practice for Protection and Coordination of Industrial and Commercial Power Systems', pp. 1–710, 2001.
[25] A. Shreyas, 'Analysis of Communication Protocols for Neighborhood Area Network for Smart Grid', Sacramento, 2010. Available: https://scholars.csus.edu/esploro/outputs/graduate/Analysis-of-communication-protocols-for-Neighborhood/99257831099701671#file-0 [accessed 22 May 2024].
[26] IEEE, 'IEEE Std. 2030 – IEEE Guide for Smart Grid Interoperability of Energy Technology and Information Technology Operation with the Electric Power System (EPS), End-Use Applications, and Loads', 2011.
[27] National Rural Electric Cooperative Association (NRECA), *MultiSpeak and the NIST Smart Grid*. 'MultiSpeak and the NIST Smart Grid': NRECA, 2014.
[28] IEC Technical Committees & Subcommittees, 'TC 57, Power Systems Management and Associated Information Exchange'.
[29] IEC Subcommittees and Working Groups, 'TC 57 – WG14, System Interfaces for Distribution Management (SIDM)'.
[30] Neplan., 'IEEE14 – AC Steady State Load Flow', 2023.
[31] International Electrotechnical Commission, 'IEC 61970 Series: Energy Management System Application Program Interface (EMS-API) – All Parts', IEC, Geneva, Switzerland, 2024.

[32] International Electrotechnical Commission 'IEC 61968-4:2019, Application integration at electric utilities – System interfaces for distribution management – Part 4: Interfaces for records and asset management', IEC, 2019.

[33] National Institute of Standards and Technology, 'NISTIR – 7628 – Guidelines for Smart Grid Cybersecurity', 2014.

[34] Department of Homeland Security (DHS), *DHS Cyber Security Procurement Language for Control Systems*, 2009.

[35] IEEE, 'C37.240-2014 – IEEE Cyber Security Requirements for Substation Automation, Protection and Control Systems', 2015.

[36] IEEE, '1686-2022 – IEEE Standard for Intelligent Electronic Devices Cyber Security Capabilities', 2023.

[37] ISO/IEC, 'ISO/IEC 27019:2017. Information Technology – Security Techniques – Information Security Controls for the Energy Utility Industry', 2017.

[38] V. Tosic and S. Dordevic-Kajan, 'The Common Information Model (CIM) Standard – An Analysis of Features and Open Issues', in *4th International Conference on Telecommunications in Modern Satellite, Cable and Broadcasting Services. TELSIKS'99*, Nis, Yugoslavia: IEEE, pp. 677–680, 1999, doi:10.1109/TELSKS.1999.806301.

[39] P. System, 'Relaying Committee Working Group H6: "Application Consideration of IEC 61850/Uca2 for Substation Ethernet Local Area Network Communication for Protection and Control"', 2005.

[40] NIST, 'Framework and Roadmap for Smart Grid Interoperability Standards, Release 4.0', NIST Special Publication 1108r4, Gaithersburg, MD, USA, 2021. [Online]. Available: https://doi.org/10.6028/NIST.SP.1108r4 [accessed 21 May 2024].

[41] N. I. O. F. Standards and Technology, 'IEC 61850 Objects/DNP3 Mapping', *NIST*, vol. 20090730, 2009.

[42] B. Kasztenny, 'IEC 61850 : A Practical Application Primer for Protection Engineer', in *60th Annual Georgia Tech Protective Relaying Conference*, pp. 1–60, 2006.

[43] D. Choi, H. Kim, D. Won and S. Kim, 'Advanced key-management architecture for secure SCADA communications', *IEEE Trans. Power Deliv.*, vol. 24, no. 3, pp. 1154–1163, 2009, doi:10.1109/TPWRD.2008.2005683

[44] R. Aguilar and J. Ariza, *Testing and Configuration of IEC 61850 Multivendor Protection Schemes*. New Orleans, 2010.

[45] IEEE, 'IEEE 802.15.4-2020 – IEEE Standard for Low-Rate Wireless Networks', 2020.

[46] IEEE, 'IEEE Std. 802.11-2020 – IEEE Standard for Information Technology – Telecommunications and Information Exchange between Systems – Local and Metropolitan Area Networks – Specific Requirements. Part 11: Wireless LAN Medium Access Control (MAC) and Physical Layer (PHY) Specifications', 2020.

[47] ISO/IEC, ISO/IEC 7498-1:1994. 'Information Technology Open Systems Interconnection Basic Reference Model: The Basic Model', 1994.

[48] A. Wright, 'Application of Fuses to Power Networks', *Power Eng. J., UK*, vol.1991, no. 4, pp. 129–134, 1991.
[49] A. Wright and P. G. Newberry, *Electric Fuses*, 3rd ed., London: IET, 2004.
[50] J. Momoh, *Electric Power Distribution, Automation, Protection, and Control*, 1st ed. Boca Ratón, FL: CRC Press, 2008.
[51] IEE Conference Publication, 'Developments in Power System Protection', in *7th International Conference*, p. 479, 2001.
[52] J. M. Gers and E. J. Holmes, *Protection of Electricity Distribution Networks*, 4th ed. London: IET, 2021.
[53] T. Davies, *Protection of Industrial Power Systems*, 2nd ed. Oxford: Butterworth-Heinemann, 1998.
[54] Cooper Power Systems, Inc., *Electrical Distribution-System Protection*, 3rd ed. Pittsburgh, PA: Cooper Power Systems, 1990.
[55] BASLER ELECTRIC, 'Protection and Control Devices Standards, Dimensions and Accessories', *Product Bulletin*, vol. SDA-5, pp. 1–24, 2001.
[56] Alstom, *Network Protection & Automation*, 1st ed. Barcelona: Cayfosa, 2002.

Chapter 7
Regulatory issues, VVC and optimal load flow for urban grids

Keeping voltages within predefined ranges is one of the main targets in the operation of power systems. To achieve this goal, reactive flow must be monitored since the two variables are closely related as mentioned in previous chapters. Therefore, it is common to treat both topics simultaneously.

In recent years, it was introduced the term 'VVC', which stands for volt/VAR control. This refers to the technique of using voltage regulating devices and reactive power controls to maintain voltage levels within the accepted ranges at all points of the distribution system under all loading conditions. Modern software techniques and the progress on communication technologies have brought improvements in the service quality in most utilities throughout the world.

Voltage regulation and reactive power control are performed by switching capacitor banks installed along the lines, substation transformers load tap changers (LTCs), shunt capacitor banks installed at substations and voltage regulators. The control of the devices is discussed later in this chapter.

The main objectives of implementing VVC are summarised as follows:

- Maintain acceptable voltages at all points along the feeder under all loading conditions.
- Increase the efficiency of distribution systems without violating any loading and voltage constraints.
- Support the reactive power needs of the bulks power system during system emergencies.
- Keep power factor within the accepted ranges which normally are above 0.9 inductive.

This term should not be confused with 'VVO', which stands for 'Volt/Var Optimisation'. While in VVC, the controls are manipulated in the direction of expected improvement (based on statistical or experiential), VVO gets an explicit representation of business objectives (reflected into a loss or demand minimisation), which requires to manipulate the controls to find the best control objective without any preconceived direction. VVC is focused on local targets for values of voltage, current, var or power factor. VVO uses available data to produce better results.

Voltage values are the nominal single-phase supply voltages. All values are given in rms and the peak AC voltage is greater by a factor of $\sqrt{2}$. In most countries if not all, they are in the range 100–240 V.

For example, the normal rated voltage is 120 V according to ANSI C84.1-2016 [1]. Figure 7.1 shows a maximum of 126 V and a minimum of 114 V since the voltage regulation is 5%. In the United Kingdom, BS 7671:2018+A2:2022 the statutory limits of the LV supply are 230 V minus 6% and 230 V plus10%. This represents a range from 216.2 to 253 V [2].

Figure 7.2 shows a portion of a typical system that illustrates the effect of the loading of a feeder on the voltage profile which risks the fulfilment of the voltage ranges specified by the standard.

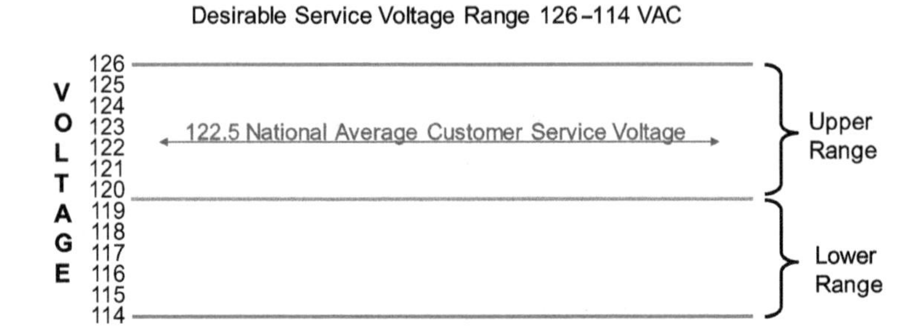

Figure 7.1 Rated voltages specified for the United States

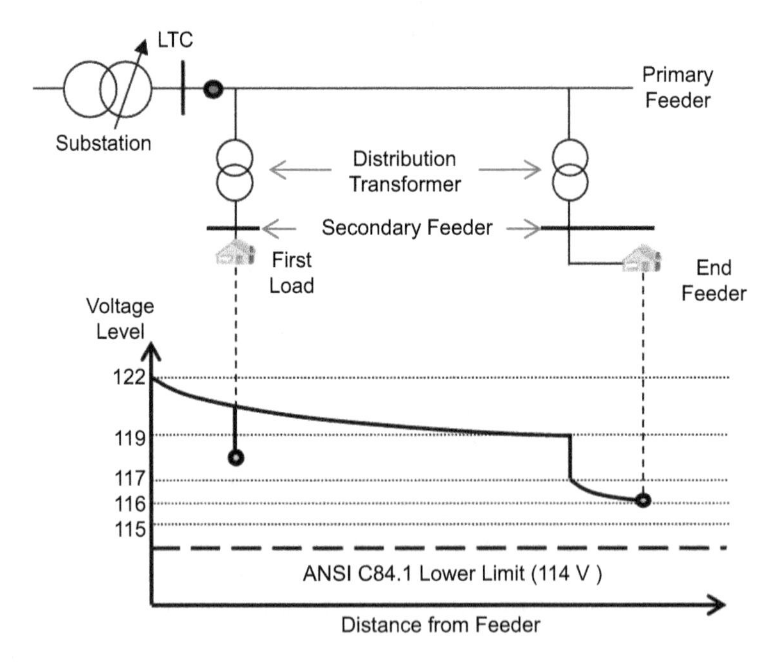

Figure 7.2 Voltage regulation limits in the United States

Different devices are required to achieve the VVC including on LTCs at the main substation transformers, voltage regulators and capacitor banks. Distributed energy resources (DER) can also help the VVC on those feeders where they are deployed. FACTS equipment (like STATCOM) and distributed generation (DG) if available should be included under the VVC analysis and be part of it.

Certain distribution utilities commonly supply voltage at the higher end of the allowed range. This poses a challenge for the implementation of conservation voltage reduction (CVR), a concept aimed at sustaining the voltage supplied to customers within the lower limit of the permissible voltage range [3,4].

CVR proposes that a minor reduction in voltage at the customer's end, while still adhering to the prescribed standards, can lead to energy preservation and enhanced efficiency. This reduction helps not only in lowering peak demand but also in minimising power losses. A drop of voltage of 1.0% at the substation level can lead to an approximate 1.0% reduction in load.

To perform voltage reduction at the substation side, the feeder must have a voltage profile sufficiently above the lower voltage level if no compensation is applied. Otherwise, the voltage at the end side will be out of the allowable limits during peak condition.

Therefore, implementing CVR requires information like load characteristics, load profiles, feeder characteristics, and if applicable, size and location of capacitor banks, voltage regulators and DER technologies. Typically, feeders based on constant impedance load models can have losses reduction. That is not the case for constant power loads, where the impact would be the opposite since the current can increase due to the voltage reduction.

A common application of CVR for heavily loaded feeders with a significant voltage drop at the feeder's end is installing capacitor banks, or other voltage regulation technologies along the distribution feeder. In this way, the voltage profile is supported within acceptable limits, allowing for substation voltage reduction when required.

7.1 Definition of voltage regulation

Voltage regulation can be defined as the percent voltage drop of a line (e.g., a feeder) with respect to the receiving-end voltage.

Therefore:

$$\% \ regulation = \frac{|V_s| - |V_r|}{|V_r|} x100 \tag{7.1}$$

where:

$|V_s|$ is the sending voltage

$|V_r|$ is the receiving-end voltage

7.2 Options to improve voltage regulation

There are numerous ways to improve the distribution system's overall voltage regulation:

- Use of generator voltage regulators
- Application of voltage-regulating equipment in the distribution substations
- Application of capacitors in the distribution substation
- Balancing of the loads on the primary feeders
- Increasing feeder conductor size
- Changing feeder sections from single-phase to multiphase
- Transferring loads to new feeders
- Installing new substations and primary feeders
- Increasing of primary voltage level
- Application of voltage regulators out on the primary feeders
- Application of shunt capacitors on the primary feeders
- Application of series capacitors on the primary feeders
- Use of DG or DER

The commonest way to regulate voltage in power systems is by means of tap changers in transformers. Normally, they are used to control the voltage level at the LV side of transformers associated with transmission or distribution substations. They could control voltage at the HV side in generator plants. However, this is not usual since in this case the voltage is controlled by the excitation of the generators.

Tap changers can be of the no-load operating type NLTC or on-load operation type OLTC. The first one is used mainly on small transformers or non-important transformers. OLTC tap changers could be mechanical type or thyristor assisted. Mechanical tap changers physically make a connection before releasing the previous connection point employing different tap selectors. They avoid the presence of elevated circulating currents through a diverter switch that provides a temporary high impedance value in series with the short-circuited winding section.

Tap changers assisted by thyristors are used to take the load current while the main contacts switch from one source to the other. This prevents arcs in the main contacts and can result in an extended life cycle in the maintenance activities. They require an additional low voltage supply for the thyristor circuit. This type of tap changers is not very popular due to their complexity and the elevated cost.

The most economical way of regulating the voltage along the feeders within the required limits is to apply both step voltage regulators and shunt capacitors. Of course, a fixed capacitor is not a voltage regulator and cannot be directly compared to regulators; however, in some cases, automatically switched capacitors can replace conventional step-type voltage regulators for voltage control on distribution feeders. The following sections make specific reference to voltage regulators and capacitors.

7.3 Voltage regulators

A step-type voltage regulator is fundamentally an autotransformer with many taps (or steps) in the series winding that adjusts itself automatically by changing the taps until the desired voltage is obtained. It can be either station-type, distribution-type.

Voltage regulators referred to in [5] are designed to correct the line voltage from 100% boost to 10% buck (i.e. 10%) in 32 steps, with a 5% or 8% voltage change per step as shown in Figure 7.3. The effect of each step is shown in Figure 7.4.

The voltage regulators can be located at the substation busbar or on the feeders. The settings are calculated such as to keep the voltage within certain ranges at a regulating point along the feeders. The procedure to determine the settings has been very well treated in the book *Electric Power Distribution System Engineering* by Turan Gönen [6], from which (7.2)–(7.7) and the corresponding explanations have been taken.

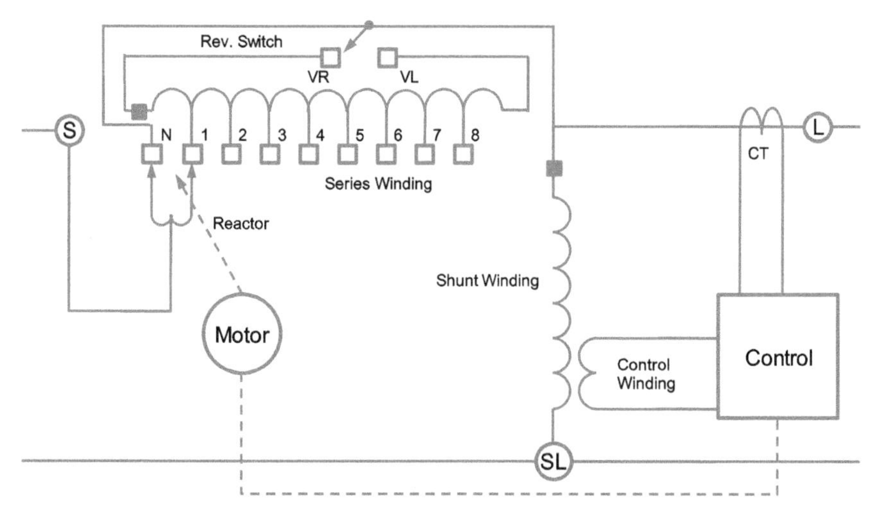

Figure 7.3 Schematic of a single-phase 32-step voltage regulator

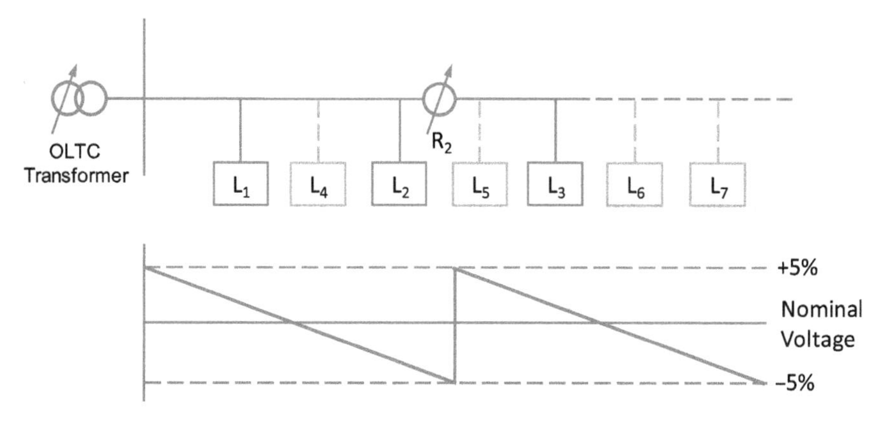

Figure 7.4 Voltage profile with step-type voltage regulators

The settings correspond to the dial of the resistance and reactance elements built in the so-called line-drop compensator (LCD) which is in the control panel of the device.

In case there is no load tapped off between the regulator and the controlled point, then the R dial setting of the LCD is determined from:

$$R_{set} = \frac{CT_P}{PT_N} x R_{eff} \tag{7.2}$$

where:

CT_P is the rating of the current transformer's primary

PT_N is the potential transformer's turns ratio ($V_{pri}/V_{\sec}$)

R_{eff} is the effective resistance of a feeder conductor from regulator station to regulation point, Ω

$$R_{eff} = r_a x \frac{l - s_1}{2} \tag{7.3}$$

where:

r_a is the resistance of a feeder conductor from regulator station to regulation point, Ω/mi per conductor

s_1 is the length of three-phase feeder between regulator station and substation, mi (multiply length by 2 in feeder is in single phase)

l is the primary feeder length, mi

Also, the Z dial setting of the LCD can be determined from:

$$X_{set} = \frac{CT_P}{PT_N} x X_{eff} \tag{7.4}$$

where:

X_{eff} is the effective reactance of a feeder conductor from regulator to regulation point,

$$X_{eff} = x_L * \frac{l - s_1}{2} \tag{7.5}$$

and,

$$x_L = x_a + x_d \ \ [\Omega/mi] \tag{7.6}$$

where:

x_a is the inductive reactance of individual phase conductor of feeder at 12-in spacing

x_d is the inductive reactance spacing

x_L is the inductive reactance of feeder conductor

The difference between the two voltage values is the total voltage drop between the regulator and the regulation point, which can also be defined as:

$$VD = |I_L| \ x \ R_{eff} x \cos\theta + |I_L| \ x \ X_{eff} \ x \sin\theta \tag{7.7}$$

from which R_{eff} and X_{eff} values can be determined easily if the load power factor of the feeder and the average R/X ratio of the feeder conductors between the regulator and the regulation point are known.

Automatic voltage regulation is provided by bus regulation at the substation, or individual feeder regulation in the substation, or supplementary regulation in the main by regulators mounted on poles or a combination of the above.

7.4 Capacitor application in distribution systems

Generally, inductive and resistive elements of the networks cause active power losses and voltage drop. Capacitor location in distribution networks is a very important option to reduce electrical losses and therefore it has been widely used. This is possible by reducing the reactive power component of the current.

Capacitors not only help to save losses but also play an important role in power factor correction and in the improvement of the voltage profile especially in long feeders. Even with the benefits of capacitor location in distribution systems, their location must be analysed carefully because of the high costs involved and the overvoltage that may be produced in networks with harmonic circulation when there are resonance conditions.

A capacitor is a device consisting essentially of two electrodes separated by a dielectric insulating material that is capable of supplying magnetising kVAR to the system. The capacitive reactance has the nature of a negative inductive reactance. This property is utilised in electrical circuits to balance the effects of inductive reactance and the lagging reactive kVA of inductive loads.

Capacitors can be classified as series or parallel according to the type of connection they have. Series capacitors are connected in series with lines to compensate for inductive reactance. Shunt capacitors are connected in parallel with lines to balance the reactive power or current required by an inductive load.

Example 7.1

Consider a simple 11 kV radial line transmitting power by an overhead system to a lagging power factor load. Figure 7.5 presents the system diagram and the equivalent circuit.

Analyse the sending and receiving-end conditions for each of the following three cases:

- Without capacitors: the line-to-line receiving-end voltage is assumed constant at 11 kV. All calculations are referred to as 11 kV base voltage.
- With shunt capacitors: three single-phase capacitors units, each having a reactance of 45.7 Ω, are connected between phase and neutral adjacent to the load. Calculate the current taken by each capacitor, I_C, I_R and E_s.
- With series capacitors: in this case, a capacitor unit is connected in series with each phase, the reactance of each unit being 5 Ω. Calculate I_R and E_s.

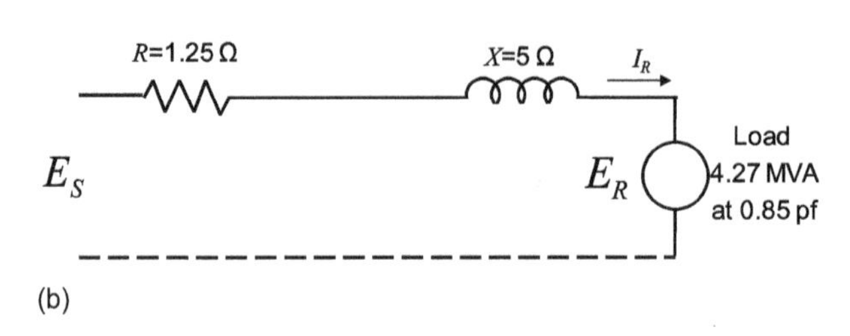

Figure 7.5 Diagrams for Example 7.1: (a) System diagram and (b) Equivalent circuit

For case (a):

The line-to-line receiving-end voltage is assumed constant at 11 kV. All calculations are referred to as 11 kV base voltage.

$$I_R = \frac{4.27\ \text{MVA}}{\sqrt{3} \cdot 11\ \text{kV} \cdot 0.85} \angle\left(-\cos^{-1}(0.85)\right) = 264\angle(-31.78^\circ)\ \text{A per phase}$$

Taking E_R as reference vector,

$$\begin{aligned} E_S &= E_R + \left[\sqrt{3}(I_R\angle(\phi_R))\ (R + jX_L) \cdot 10^{-3}\right]\ \text{kV} \\ &= 11 + \left[\sqrt{3}(264\angle(-31.78^\circ)) \cdot (1.25 + j5) \cdot 10^{-3}\right]\ \text{kV} \\ &= 11 + [2.36\angle(44.18^\circ)]\ \text{kV} \\ &= 12.79\angle 7.37^\circ\ \text{kV} \end{aligned}$$

Figures 7.6 and 7.7 show the vector diagram for the current case. That is the base case for the other points.

For case (b):

Three single-phase capacitors units, each having a reactance of 45.7 Ω, are connected between phase and neutral adjacent to the load.

Then, current taken by each capacitor:

$$I_C = \frac{11,000}{\sqrt{3} \cdot 45.7} \angle(90^\circ)\ \text{A.}$$

$$I_C = 139\angle(90^\circ)\ \text{A. leading the applied voltage by } 90^\circ$$

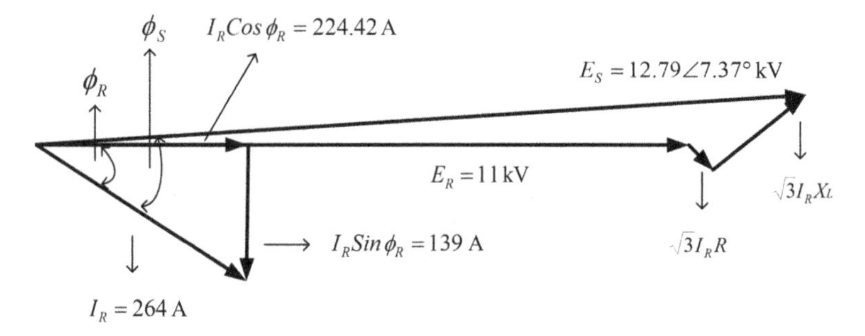

Figure 7.6 Voltage and current vector diagram for the base case

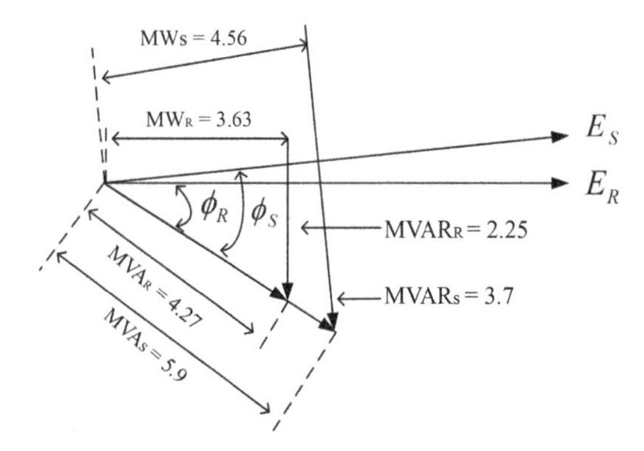

Figure 7.7 Power diagram for the base case

The inductive component of the load current:

$$= 264 \sin(31.78°)\angle(-90°)$$
$$= 139\angle(-90°) \text{ lagging the applied voltage by } 90°$$

Therefore, the current taken by the capacitor will neutralise the inductive component of the load current and the actual current taken from the line will be only 224.42 A, in phase with the applied voltage:

$$I_R = 224.42\angle(0°)$$
$$E_S = 11 + \left[\sqrt{3} \cdot 224.42 \cdot (1.25 + j5) \cdot 10^{-3}\right] \text{ kV}$$
$$E_S = 11.65\angle(9.6°) \text{ kV}$$

For case (c):

In this case, a capacitor unit is connected in series with each phase, the reactance of each unit being 5Ω.

$$I_R = 264\angle(-31.78°)$$

$$
\begin{aligned}
E_S &= E_R + \left[\sqrt{3}(I_R\angle(\phi_R))(R + j(X_L - X_C)) \cdot 10^{-3}\right] \text{kV} \\
&= 11 + \left[\sqrt{3}(264\angle(-31.78^\circ)) \cdot (1.25 + j(5 - 5)) \cdot 10^{-3}\right] \text{kV} \\
&= 11 + [0.57\angle(-31.78^\circ)] \text{ kV} \\
&= 11.48\angle(-1.5^\circ) \text{ kV}
\end{aligned}
$$

For the purpose of comparison, the relevant sending and receiving-end conditions for the three cases are tabulated in Table 7.1.

After the information shown in Table 7.1, the following results can be highlighted:

1. Both shunt and series capacitors reduce the voltage drop, and the MVAR demand at source of supply.
2. The MVAR rating or the shunt capacitor bank for the three phases.

$$
\begin{aligned}
&= 3 \cdot I_C{}^2 \cdot X_C \cdot 10^{-6} \text{ MVAR} \\
&= 3 \cdot (224.42)^2 \cdot 45.7 \cdot 10^{-6} \text{ MVAR} \\
&= 6.9 \text{ MVAR}
\end{aligned}
$$

   The MVAR rating of the series capacitor bank

$$
\begin{aligned}
&= 3 \cdot (264)^2 \cdot 5 \cdot 10^{-6} \text{ MVAR} \\
&= 1.05 \text{ MVAR}
\end{aligned}
$$

   Thus, for the same improvement in voltage regulation, the series capacitor bank is much smaller than the shunt capacitor bank. At the same time, it should be noted that series capacitors reduce MVAR demand at source far less than shunt capacitors do.
3. Shunt capacitors reduce the line power losses by reducing the receiving-end current.

Table 7.2 shows a comparison between the shunt capacitor and the series capacitor applications.

7.4.1 Feeder model

A general model is considered here, which corresponds to a distributed load along the feeder with the possibility of having a concentrated load at the end whose value can also be set to zero, as proposed by Neagle and Samson in [7]. If the total feeder current at the substation end is I_1 and the concentrated load at the other end takes a current I_2, the expression for the current along the feeder depends on the distance from the substation according to the following equation:

$$i(x) = I_1 - (I_1 - I_2)x \tag{7.8}$$

If a factor $p = \frac{I_2}{I_1}$ is introduced, the equation becomes:

$$i(x) = I_1[(p - 1)x + 1] \tag{7.9}$$

The profile corresponding to that expression is shown in Figure 7.8 If the load is uniformly distributed along the feeder and there is not concentrated load at the

Table 7.1 Results comparison for Example 7.1

Type of system	Line voltage		Voltage drop	Power factor		Sending-end power			Receiving-end power			Active power loss
	Sending end kV	Receiving end kV		Sending end	Receiving end	Active power MW	Reactive power MVAR	Apparent power MVA	Active power MW	Reactive power MVAR	Apparent power MVA	
Case A without capacitors	12.79	11	1.8 kV (14.0%)	0.78 lag	0.85 lag	4.56	3.7 lag	5.9	4.3	2.7 lag	5.1	0.8 MW (13.6%)
Case B with shunt capacitors	11.65	11	1.65 kV (5.9%)	0.986 lag	1	4.47	0.76 lag	4.53	4.3	Nil	4.3	0.17 MW (3.9%)
Case C with series capacitors	11.48	11	0.48 kV (4.4%)	0.864 lag	0.85 lag	4.53	2.65 lag	5.3	4.3	2.7 lag	5.1	0.23 MW (5.35%)

Table 7.2 Comparison between shunt capacitor and series capacitor

	Shunt capacitor	**Series capacitor**
Size	45.7Ω	5Ω
Voltage regulation	Good	Good
Losses reduction	Very good	Nothing
Power factor	Very good	Fair
Stability	Good	Very good
Cost	Reasonable	High

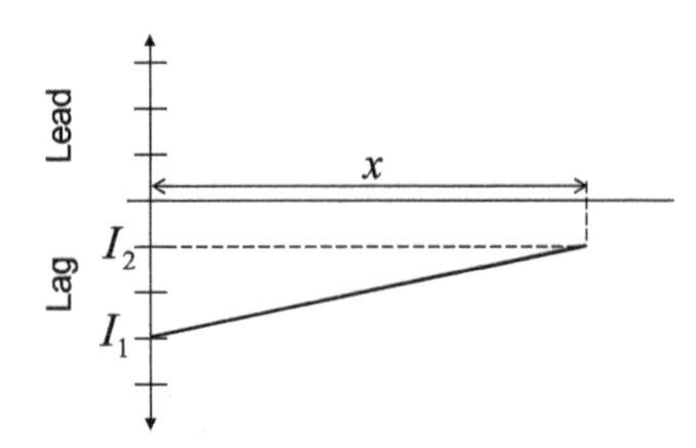

Figure 7.8 Current profiles for feeders with uniformly distributed loads

end, I_2 and therefore p is equal to zero, and (7.9) becomes

$$i(x) = I_1(1 - x) \tag{7.10}$$

Likewise, if the load is concentrated at the end of the feeder, $p = 1$ and therefore the expression for the current is:

$$i(x) = I_1 = I_2 \tag{7.11}$$

7.4.2 Capacitor location and sizing

The reduction in losses by using network reconfiguration can be further enhanced by placing capacitors along the feeders [8,9]. The following sections will develop theories for such applications which have been proposed in several works. The different magnitudes, i.e., power, energy, impedance and current, will be dealt with in pu values.

The power loss dissipated in a circuit with a total current I, a resistance value R and a power factor angle ϕ can be expressed as:

$$P_1 = I^2R = (I\cos\phi)^2R + (I\sin\phi)^2R \tag{7.12}$$

If a capacitor with a current I_c is installed, the reactive part of the current will be compensated and therefore the new total losses (P_2) can be expressed as:

$$P_2 = I^2R = (I\cos\phi)^2R + (I\sin\phi - I_C)^2R \tag{7.13}$$

Therefore, the losses reduction can be found as:

$$\Delta P = P_1 - P_2 = 2IRI_C \sin\phi - I_C{}^2R \tag{7.14}$$

Equation 7.14 shows that only the reactive component of the load current is required for power loss reduction studies. Therefore, in the rest of this section, that component will be referred to as I_{1R}.

If a feeder has a uniformly distributed loading whose current is that given by (7.9), the total losses due to the *reactive* component without capacitors will again be referred to as P_1 and can be found with the following equation:

$$P_1 = 3\int_0^1 [I_{1R}(1 + (p-1)x)]^2 R dx \tag{7.15}$$

where:

$p = I_{2R}/I_{1R}$

I_{1R} is the reactive current at the substation end

I_{2R} is the reactive current at the feeder end

x is the distance along the feeder ranging from 0 to 1

The result is given in the following equation:

$$P_1 = RI_{1R}{}^2\left(p^2 + p + 1\right) \tag{7.16}$$

7.4.3 Reduction in power losses with one capacitor bank

Figure 7.9 shows a distribution feeder with one capacitor bank installed at a distance a from the source.

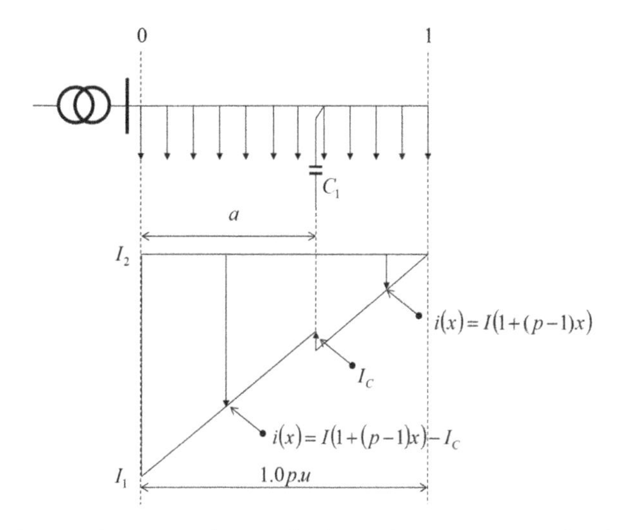

Figure 7.9 Distribution feeder with one capacitor bank

If the capacitor bank takes a current I_c as shown in Figure 7.9, the new total power losses P_2, due to the reactive current including the capacitor bank, are calculated as:

$$P_2 = 3\int_0^a [I_{1R}(1 + (p-1)x) - I_C]^2 R dx + 3\int_a^1 [I_{1R}(1 + (p-1)x)]^2 R dx \tag{7.17}$$

The result of this expression is:

$$P_2 = RI_{1R}^2(p^2 + p + 1) - 3RI_{1R}^2\left[(p-1)\frac{I_C}{I_{1R}}a^2 + 2\frac{I_C}{I_{1R}}a - \left(\frac{I_C}{I_{1R}}\right)^2 a\right] \tag{7.18}$$

The power loss reduction due to the capacitor installation is found by subtracting (7.18) from (7.16) as follows:

$$\Delta P = P_1 - P_2 = 3RI_{1R}^2\left[(p-1)\frac{I_C}{I_{1R}}a^2 + 2\frac{I_C}{I_{1R}}a - \left(\frac{I_C}{I_{1R}}\right)^2 a\right] \tag{7.19}$$

The optimum sizing and location of one capacitor bank to reduce power losses is found by taking partial derivatives of (7.19) with respect to I_c and the distance a, as follows:

$$\frac{\partial \Delta P}{\partial I_C} = 3RI_{1R}^2\left[(p-1)\frac{1}{I_{1R}}a^2 + 2\frac{1}{I_{1R}}a - 2\frac{I_C}{I_{1R}^2}a\right] \tag{7.20}$$

$$\frac{\partial \Delta P}{\partial a} = 3RI_{1R}^2\left[2(p-1)\frac{I_C}{I_{1R}}a + 2\frac{I_C}{I_{1R}} - \left(\frac{I_C}{I_{1R}}\right)^2\right] \tag{7.21}$$

By equating (7.20) and (7.21) to zero and solving simultaneously for I_c and a, the following values are obtained:

$$I_C = \frac{2}{3}I_{1R} \tag{7.22}$$

$$a = \frac{2}{3(1-p)} \tag{7.23}$$

The maximum power loss reduction with one capacitor is obtained by replacing (7.22) and (7.23) into (7.19). The corresponding expression is:

$$\Delta P_{MAX} = R\,I_{1R}^2\frac{8}{9(1-p)} \tag{7.24}$$

Equations 7.23 and 7.24 are valid only if $p \leq 1/3$. For larger values of p, the magnitude of a becomes greater than 1 or even negative, which does not have physical meaning. Therefore, for p greater than 1/3, the capacitor bank should be located at the end of the feeder which corresponds to a value of the variable a equal to 1. If this value is introduced in (7.19), the value of the capacitor current is:

$$I_c = \frac{1}{2} I_{1R}(p + 1) \quad \text{for } p > 1/3 \tag{7.25}$$

Likewise, if the value of I_c given in (7.23) and $a = 1$ are introduced in (7.19), the reduction in losses when a capacitor bank is installed and $p>1/3$ is given by the following equation:

$$\Delta P = \frac{3}{4} R I_{1R}^2 (p + 1)^2 \quad \text{for } p > 1/3 \tag{7.26}$$

7.4.4 Reduction in power losses with two capacitor banks

Figure 7.10 shows a distribution feeder with two capacitor banks, one installed at a distance a and the other a distance b from the source.

Initial losses:

$$P_1 = 3\int_0^1 [I(1-x)]^2 R dx = I^2 R \tag{7.27}$$

Losses if capacitors are connected:

$$\begin{aligned} P_2 &= 3\int_0^a R[I(1-x) - 2I_C]^2 dx + 3\int_a^b R[I(1-x) - I_C]^2 dx + 3\int_b^1 R[I(1-x)]^2 dx \\ &= 3R\left(II_C a^2 - 2II_C a + 3{I_C}^2 a - 2II_C b + II_C b^2 + {I_C}^2 b + \frac{I^2}{3}\right) \end{aligned} \tag{7.28}$$

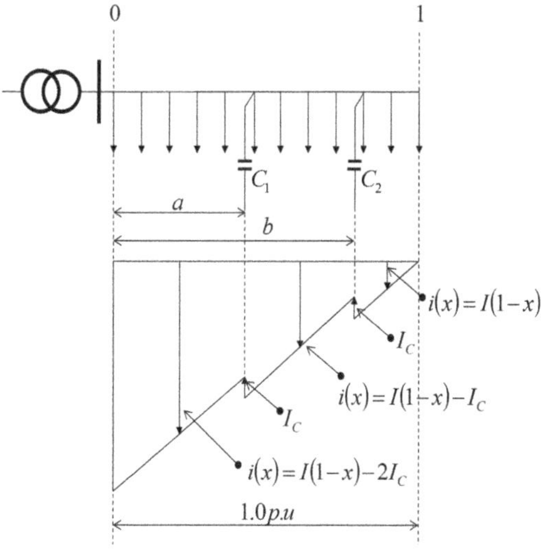

Figure 7.10 Distribution feeder with two capacitor banks

Therefore, the losses reductions are expressed as:

$$\Delta P = 3R\left(a\left(2II_C - II_C a - 3I_C{}^2\right) + b\left(2II_C - II_C b - I_C{}^2\right)\right) \tag{7.29}$$

It can be demonstrated that the optimal values of a and b are:

$$a = 1 - \frac{3I_C}{2I} \tag{7.30}$$

$$b = 1 - \frac{I_C}{2I} \tag{7.31}$$

7.4.5 Losses reduction with three capacitor banks

$$\Delta P = 3R\left(a\left(2II_C - II_C a - 5I_C{}^2\right) + b\left(2II_C - II_C b - 3I_C{}^2\right) + c\left(2II_C - II_C c - I_C{}^2\right)\right) \tag{7.32}$$

The optimum localisation of capacitors:

$$a = 1 - \frac{5I_C}{2I} \tag{7.33}$$

$$b = 1 - \frac{3I_C}{2I} \tag{7.34}$$

$$c = 1 - \frac{I_C}{2I} \tag{7.35}$$

7.4.6 Consideration of several capacitor banks

The expressions presented in (7.19), (7.29) and (7.32) can be generalised for any feeder with n sections to find the power and energy loss reductions when capacitors are located in different places.

For more than one segment on the feeder, expressions (7.19), (7.29) and (7.32) are still valid, although new techniques that consider heuristic strategies have been proposed. Under these circumstances, the expressions can be used by putting $a = 1$, which is then directly applicable to one segment at a time. Therefore, when a capacitor is installed downstream at a segment n, the reduction in the power losses is given as a function of that capacitor current, and of currents at the beginning and at the end of the segment, respectively, as follows:

$$\Delta P_m = P_{1m} - P_{2m} = 3\,R_m\left[\left(I_{1Rm} + I_{2Rm}\right) I_{cn} - I_{cn}^2\right] \tag{7.36}$$

For the segment where a capacitor is located, the expressions for power and energy have already been developed. General expressions are obtained by considering the effect of one capacitor at a time for every section towards the source as follows:

$$\Delta P = \Delta P_n + \sum_{m=1}^{n-1} \Delta P_m \tag{7.37}$$

That equation means that for a capacitor installed in a segment n, the power loss reduction is:

$$\Delta P = 3\,R_n\,I_{1Rn}^2\left[(p-1)\frac{I_{cn}}{I_{1Rn}}\,a^2 + 2\,\frac{I_{cn}}{I_{1Rn}}\,a - \left(\frac{I_{cn}}{I_{1Rn}}\right)^2 a\right] + \sum_{m=1}^{n-1} 3\left[(I_{1Rm} + I_{2Rm})\,I_{cn} - I_{cn}^2\right] R_m \tag{7.38}$$

Likewise, if a number of k banks are considered:

$$\Delta P = \sum_{n=1}^{k}\left\{3\,R_n\,I_{1Rn}^2\left[(p-1)\frac{I_{cn}}{I_{1Rn}}\,a^2 + 2\,\frac{I_{cn}}{I_{1Rn}}\,a - \left(\frac{I_{cn}}{I_{1Rn}}\right)^2 a\right] + \sum_{m=1}^{n-1} 3\left[(I_{1Rm} + I_{2Rm})\,I_{cn} - I_{cn}^2\right] R_m\right\}$$

7.4.7 Modern techniques in capacitor placement and sizing

Regarding the motivation and philosophy behind Volt/VAR control and capacitor placement, the issue under analysis and of particular importance is the installation of these devices. These analyses lead to energy loss reductions, increase of feeder transport capacities, improvement of voltage profiles and network stability [10–13].

Hence, the problem of optimal capacitor placement involves identifying suitable installation points, the dimension of each capacitor bank to be installed and a proper utilisation and operation of existing capacitor banks at different voltage levels, which turns the capacitor banks placement issue into a matter of capacitor banks management.

For many years, these studies have been carried out with different analytical methodologies as the one described above with satisfactory results. Nonetheless, due to the variation, modern equipment's and load's behaviour of urban distribution networks, the problem of determining the location and optimal size of the capacitor becomes a nonlinear and high-order optimisation problem [14–17]. Thus, non-linear methods, also known as meta-heuristic algorithms, represent an adequate approach.

As in every real-life application problem, several evaluation aspects criteria need to be considered simultaneously to appraise the merit of potential solutions. In this matter, multi-objective models have been developed incorporating different objectives that must be weighed by planning engineers as well as operational constraints that in the first applications, some approaches used mathematical models with unrealistic assumptions for the sake of computer tractability.

In simple terms, as expressed in [18], an objective function can be seen as the minimisation of both investment costs and active losses considering the injection of reactive power and specifying suitable bus positions for their integration.

$$f_{obj} = \min\left\{k^P \cdot P_{loss}^{Tot} + \sum_{j=1}^{NB} k_j^C \cdot Q_j^C\right\}$$

where k^P represents the annual cost of power losses in \$/kW, P_{loss}^{Tot} the overall power losses in kW, k_j^C the annual cost of a capacitor bank in \$/kVAR; Q_j^C the capacitor size installed at bus j, evaluated along all the number of candidate buses (NB).

The objective function is subjected to the following equalities and inequalities that constrain the operation of the capacitor banks. It's worth noting that variations in the objective function and constraints may take place bearing in mind the main objective and different network considerations.

$$\sum_{i=1}^{NG} P_{G,i} = \sum_{i=1}^{NL} P_{D,i} - \sum_{i=1}^{NB} P_{loss,i}$$

$$\sum_{i=1}^{NG} Q_{G,i} = \sum_{i=1}^{NL} Q_{D,i} - \sum_{i=1}^{NB} Q_{loss,i} - \sum_{i=1}^{NC} Q_{C,i}$$

Evaluated by taking into account the generated power, demand, losses and shunt capacitors balance.

And for the voltage limits, transmitted power, real power generation and shunt capacitor bank limits, respectively:

$$V_{i,\min} \leq V_i \leq V_{i,\max}, \quad for\ i = 1 \ldots N$$

$$S_k \leq S_{k,\max}, \quad for\ i = 1 \ldots N$$

$$P_{G,i,\min} \leq P_{G,i} \leq P_{G,i,\max}, \quad for\ i = 1 \ldots NG$$

$$\sum_{j=1}^{NC} Q_j^C \leq Q_{\max}^C, \quad for\ i = 1 \ldots NC$$

In the literature, different robust optimisation approaches encompassing classical and metaheuristic methods can be found and well-referenced with their applications in power and energy systems, being their performance rigorously evaluated under IEEE networks for instance. Some references in this matter can be found in [18–22].

Some of the algorithms reported in the literature are:

- Simulated annealing
- Ant colony optimisation
- Harmony algorithm
- Genetic algorithm (GA)
- Particle swarm optimisation (PSO) algorithm
- Gravitational search algorithm
- Fuzzy harmony search algorithm
- Modified honeybee mating optimisation evolutionary algorithm
- Hybrid differential evolutionary and PSO algorithm

Some of them are assessed to cope with distribution network applications under different load conditions and even harmonic flows due to the widespread expansion of nonlinear loads.

7.5 Modelling of distribution feeders including VVC equipment

Modelling of distribution feeders involving the elements mentioned above can be a comprehensive task due, particularly to the non-symmetry that distribution systems have. For example, models for load flow in distribution circuits allow identifying points with low or high voltage levels, to take appropriate corrective actions, which may include the operation of VVC devices. In addition, load flow contributes to the optimal feeder's reconfiguration in order to reduce losses, which is a constraint in switching optimisation algorithms [23–31].

A small system taken from IEEE will be considered to illustrate the modelling [32]. The data from the model is introduced in the illustration presented below. The system has the following characteristics:

1. Short and relatively highly loaded for a 4.16 kV feeder
2. One substation voltage regulator consisting of three single-phase units connected in wye-configuration
3. Overhead and underground lines with variety of phasing
4. Shunt capacitor banks
5. In-line transformer
6. Unbalanced spot and distributed loads

Figure 7.11 presents a radial feeder taken from IEEE 13 Node Test Feeder that will be used to illustrate the modelling and the type of analysis that can be performed.

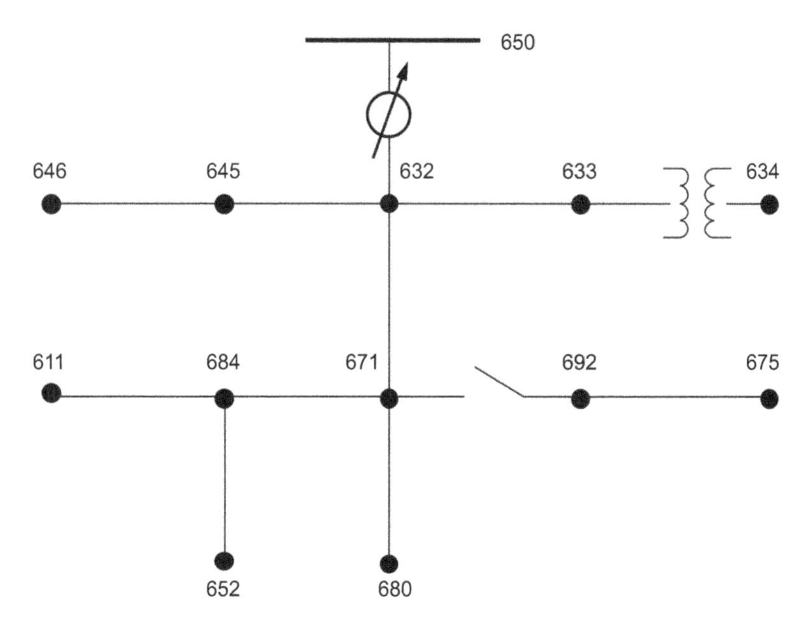

Figure 7.11 Layout of the feeder used in the modelling [32]

Figure 7.12 shows the modelling of the voltage regulator installed in the system considered, taken from NEPLAN Resources [33]. The modelling requires a previous analysis of the device used and the setting of the taps, based on the system requirements.

Figure 7.13 shows the load flow results for the system modelled, including voltage levels at the different nodes and active and reactive flow through all the elements.

Example 7.2

A distribution network composed of five feeders associated with the substations Sub 1, Sub 2 and Sub 3. Determine the required capacitors and their locations considering the topology changes to improve the losses and the voltage profile. Figure 7.14 illustrates the georeferenced feeders.

The network has the following number of elements: 5295 nodes, 3916 lines, 1069 transformers, 1078 loads and 3 network equivalents. The Substation Sub 1 feeds the FEEDER 5 and the Substation Sub 2 feeds the FEEDER 4. Three feeders are connected to the Substation Sub 3: FEEDER 1, FEEDER 2 and FEEDER 3. The network corresponds to a symmetrical distribution system. All the loads are modelled as PQ elements, and they are connected to transformers of 13.2/0.208 kV. The Slack nodes are set in the main substations and the topologies are described in Table 7.3.

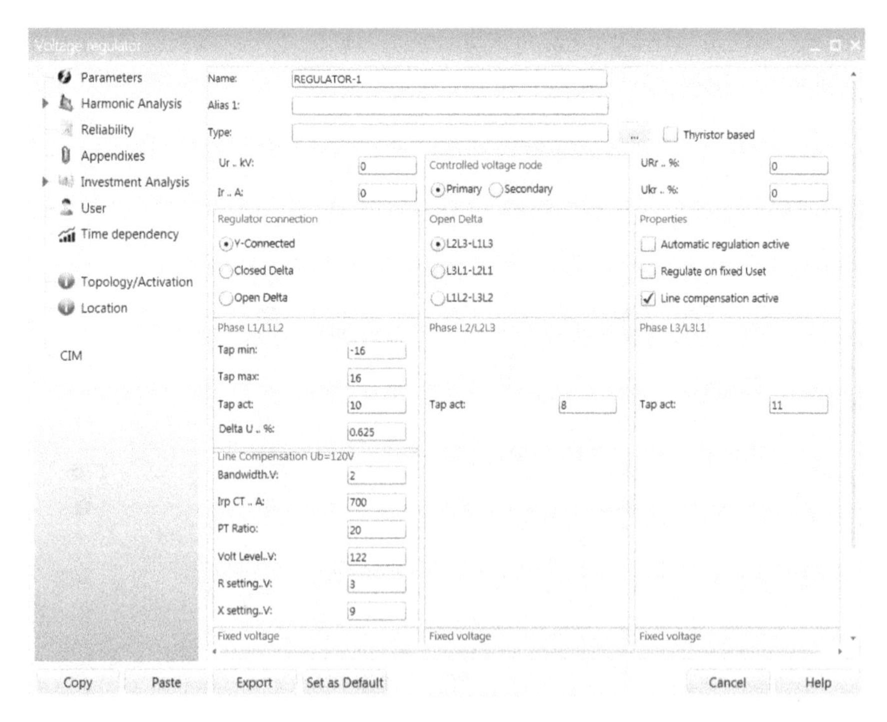

Figure 7.12 Typical voltage regulator modelling input data [33]

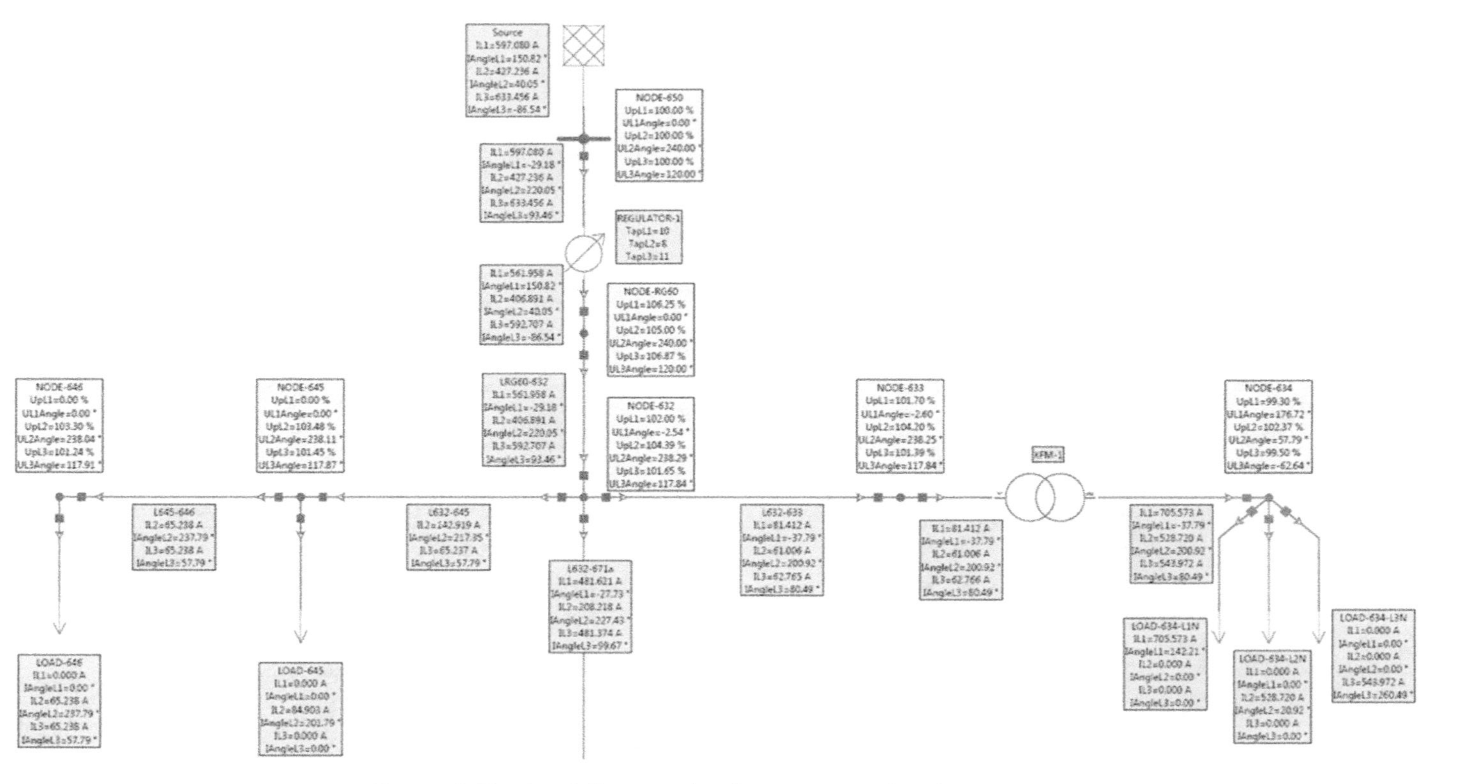

Figure 7.13   Load flow run for the system used with VVC [33]

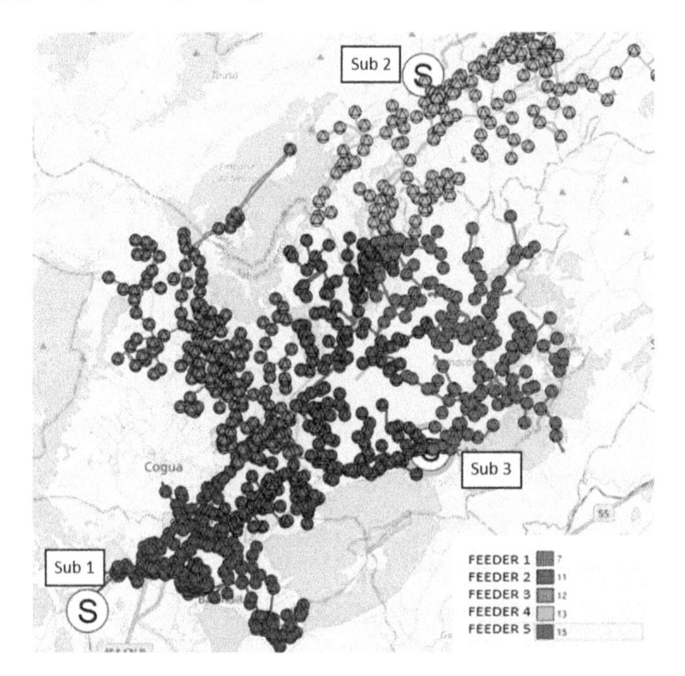

Figure 7.14 Study case in the georeferenced system

Table 7.3 Topologies in the study case

Topology	Feeders	Type
1	FEEDER 5 – FEEDER 2	PARTIAL
2	FEEDER 4 – FEEDER 2	COMPLETE
3	FEEDER 1 – FEEDER 2	PARTIAL
4	FEEDER 3 – FEEDER 2	COMPLETE

As presented in Table 7.3, FEEDER 2 is connected to all the other feeders, which allows it to change the topology in partial or complete configurations. For partial configuration, it is assumed that the load is shared with both feeders. For complete configuration, all the load of one feeder can be assumed by the other.

In the topology 1, part of the load is assumed by FEEDER 5 and the other is assumed by FEEDER 2. In topology 3, part of the load is assumed by FEEDER 1 and the other is assumed by FEEDER 2. Finally, the load of FEEDER 2 will be completely assumed by FEEDER 2 and FEEDER 3 in the topologies 2 and 4, respectively. Figure 7.15 presents the initial power flow results considering a voltage of 100% in the substations for the four topologies considered.

From Tables 7.4–7.7, the power flow results are shown, where $V_{\min}$ indicates the minimum voltage on each feeder, while Losses FEEDER indicates the total losses per feeder.

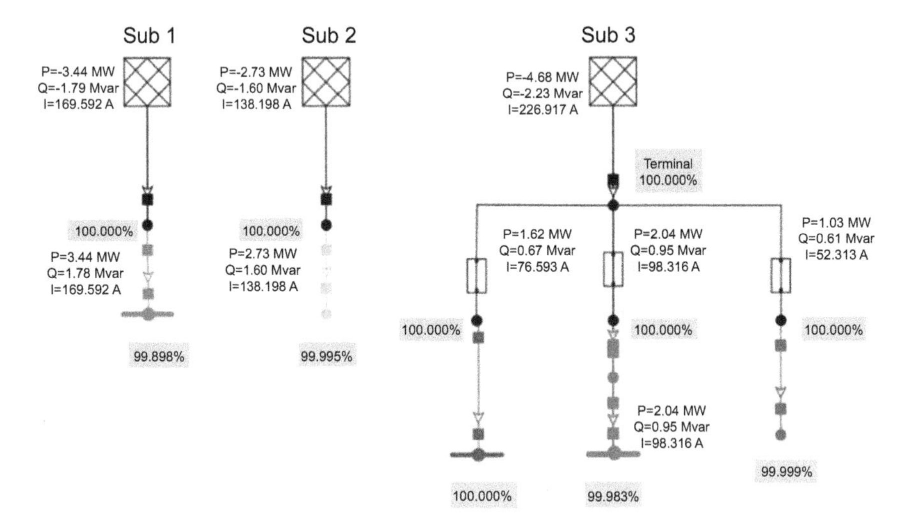

(a)

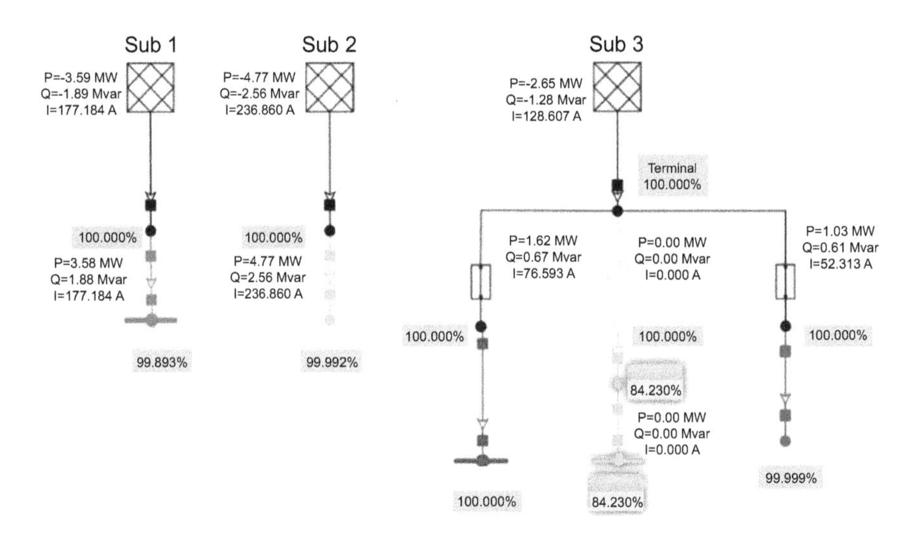

(b)

Figure 7.15 Power flow results for each topology: (a) Topology 1; (b) Topology 2; (c) Topology 3 and (d) Topology 4

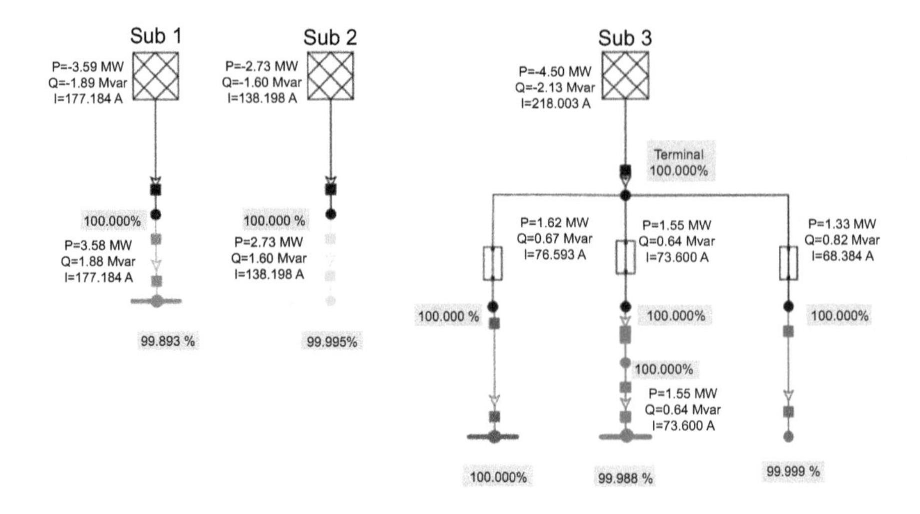

(c)

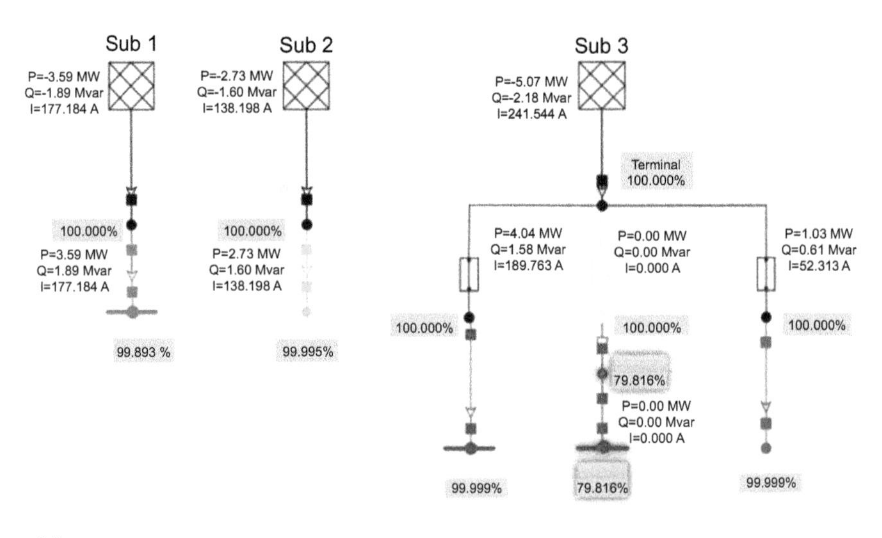

(d)

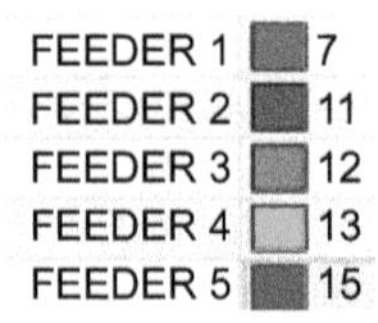

Figure 7.15 (Continued)

Table 7.4 Summary of power flow results for Topology 1: partial feeder 5–feeder 2

FEEDER	INITIAL		
	V_{min} (%) – 13.2 kV	Losses FEEDER (kW)	Losses NETWORK (kW)
FEEDER 1	96.37	57.13	838.81
FEEDER 3	89.81	195.68	
FEEDER 2	90.93	220.18	
FEEDER 4	93.61	170.81	
FEEDER 5	95.67	195.01	

Table 7.5 Summary of power flow results for Topology 2: complete feeder 4–feeder 2

FEEDER	INITIAL		
	V_{min} (%) – 13.2 kV	Losses FEEDER (kW)	Losses NETWORK (kW)
FEEDER 1	96.37	57.13	994.38
FEEDER 3	89.81	195.68	
FEEDER 2	CONNECTED TO FEEDER 4		
FEEDER 4	83.58	529.33	
FEEDER 5	95.28	212.24	

Table 7.6 Summary of power flow results for Topology 3: partial feeder 1–feeder 2

FEEDER	INITIAL		
	V_{min} (%) – 13.2 kV	Losses FEEDER (kW)	Losses NETWORK (kW)
FEEDER 1	94.28	98.96	803.27
FEEDER 3	89.81	195.68	
FEEDER 2	93.63	125.58	
FEEDER 4	93.61	170.81	
FEEDER 5	95.28	212.24	

Table 7.7 Summary of power flow results for Topology 4: complete feeder 3–feeder 2

FEEDER	INITIAL		
	V_{min} (%) – 13.2 kV	Losses FEEDER (kW)	Losses NETWORK (kW)
FEEDER 1	96.37	57.13	1371.28
FEEDER 3	78.27	931.1	
FEEDER 2	Connected to FEEDER 3		
FEEDER 4	93.61	170.81	
FEEDER 5	92.95	212.24	

If it is desired to maintain voltage regulation within a range of 90%–105%, the previous tables demonstrate that the minimum voltage in FEEDER 3 is 89.81%. This can be improved through remedial actions such as raising the substation voltage using the main substation transformer. Based on the power flow analysis, the system achieves a minimum voltage of 90.92% with a voltage substation output of 101%. However, in the case of complete Topologies 2 and 4, voltage regulation becomes more complex. According to power flow simulations, Topology 2 requires a voltage substation of 105% to achieve a minimum voltage of 90%, while Topology 4 requires 109%. Subsequently, an optimal capacitor placement strategy is proposed to ensure that voltage remains within the prescribed limits (90%–105%) for all topologies.

Considering that this system presents different topologies to supply five feeders, performing optimal capacitor simulations is affected by the operational state. For example, the location of two capacitors in FEEDER 2 for the operational state defined as the Topology 1 could change the associated feeder of one (or two) of the capacitor banks once the operational state changes. In this case, two ways for the optimal capacitor placements are analysed: optimisation with partial configurations and optimisation with complete configurations.

According to the results presented in the previous tables, higher losses in partial configuration are obtained in Topology 1. Based on this, an optimal capacitor placement analysis is performed by each feeder using Topology 1 as the operational state and the summary of capacitor banks is shown in Table 7.8.

Table 7.9 presents a description of the different optimisations performed for Topologies 2 and 4. The value N in table denotes that the optimisation module will calculate the number of banks based on the maximum losses reduction capacity of the system. Table 7.10 presents the capacitor banks obtained after the analysis.

Four capacitor banks were obtained in both optimisations. Nevertheless, remedial actions are required to get a voltage regulation in the range 90%–105%.

Table 7.8 Capacitor placement in partial topology

Optimisation name	**Feeder**	**Obtained bank (kVAR)**	
		First	**Second**
OPT1	FEEDER 1	396.51	
OPT2	FEEDER 1	349.1	122.698
OPT3	FEEDER 3	729.99	
OPT4	FEEDER 3	196.35	535.21
OPT5	FEEDER 2	842.78	
OPT6	FEEDER 2	578.48	344.35
OPT7	FEEDER 4	916.92	
OPT8	FEEDER 4	328.9	897.01
OPT9	FEEDER 5	1767.81	
OPT10	FEEDER 5	414.38	1359.19

Different combinations were determined using power flow simulations and they include the capacitors obtained in the optimisations OPT1–OPT12. Table 7.11 presents a summary for the representative combinations in Topology 4, in which a minimum voltage of 90% was obtained.

Table 7.9 Optimisations performed in complete topology

Optimisation	Feeder	Topology	Number of capacitors
OPT11	FEEDER 3	4	N
OPT12	FEEDER 4	2	N

Table 7.10 Capacitor placement in complete topology

Optimisation	Feeder	Obtained banks (kVAR)			
OPT11	FEEDER 3	982.5	506.8	150.0	335.9
OPT12	FEEDER 4	1060.0	325	321	811

Table 7.11 Best combinations for all the optimisation functions to get a voltage of 90%

OPT11+OPT3		**OPT11+OPT12+OPT4**	
Five selected capacitor banks		**Six selected capacitor banks**	
Vo = 105%		**Vo = 105%**	
	Plosses kW		**Plosses kW**
FEEDER 1	57.52	FEEDER 1	57.52
FEEDER 3	749.66	FEEDER 3	810.57
FEEDER 4	170.81	FEEDER 4	170.81
FEEDER 5	212.24	FEEDER 5	212.24
losses kW	1,190.23	losses kW	1,251.15
V_{min}	89.39	V_{min}	89.99
PT11+OPT4+OPT5		**OPT11+OPT3 + OPT6**	
Six selected capacitor banks		**Six selected capacitor banks**	
Vo = 105%		**Vo = 105%**	
	Plosses kW		**Plosses kW**
FEEDER 1	51.52	FEEDER 1	51.52
FEEDER 3	752.63	FEEDER 3	740.95
FEEDER 4	170.81	FEEDER 4	170.81
FEEDER 5	212.24	FEEDER 5	212.24
losses kW	1,193.20	losses kW	1,181.52
V_{min}	89.57	V_{min}	89.62

Table 7.12 Selected capacitors after analysis

Bank name	kVAR	Distance [km]
CAP_16	982.5	5.59
CAP_17	506.8	15.34
CAP_18	150	14.3
CAP_19	335.9	13.16
CAP_4	730	8.7
CAP_8	578.5	10.48

According to Table 7.11, it is required at least five capacitors using OPT11+OPT3 combination to get a minimum voltage of 89.39%. An optimal solution is proposed using the combination of OPT11+OPT3+OPT6, which offers the best results in losses reduction and the minimum voltage is close to 90%. The selected capacitors are indicated in Table 7.12, and the distance shown is based on FEEDER 3 for Topology 4.

According to previous results, at least six capacitor banks are suggested to reduce technical losses (kW) in all the topologies and ensure a voltage regulation within the desired range. The capacitor sizing and location used different combinations of the performed optimisations and it considered the impact of the capacitor banks in two ways: optimisation in partial configuration and optimisation in complete configuration. Table 7.13 presents a summary with the initial conditions (using remedial actions to get V_{min} = 90%) and the final conditions (to get voltage in the range of 90%–105% and switching on/off the capacitor banks to get the minimum losses). Redistribution of capacitor banks with the different topologies is illustrated in Figure 7.16.

7.6 Voltage/VAR control

Previously, the information to monitor and command equipment in charge of the control of voltage and reactive power came from the current and voltage transformers. Nowadays, better results are achieved by employing software packages interacting with SCADA systems [34–43]. Figure 7.17 illustrates various elements of a VVC which are controlled and monitored by a SCADA system.

Example 7.3

The following system is a reference to illustrate how Volt/VAR control is done. The voltage profile across the complete feeder is shown in Figure 7.18. It is shown how the performance and losses can be increased and reduced, respectively, by implementing the voltage control processor and the VAR dispatch processor. The first one is used to implement the CVR. The second helps to improve the reactive power requirements, and therefore, the voltage profile becomes more uniformly distributed across the feeder. In this case, the following sample rules are defined:

1. The 'candidate' capacitor bank for switching is identified considering that it is currently 'off' and the rating value is less than the measured reactive power flow at the head end of the feeder.

Table 7.13 Summary of benefits for the selected capacitor banks

Topology	Feeder	Initial				Final				Saving kW		Cap on	Cap off
		V_{max} (%)	V_{min} (%)	FEEDER Losses (kW)	NETWORK Losses (kW)	V_{max} (%)	V_{min} (%)	FEEDER Losses (kW)	NETWORK Losses (kW)	kW per feeder	kW in the network		
1	FEEDER 1	100	96.37	57.13	838.81	100	96.37	57.13	795.31	0	43.5	CAP_16	CAP_19
	FEEDER 3	100	89.81	195.8		100	91.08	169.39		26.41		CAP_17	CAP_4
	FEEDER 2	100	90.93	220.18		100	92.48	193.8		26.38		CAP_18	CAP_8
	FEEDER 4	100	93.61	170.81		100	93.61	170.81		0			
	FEEDER 5	100	95.67	195.01		100	95.67	195.01		0			
3	FEEDER 1	100	94.28	98.96	803.27	100	95.58	84.11	751.02	14.85	52.25	CAP_16	CAP_19
	FEEDER 3	100	89.81	195.68		100	91.08	169.39		26.29		CAP_17	CAP_4
	FEEDER 2	100	93.63	125.58		100	94.18	114.46		11.12		CAP_18	CAP_8
	FEEDER 4	100	93.61	170.81		100	93.61	170.81		0			
	FEEDER 5	100	95.28	212.24		100	95.28	212.24		0			
4	FEEDER 1	109	107.39	58.21	1195.87	105	101.52	57.52	1181.52	0.69	14.35	CAP_16	CAP_19
	FEEDER 3	109	89.83	754.61		105	89.62	740.95		13.66		CAP_17	CAP_4
	FEEDER 2									0		CAP_18	CAP_8
	FEEDER 4	100	93.63	170.81		100	93.61	170.81		0			
	FEEDER 5	100	95.4	212.24		100	95.28	212.24		0			
2	FEEDER 1	100	96.55	57.13	953.74	100	96.37	57.13	915.8	0	37.94	CAP_16	CAP_19
	FEEDER 3	100	89.88	195.68		100	91.06	173.06		22.62		CAP_17	CAP_4
	FEEDER 2									0		CAP_18	CAP_8
	FEEDER 4	105	92.3	488.69		103	93.38	473.37		15.32			
	FEEDER 5	100	95.89	212.24		100	95.28	212.24		0			

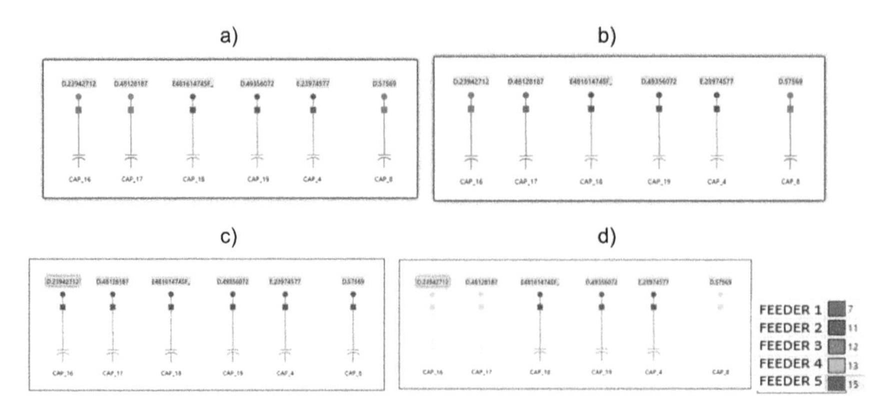

Figure 7.16 Capacitor banks on the feeder according to the topology: (a) Topology 1; (b) Topology 3; (c) Topology 4 and (d) Topology 2

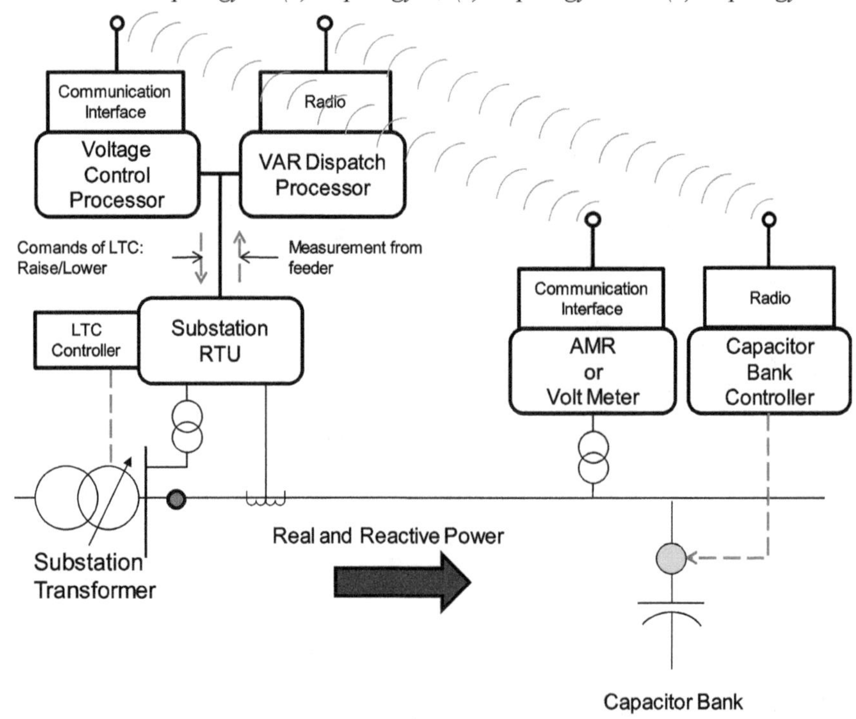

Figure 7.17 Components of a VVC assisted with SCADA

2. Once all the 'candidate' capacitor banks are checked, the one that has the lowest measured local voltage.
3. The switch of the chosen bank is changed to the 'ON' position.

The first module control for VAR dispatch is used. It is shown how the voltage profile increases by the end of the feeder and how the losses and the reactive power

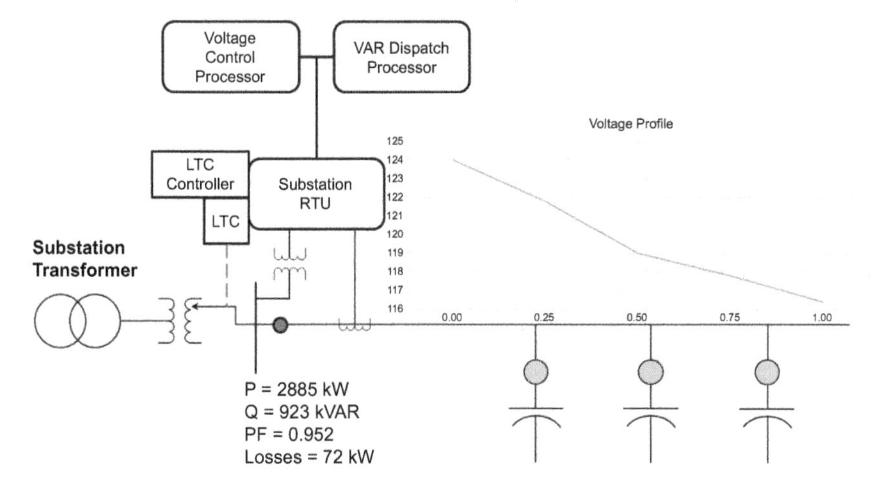

Figure 7.18 System for Example 7.3

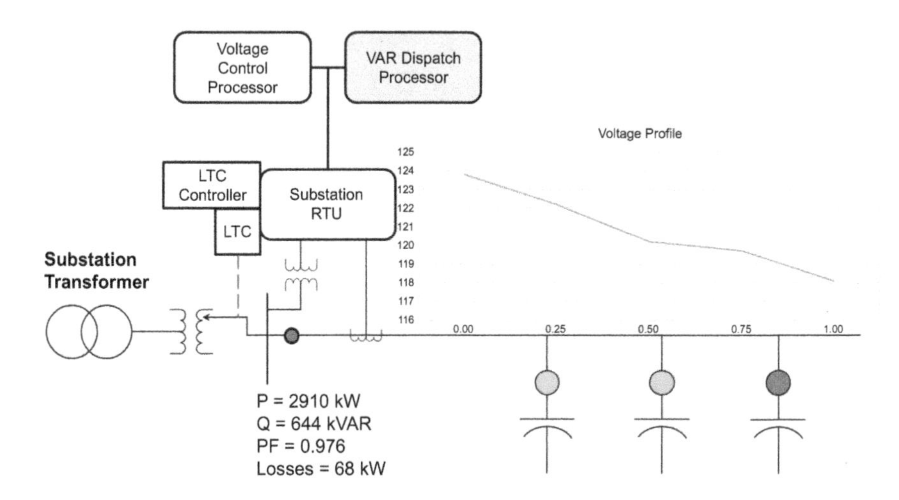

Figure 7.19 VAR dispatch processor control module for Example 7.3

requirements improve. Also, the power factor improves considerably. This is shown in Figure 7.19.

Then, all the capacitor banks are connected; it is shown in Figure 7.20. A comparison of the system with and without the VAR dispatch processor module is illustrated, in which an important improvement is obtained for the overall power losses and the reactive power requirements.

It is noticed that a uniformed voltage profile distribution across the feeder improves considerable the overall efficiency of the system since the transferred active power increases while the reactive power reduces, which improves the power factor of the system and reduces the losses.

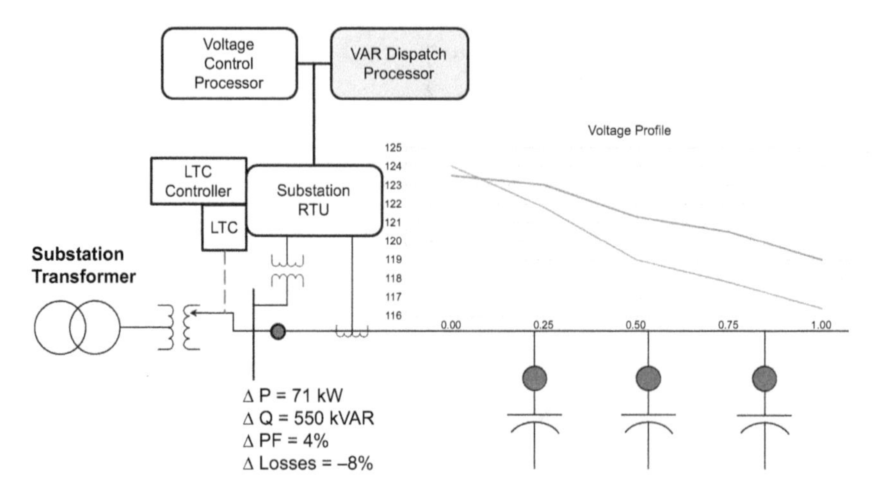

Figure 7.20 VAR dispatch processor control module for all capacitor banks in Example 7.3

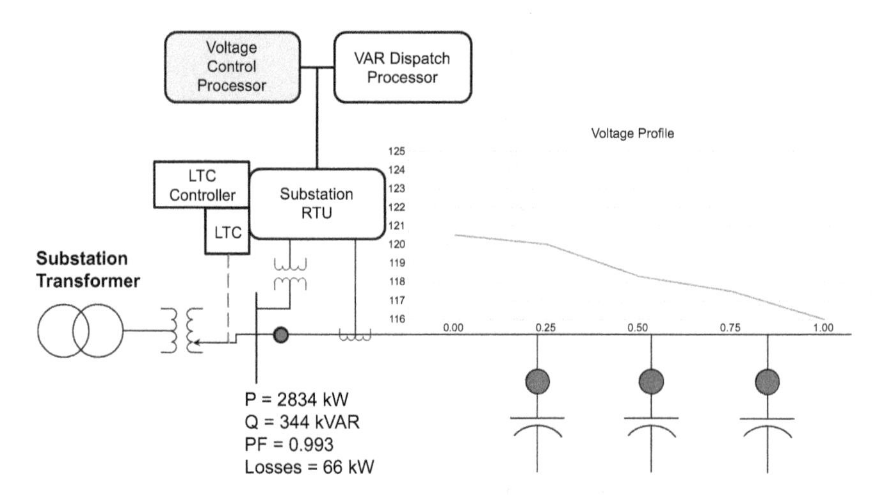

Figure 7.21 Voltage control processor module for Example 7.3

The next step is using the voltage control processor module to implement the CVR mentioned previously; it is illustrated in Figure 7.21. For this case, the following sample rule for voltage reduction is implemented: if voltage at head end of the feeder exceeds LTC set point, then lower the voltage. As it was stated before, this is done to improve the efficiency of the transferred power.

For the first moment, the LTC changes its position and reduces the overall voltage profiles across the distribution system. The voltage remains in the allowed limits and there is a reduction in the transferred active and reactive power supplied by the system.

The idea is to implement the CVR scheme. Therefore, it is expected to reduce the voltage as much as possible (that means, the lower limit is the voltage at the end of the feeder). Again, there is an important reduction in the transferred active and reactive power supplied by the system.

Comparing this with the previous condition (with the maximum position in the LTC and all the capacitor bank used), it is shown how there is an increase of efficiency for the required power in the overall system. This is illustrated in Figure 7.22.

The effect of both modules, as shown here, is to considerably improve the voltage profile and the efficiency of the whole distribution system. The SCADA control is based on the corresponding regulation, as illustrated in Figure 7.23.

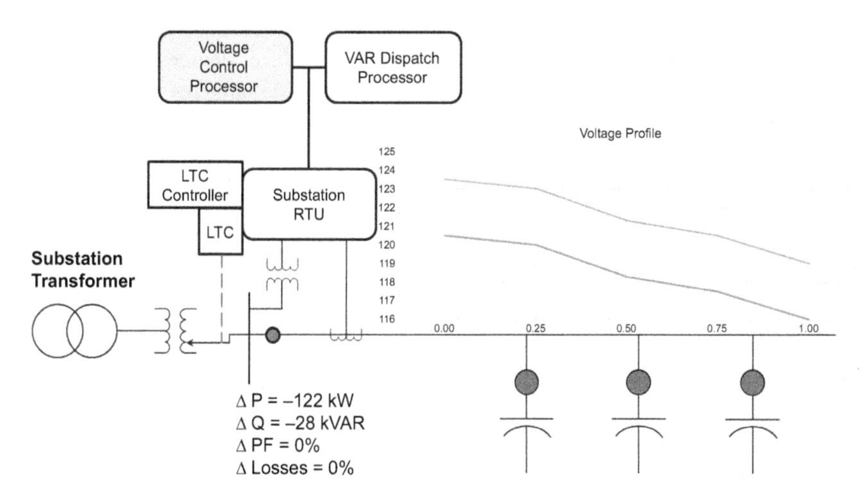

Figure 7.22 Voltage control processor results comparison for Example 7.3

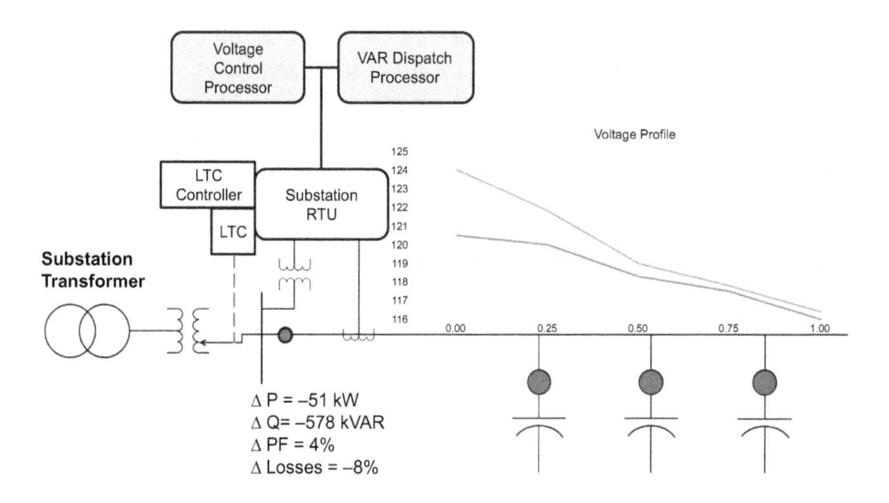

Figure 7.23 Volt/VAR modules applied for Example 7.3

VCC implementation is getting more popular due to the favourable experiences in service quality and increased efficiency. The system should be able to maintain the range of voltages specified by the user on feeders and branches. Normally, power factors above 0.9 are required although some utilities request 0.95 and above.

It is important to enable the operator to override in the event of topology changes because of a FLISR sequence or reconfiguration to reduce losses under normal conditions. In addition, various setting conditions must be developed for the different operational scenarios. This helps the operator overview the actions, especially in the event of emergencies where time is crucial.

The current number of tap changers, their location in the distribution system and capacitor bank controls is an initial potential information contribution to the database generated in the IEDs control systems. Therefore, devices must operate with various communication protocols that provide a wide range of possibilities to transmit information.

An important coordination between tap changers connected in series is required in a distribution system. Various commutator control functions that could be considered for the coordination such as types and time settings, line drop compensations, R+X and Z type, inverse power configurations, reference points and voltage ranges, and voltage reduction operation.

A coordination example is the timed control configuration. Typically, the voltage control manipulates the tap changer when voltage exceeds the voltage ranges for a period longer than the timer setting. Various tap changers allow the voltage to return within the permissible range. Normalised distribution loads cause major voltage changes the further they are from the circuit due to increase of the source/line impedance ratio. As a result, the first line tap changer upstream of the load (further from the circuit) operates first and controls voltages accordingly. However, assuming a voltage fluctuation at the source, all tap changer controls would experience the same change simultaneously. Aimed to avoid multiple or unnecessary tap changer operations, the time in each one (upstream) should be faster than the following control (upstream), where a time of 10–15 seconds between elements will allow an adequate coordination.

Another control approach to be considered is the inverse power configuration control. If a reverse power situation is presented in any circuit with multiple tap changers connected in series, attention must be focused on the appropriate control actions and coordination times. These operations and the coordination are very different for the radial topology or DG schemes. If the voltage reduction techniques are applied to a distribution system, it should be applied simultaneously in all tap changer controls in series. Some characteristics have been implemented in the most common commutator controls. These characteristics aid in the tap changers coordination and voltage capacitor controls for more efficient Volt/VAR management (VVM). The characteristics are designed to utilise local intelligence, providing a better managed Volt/VAR.

Finally, it could be set a voltage range control configuration. If the upstream range control is set to reduce the tap changers, these could be changed from the upstream control to the downstream control. Since the upstream tap changer voltage range affects downstream voltages, the tap changer general maintenance could be reduced in one to increase the next one. The lower the range in the upper section of voltage variation, the more stable will be the voltage source downstream, which results in a lower quantity of tap changers.

Distributed resources could present a challenge in the implementation of VVC, especially those that are inverter-based resource (IBR). The impact on feeders can be remarkable based on their position in the system and the generator size; consequently, corresponding coordination with the VVC devices must be accomplished. To achieve coordination with the electric power system (EPS) operator, DER operators shall be required to actively participate in the voltage regulation by controlling active and reactive power. According to IEEE 1547A-2020 [44], DERs shall not make that voltage of the EPS service area go beyond the limits established by required regulation standards such as ANSI C84.1-2020, Range A [1].

In the next sections, some control considerations can be applied for systems with high penetration of DERs.

7.6.1 Line drop compensation principle

An example of commutator coordination control is LCD configuration. The line drop compensation controls R and X characteristics and regulates voltage at the end of the defined line section (corresponding to load centre) [41].

This compensation is conducted by calculating the line voltage drop between the busbar to be controlled and the load centre using the incoming current (including angle) and the R and X values. The voltage control configuration changes to maintain destination voltage at load centre. The value of impedance Z is adjusted as a function of load magnitude with the impedance values for the defined section.

In the real implementation, the impedance values utilised in the first compensator type are very precise and the load centre is not defined accurately. Both issues cause unexpected voltages during operation. It is not necessary to coordinate the tap changer that controls the compensator configuration. However, the upstream settings could have a greater effect over the location requirements for the next downstream installation.

When there are DERs units connected into the feeder, the line current seen at substation is reduced at substation and the line drop compensation cannot be calculated properly. Therefore, the voltage calculated at the end of the line is higher than the actual value, causing lower voltage to the customers. Figure 7.24 represents an example of this situation.

7.6.2 VAR bias control

OLTC components ideally should be employed to rectify voltage deviations arising from fluctuations in real power consumption [42–44]. Conversely, line capacitors (a type of reactive element) should be deployed to address voltage deviations caused by changes in reactive power demand. The conventional approach of

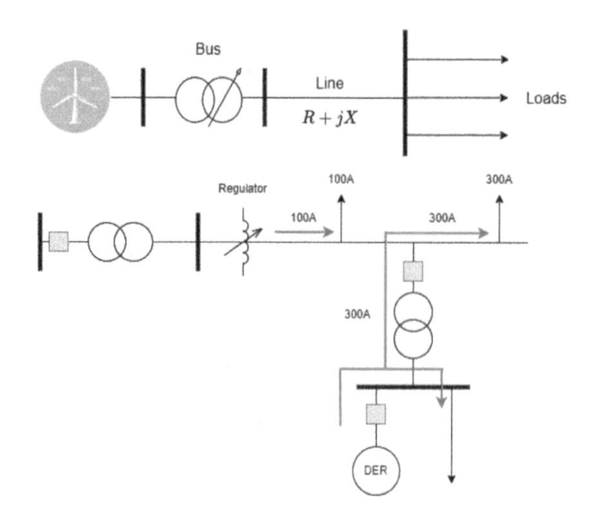

Figure 7.24 Line drop compensation calculation issues with DER

utilising line drop compensation to regulate OLTC elements doesn't achieve the most effective synchronisation with voltage-triggered line capacitors.

A technique that enhances the coordination between OLTC elements and voltage-triggered line capacitors is VAR-Bias. This method involves monitoring voltage levels, the direction of VAR (Volt-Ampere Reactive) flow and the magnitude of VAR flow in OLTC control processes. VAR-Bias considers the operating scenarios and the direction of reactive power flows. An adequate coordination between the control times along the distribution circuits eases the approach to the voltage and power factor regulation.

For example, the following cases can occur:

1. If the voltage moves away from the desired point and has a value close to the unity power factor (PF), the control action will be assumed by the OLTC.
2. If the voltage moves away from the desired point and there is a leading PF, the bank control must reduce the injection of reactive power, which is possible by turning off the banks or reducing their steps.
3. If the voltage moves away from the desired point and there is a lagging PF, the bank control must increase the injection of reactive power, which is possible by turning on banks or increasing their steps.

Based on it, a detailed explanation for these types of scenarios is presented as follows:

When the voltage of the feeder strays from the desired level and the quantity of VAR flow remains low (indicating the feeder operating close to unity power factor), the OLTC elements will make tapping adjustments to rectify the voltage discrepancy. During this scenario, the line capacitors will remain inactive. In cases where the feeder voltage deviates from the intended value and the VAR flow quantity becomes highly leading (indicating an excess of reactive power), the OLTC elements must refrain from tapping adjustments (instead, dynamically altering their setpoints upwards). Meanwhile, the voltage-sensitive controls of the line capacitors will detect the rise in voltage and disengage to counter the excess reactive power condition, ultimately decreasing the voltage. If the feeder voltage undergoes deviation from the target level and the VAR flow quantity becomes highly lagging (indicating an insufficiency of reactive power), the OLTC elements must avoid tapping adjustments (opting to dynamically modify their setpoints downwards). In this case, the voltage-sensitive controls of the line capacitors will perceive the voltage decrease and activate to address the insufficient reactive power condition, thereby raising the voltage.

If the deviation in feeder voltage arises from changes in real power loading, the OLTC elements come into play through tapping adjustments. Conversely, if the deviation originates from variations in reactive power loading, the reactive elements adjust their VAR output accordingly. Communications to the line capacitors are not required in this scheme, which promotes considerable savings in CAPEX and OPEX.

For VVM, ideally, OLTC elements should be used to correct voltage excursions due to real power loading changes; reactive elements such as line capacitors and active VAR regulating DER should be used to correct voltage excursions due to reactive loading changes. OLTC element control using conventional line drop compensation does not optimally coordinate with voltage-switched line capacitors.

Additionally, active VAR regulating DER can be embraced and used as part of a VVM strategy. Advantages of using active VAR regulating DER include granular and analogue control, no switching transients due to the switching of reactive devices and the capability to tune feeder VAR conditions to very close to unity since they can absorb/output small amounts of VAR between the larger steps of line capacitor banks.

To integrate active VAR regulating DER with voltage-sensing line capacitors effectively, the DER should utilise Volt/VAR droop local control, a prevalent feature in many inverter-based regulation technologies today. With this approach, as the voltage level increases, the DER will autonomously reduce its VAR output, and if necessary, can even absorb VAR. Conversely, when the voltage level decreases, the DER will enhance its VAR output to support the system. A comparison of this scheme with a traditional capacitor bank control is shown in Figure 7.25.

Considering the action of both OLTC and DER devices, when the feeder voltage strays from the desired level and the VAR flow quantity remains low (indicating the feeder operating close to unity power factor), the OLTC elements will adjust by tapping to rectify the voltage deviation. Meanwhile, the line capacitors and DER will maintain their VAR output without alteration. In situations where the feeder voltage deviates from the intended value and the VAR flow quantity becomes substantially leading, the OLTC elements will abstain from tapping adjustments. Instead, the voltage-sensitive controls of the line capacitors and DER will detect the voltage rise and initiate a switch-off or reduction in VAR

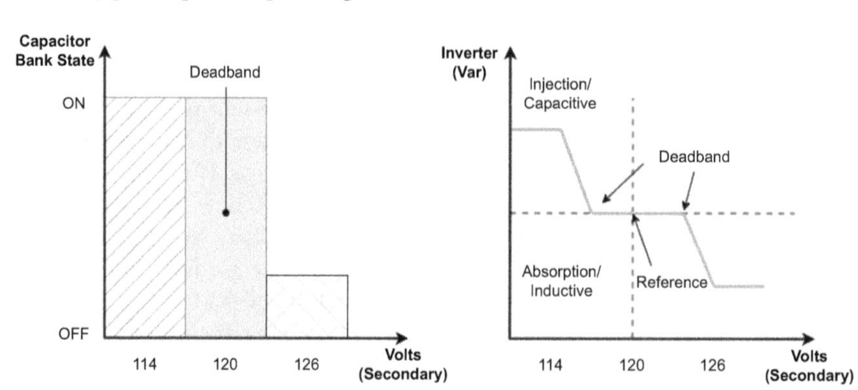

Figure 7.25 Comparison of voltage sensing line capacitors and active VAR DER (Volt/VAR droop)

output. This action aims to mitigate the over VAR condition, thus reducing the voltage. If the feeder voltage deviates from the desired value and the VAR flow quantity becomes significantly lagging, the OLTC elements will refrain from tapping adjustments. Instead, the voltage-sensitive controls of the line capacitors and DER will perceive the voltage drop and trigger a switch-on or increase in VAR output. This effort aims to alleviate the under-VAR condition, resulting in a voltage increase. In this manner, both capabilities are used efficiently. If the feeder voltage deviation is from real power loading change, the OLTC elements tap. If the feeder voltage deviation is from reactive power loading change, the reactive elements change their VAR output.

To ensure effective operational harmony, several guiding principles should be implemented for the coordination of OLTC elements, line capacitors and active VAR regulating DER. Reactive support elements positioned at the far end of the feeder should have shorter response delays, gradually increasing as they approach the feeder's origin. OLTC elements at the feeder origin should tap first, with downstream elements having progressively longer time delays. The shortest OLTC time delay must surpass the longest delay of any reactive support element. Adhering to these principles establishes a coordinated system optimising the interplay among these components.

Reactive support elements' voltage setpoints should lie beyond the setpoints of OLTC elements without VAR flow, while OLTC elements' voltage setpoints should be within reactive support elements' range when no VAR flow is present, and outside that range when VAR flow is active. An illustration in Figure 7.26 demonstrates voltage setpoints for OLTC and reactive power in line with this approach. This assumes that the DER contributes to reactive output, prompting line capacitors to react before the DER. The DER could potentially exhibit a similar response to line capacitors, allowing both assets to collaborate seamlessly.

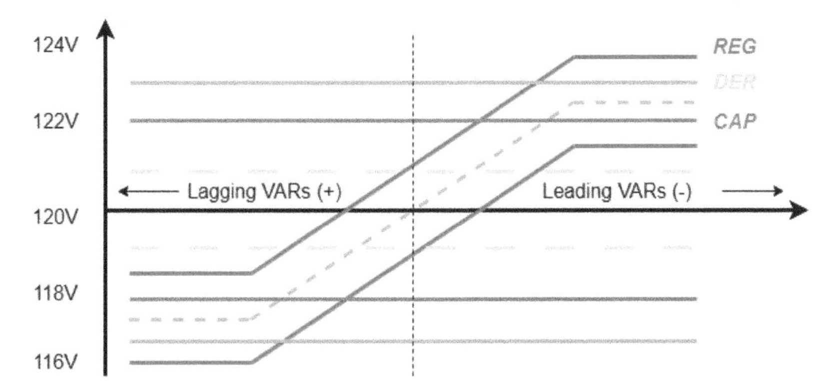

Figure 7.26 Example voltage setpoint coordination for OLTCs and reactive support elements

7.7 Volt-VAR schedule

The integrated VCC (IVVC) is considered an exceptional function of distributed automation that helps identify the most suitable control procedures for the devices involved in voltage regulation and VAR control to help achieve the specified operational goals of utilities avoiding violations to the fundamental operational constraints like high/low voltage limits and load limits.

The established VVC considers operating goal conditions such as minimal electrical losses, minimal electrical demand and reduced energy consumption. The decision criterion of IVVC requires the employment of an online power flow (OLPF).

The IVVC offers improvements compared to VVC since it can accommodate scenarios resulting from FLISR or feeder reconfiguration. It could include modelling the dynamic effects of DER. Figure 7.27 illustrates the system employing IVVC along with DER.

IVVC control can be achieved similar to feeders, i.e. in a centralised or decentralised controller way.

For the case of a centralised Volt/VAR controlled system, the distribution system SCADA integrates the IVVC which determines how to proceed and provides information for the recording system. Figure 7.28 presents the aforementioned case. The solution is considered very reliable and offers an overall illustration of the system. However, it is more expensive and could possibly congest the SCADA system.

On the other hand, the decentralised Volt/VAR controller is a standalone system that does not involve the overall system SCADA. It relies on local interaction with various devices associated with IVVC as observed in Figure 7.29. The recording takes place once all operations have been completed.

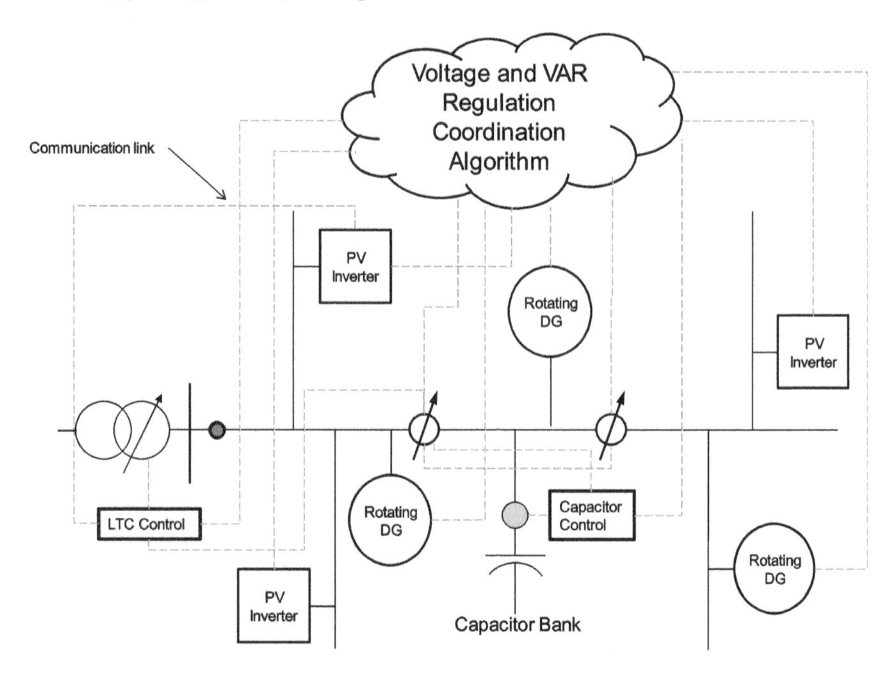

Figure 7.27 Integrated volt/VAR control including DG

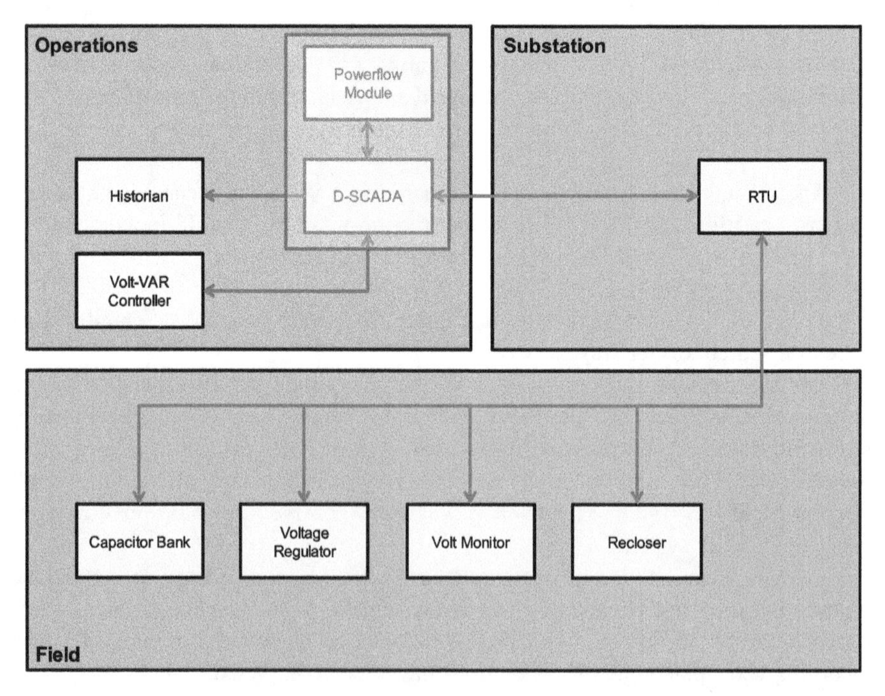

Figure 7.28 Context diagram for centralised integrated volt/VAR control

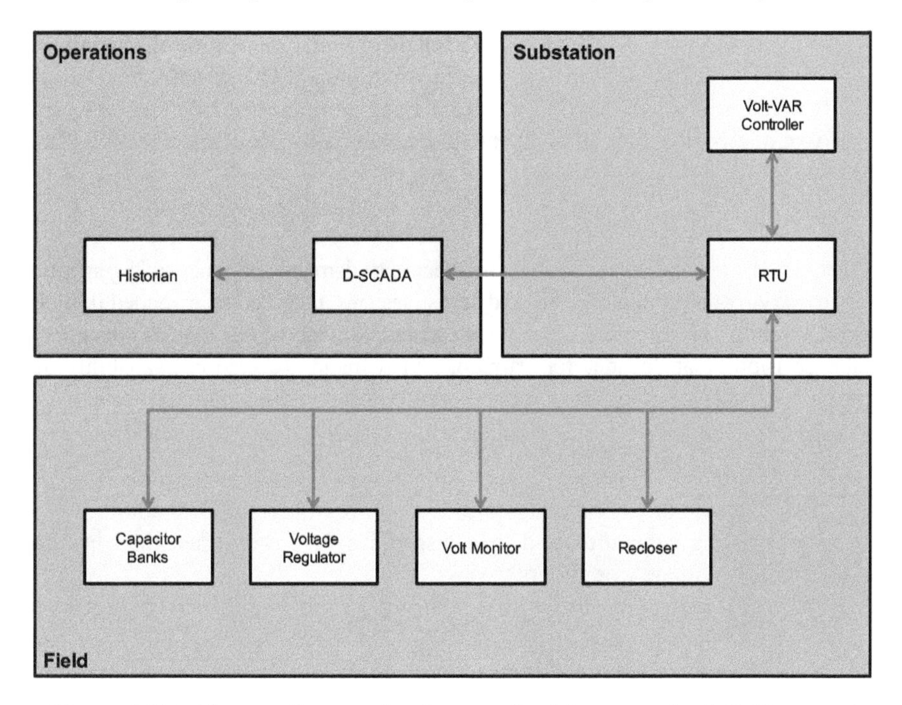

Figure 7.29 Context diagram for decentralised integrated volt/VAR control

7.7.1 Optimal power flow

The optimal power flow (OPF) is formulated as a nonlinear function to solve an optimisation problem with constraints. Some objective functions described in the literature are minimisation of active losses, minimisation of reactive losses, generation costs and reactive power flow [44].

For Volt-VAR control applications, the main objective of the OPF is to optimise the operation of the reactive power (VAR) and manage the output of all types of generators and the voltage levels at various buses in a way that system losses are minimised, and voltage stability is maintained within acceptable limits. The latter involves voltage control devices such as transformers, voltage regulators, capacitor banks and shunt reactors to maintain voltage levels across the network within specified bounds. OPF algorithms for Volt-VAR control consider a set of optimisation objectives and constraints to achieve these goals.

Some of the optimisation techniques used for this are nonlinear programming (NLP), which is used in many OPF algorithms and handle nonlinear constraints and objectives efficiently; interior point methods, which is able to handle large-scale nonlinear optimisation problems efficiently; and decomposition techniques, which are used in large and complex power systems, these techniques decompose the system into subproblems for faster convergence (these subproblems may involve optimising different areas or regions of the network).

OPF has been used for many years for the optimal generation dispatching in transmission systems. However, with the incorporation of DG systems and DERs in distribution systems, OPF algorithms are playing an important role for the tertiary control of microgrids that can be operated in connected or disconnected mode from the external grid.

7.7.2 Optimal power flow multi-period

Power systems operate dynamically, with electricity demand and renewable generation sources varying throughout the day and across seasons. OPF can be extended in multi-period scenarios to become even more versatile, allowing power system operators to make decisions that account for variations in demand, generation availability and system conditions over different time intervals. Multi-period OPF takes into consideration these dynamic changes, posing unique challenges and opportunities:

- Time-varying demand: Electricity demand can vary significantly between different times of day and days of the week. Multi-period OPF addresses this by optimising generation and load dispatch over a predefined time horizon, which could range from hours to days.
- Renewable generation integration: Renewable energy sources like solar and wind exhibit intermittent and uncertain generation patterns. Multi-period OPF can manage the integration of these resources by optimising their utilisation in conjunction with conventional generators.
- Energy storage: Energy storage systems, such as batteries, play a crucial role in smoothing out fluctuations in supply and demand. Multi-period OPF coordinates the charging and discharging of energy storage systems to enhance system stability and economic efficiency.
- Transmission constraints: Over longer time horizons, transmission network constraints become more pronounced. Multi-period OPF ensures that power flows remain within transmission limits, accounting for varying conditions.

Solving multi-period OPF problems involves handling larger-scale optimisation challenges compared to single-period OPF. Various techniques can be employed such as dynamic programming, which can be used for short time horizons where discretising time intervals is feasible and breaks down the optimisation problem into smaller subproblems that are solved recursively; rolling horizon optimisation for longer time horizons, in which the optimisation problem can be divided into shorter periods with the solution updated periodically as new information becomes available; stochastic optimisation, in which uncertainty in renewable generation or demand is considered in different scenarios and their probabilities; finally the mixed-integer linear programming (MILP), in which discrete decisions are taken such as generator start-up and shut-down.

Proposed exercises

1. A 13.2 kV feeder has a total load of 5.5 MW and 3.2 MVAR. The load is uniformly distributed. Perform the following calculations:

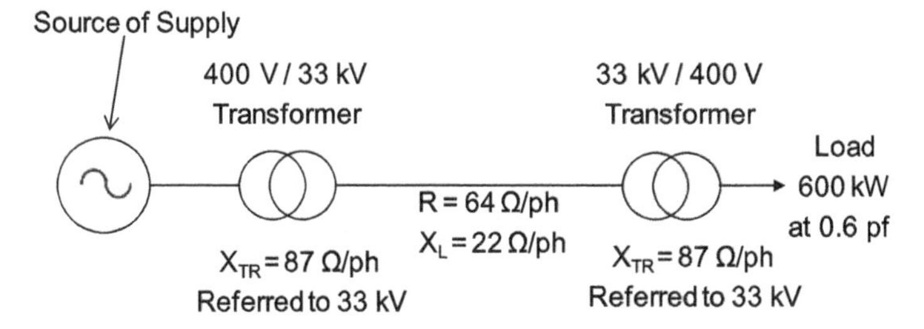

Figure 7.30 Diagrams for Exercise 7.2

(a) Location and size of the optimal capacitor bank to achieve the maximum losses reduction for load factors of 0.9 and 0.6, respectively
(b) Consider the same feeder if two capacitor banks were to be installed.

2. Consider a simple 33 kV radial line transmitting power by an overhead system to a lagging power factor load. Figure 7.30 presents the system diagram and the equivalent circuit.

Calculate shunt and series capacitors (size and impedance) to improve the power factor to 0.85 for each case. The line-to-line receiving-end voltage is assumed constant at 33 kV. All calculations are referred to 33 kV base voltage.

References

[1] ANSI, 'ANSI C84.1-2020. American National Standard for Electric Power Systems and Equipment – Voltage Ratings (60 Hz)', 2020.
[2] British Standard, 'BS 7671:2018+A2:2022. Requirements for Electrical Installations. IET Wiring Regulations', British Standard, 2022.
[3] Z. Wang and J. Wang, 'Review on implementation and assessment of conservation voltage reduction', *IEEE Trans. Power Syst.*, vol. 29, no. 3, pp. 1306–1315, 2014.
[4] P. K. Sen and K. Lee, 'Conservation voltage reduction technique: an application guideline for smarter grid', *IEEE Trans. Ind. Appl.*, vol. 52, no. 3, pp. 2122–2128, 2016.
[5] W. H. Kersting, 'Distribution feeder voltage regulation control', *IEEE Trans. Ind. Appl.*, vol. 46, no. 2, pp. 620–626, 2010.
[6] T. Gonen, *Electric Power Distribution System Engineering*, 3rd ed. Boca Raton, FL: CRC Press, 2014.
[7] N. M. Neagle and D. R. Samson, 'Loss reduction from capacitor installed on primary feeders', *Power Apparatus Syst. III. Trans. Am. Inst. Electr. Eng.*, vol. 75, no. 3, pp. 950–959, 1956.
[8] M. Ayoubi, R.-A. Hooshmand, and M. T. Esfahani, 'Optimal capacitor placement in distorted distribution systems considering resonance constraint

using multi-swarm particle swarm optimisation algorithm', *IET Gener. Transm. Distrib.*, vol. 11, no. 13, pp. 3210–3221, 2017.
[9] A. Askarzadeh, 'Capacitor placement in distribution systems for power loss reduction and voltage improvement: a new methodology', *IET Gener. Transm. Distrib.*, vol. 10, no. 14, pp. 3631–3638, 2016.
[10] J. J. Grainger and S. Civanlar, 'Volt/Var control on distribution systems with lateral branches using shunt capacitors and voltage regulators, Part I: the overall problem', *IEEE Trans. Power Apparatus Syst.*, vol. 104, no. 11, pp. 3278–3283, 1985.
[11] S. Civanlar and J. J. Grainger, 'Volt/Var control on distribution systems with lateral branches using shunt capacitors and voltage regulators, Part II: the solution method', *IEEE Trans. Power Apparatus Syst.*, vol. 104, no. 11, pp. 3284–3290, 1985.
[12] S. Civanlar and J. J. Grainger, 'Volt/Var control on distribution systems with lateral branches using shunt capacitors and voltage regulators, Part III: the numerical results', *IEEE Trans. Power Apparatus Syst.*, vol. 104, no. 11, pp. 3291–3297, 1985.
[13] IEEE, 'IEEE Guide for the Application of Shunt Power Capacitors', IEEE Std 1036-2020 (Revision of IEEE Std 1036-2010), pp. 1–96, 2021, doi:10.1109/IEEESTD.2021.9373058.
[14] G. Gutiérrez-alcaraz and J. H. Tovar-hernández, 'Two-stage heuristic methodology for optimal reconfiguration and Volt/VAR control in the operation of electrical distribution systems', *IET Gener. Transm. Distrib.*, vol. 11, no. 16, pp. 3946–3954, 2017.
[15] M. M. Aman, G. B. Jasmon, A. H. A. Bakar, H. Mokhlis, and M. Karimi, 'Optimum shunt capacitor placement in distribution system – a review and comparative study', *Renew. Sustain. Energy Rev.*, vol. 30, p. 429, 2014.
[16] R. W. Uluski, 'VVC in the smart grid era', in *Power and Energy Society General Meeting*, vol. 2010. Piscataway, NJ: IEEE, pp. 25–29, 2010.
[17] M. J. Krok and S. Genc, 'A coordinated optimization approach to Volt/VAR control for large power distribution networks', in *American Control Conference (ACC)*, vol. 2011, pp. 1145–1150, 2011.
[18] M. Mustafa, K. Abdelouahab, Z. Rabie, M. R. Djamel, O. S. Maawiya, and S. Yousef, 'Improved-GWO for shunt capacitors placements and dimensions in distribution networks', in *International Conference on Electrical Engineering and Advanced Technology (ICEEAT)*, Piscataway, NJ: IEEE, pp. 1–6, 2023, doi:10.1109/ICEEAT60471.2023.10426166.
[19] P. Díaz, M. Pérez-Cisneros, E. Cuevas, O. Camarena, F. A. F. Martinez, and A. González, 'A swarm approach for improving voltage profiles and reduce power loss on electrical distribution networks', *IEEE Access*, vol. 6, pp. 49498–49512, 2018, doi: 10.1109/ACCESS.2018.2868814.
[20] O. Saad and C. Abdeljebbar, 'Historical literature review of optimal placement of electrical devices in power systems: critical analysis of renewable distributed generation efforts', *IEEE Syst. J.*, vol. 15, no. 3, pp. 3820–3831, 2021, doi: 10.1109/JSYST.2020.3023076.

[21] D. F. Pires, C. H. Antunes, and A. G. Martins, 'Multi-objective evolutionary approaches for reactive power planning in electrical networks – an overview', in *International Conference on Power Engineering, Energy and Electrical Drives*, Piscataway, NJ: IEEE, pp. 539–544, 2007, doi:10.1109/POWERENG.2007.4380202.
[22] M. Asadi, H. Shokouhandeh, F. Rahmani, *et al.*, 'Optimal placement and sizing of capacitor banks in harmonic polluted distribution network', in *IEEE Texas Power and Energy Conference (TPEC)*, College Station, TX, USA, pp. 1–6, 2021, doi:10.1109/TPEC51183.2021.9384992.
[23] R. Anilkumar, G. Devriese, and A. K. Srivastava, 'Intelligent volt/VAR control algorithm for active power distribution system to maximize the energy savings', in *IEEE Industry Applications Society Annual Meeting*, Addison, TX, pp. 1–8, 2015.
[24] R. Anilkumar, G. Devriese, and A. K. Srivastava, 'Voltage and reactive power control to maximize the energy savings in power distribution system with wind energy', *IEEE Trans. Ind. Appl.*, vol. 54, no. 1, pp. 656–664, 2018.
[25] J. Carpentier, 'Contribution to the economic dispatch problem', *Bull. De La Societe Francoise Des Electriciens*, vol. 3, no. 8, pp. 431–447, 1962.
[26] W. Wang and N. Yu, 'Chordal conversion based convex iteration algorithm for three-phase optimal power flow problems', *IEEE Trans. Power Syst.*, vol. 33, no. 2, pp. 1603–1613, 2018.
[27] J. Lavaei and S. H. Low, 'Zero duality gap in optimal power flow problem', *IEEE Trans. Power Syst.*, vol. 27, no. 1, pp. 92–107, 2012.
[28] A. Vaccaro and A. F. Zobaa, 'Voltage regulation in active networks by distributed and cooperative meta-heuristic optimizers', *Electr. Power Syst. Res.*, vol. 99, pp. 9–17, 2013.
[29] A. Mohapatra, P. R. Bijwe, and B. K. Panigrahi, 'An efficient hybrid approach for volt/var control in distribution systems', *IEEE Trans. Power Deliv.*, vol. 29, pp. 1780–1788, 2014.
[30] M. Z. Degefa, M. Lehtonen, R. J. Millar, A. Alahäivälä, and E. Saarijärvi, 'Optimal voltage control strategies for day-ahead active distribution network operation', *Electr. Power Syst. Res.*, vol. 127, pp. 41–52, 2015.
[31] A. G. P. Li, 'Active-reactive optimal power flow in distribution networks with embedded generation and battery storage', *IEEE Trans. Power Syst.*, vol. 27, pp. 2026–2035, 2012.
[32] IEEE, 'IEEE 13 Node Test Feeder. Distribution System Analysis Subcommittee', 1992.
[33] Neplan, 'IEEE 13 Node Test Feeder', 2018.
[34] J. Gers, *Distribution System Analysis and Automation*, 2nd ed. London: The Institution of Engineering and Technology, 2020.
[35] A. A. Eajal, E. F. El-Saadany, and M. F. AlHajri, 'Distributed generation planning in smart distribution grids via a meta-heuristic approach', in *2014 IEEE 27th Canadian Conference on Electrical and Computer Engineering (CCECE)*, Toronto, ON, Canada, 2014, pp. 1–6, doi:10.1109/CCECE.2014.6900930.

[36] L. Du, L. He, and R. G. Harley, 'A survey of methods for placing shunt capacitor banks in power network with harmonic distortion', in *IECON 2012 – 38th Annual Conference on IEEE Industrial Electronics Society*, Montreal, QC, Canada, 2012, pp. 1198–1203, doi: 10.1109/IECON.2012.6388600.

[37] O. Ceylan and S. Paudyal, 'Optimal capacitor placement and sizing considering load profile variations using moth-flame optimization algorithm', in *2017 International Conference on Modern Power Systems (MPS)*, Cluj-Napoca, Romania, 2017, pp. 1–6, doi:10.1109/MPS.2017.7974468.

[38] S. Rahimi, M. Marinelli, and F. Silvestro, 'Evaluation of requirements for Volt/Var control and optimization function in distribution management systems', in *Proceedings of IEEE International Energy Conference and Exhibition*, Piscataway, NJ: IEEE, pp. 331–336, 2012.

[39] Vanja Švenda, 'Practical load reduction in distribution network', in *Power and Energy Society General Meeting (PESGM) 2016*, pp. 1–6, 2016.

[40] F. Zavoda, C. Perreault, and A. Lemire, 'The impact of a volt & VAR control system (VVC) on PQ and customer's equipment', in *IEEE/PES Transmission and Distribution Conference 2010*, New Orleans, LA, USA, 2010, pp. 1–6, doi:10.1109/TDC.2010.5484574.

[41] M. A. Azzouz and E. F. El-Saadany, 'Optimal coordinated volt/VAR control in active distribution networks', in *IEEE PES General Meeting/Conference & Exposition*, National Harbor, MD, USA, 2014, pp. 1–5, doi:10.1109/PESGM.2014.6939137.

[42] S. Karagiannopoulos, P. Aristidou, and G. Hug, 'A centralised control method for tackling unbalances in active distribution grids', in *IEEE Power System Computation Conference*, pp. 1–7, 2018.

[43] W. F. Najar, M. D. Mufti, R. Owais, and N. Rehman, 'Optimised Volt/Var control in distribution system with capacitor banks using genetic algorithm', in *1st International Conference on Sustainable Technology for Power and Energy Systems (STPES)*, Piscataway, NJ: IEEE, pp. 1–6, 2022, doi:10.1109/STPES54845.2022.10006461.

[44] IEEE, 'IEEE Standard for Interconnection and Interoperability of Distributed Energy Resources with Associated Electric Power Systems Interfaces – Amendment 1: To Provide More Flexibility for Adoption of Abnormal Operating Performance Category III', IEEE Std 1547a-2020 (Amendment to IEEE Std 1547-2018), pp. 1–16, 2020.

Chapter 8

Open challenges and future developments

With the ongoing evolution of the energy system and particularly the electric energy system, distribution system operators (DSOs) play a pivotal role in shifting towards a modern model marked by more energy demand and more variable behaviour of a bi-directional power flow. This is a result of extensive renewable energy generation at the consumer level. Despite these changes, it remains as crucial as the principles of delivering energy in an efficient, cost-effective and environmentally sustainable manner.

Vast amounts of data are accessible due to the current developments of the low and medium voltage networks and newest technologies, presenting both transformative possibilities and potential challenges. Recent trends present energy management systems (EMS), sustainable mobility and renewable energy sources as the basis for new market models. A revaluation of the existing principles, guidelines and frameworks is needed, placing an emphasis on innovation and comprehensive structure planning, deployment and integration.

In order to maintain and ensure a secure, reliable, efficient and economic performance of the electric networks, the main aspects to consider are:

- Identifying vulnerabilities and potential weaknesses in the system.
- Formulating the basis and an optimal network framework.
- Defining key performance indicators for the desired system performance.
- Strategic planning towards an enhanced network to achieve investment efficiency.
- Standardising and managing assets to reduce costs and improve performance.

Well-executed planning and operation strategies set the successful integration of generation sources connected to the distribution systems called distributed or distribution generation (DG) and distributed energy resources (DER) into the distribution network. In order to have the terminology clear, it is convenient to present the definitions given by NERC [1] as follows:

'A Distributed Energy Resource (DER) is any resource on the distribution system that produces electricity and is not otherwise included in the formal NERC definition of the Bulk Electric System (BES)'. 'Distribution Generation (DG): Any non-BES generating unit or multiple generating units at a single location owned and/or operated by (1) the distribution utility, or (2) a merchant entity'.

In the content of this book, more reference is given to DER. In any case, a decentralised energy management and proactive analysis for optimised interconnection and operational strategies involve the following:

- Techno-economic analysis for the optimal placement of DER technologies.
- Distribution system analyses for a reliable and secure power supply.
- Grid code compliance through interconnection studies.
- Exploring business case studies to harness opportunities for new businesses, roles and applications.

The progress and widespread adoption of information and communication technologies (ICT) have significantly elevated the urban sectors' intelligence level. This has led to the emergence of the smart city concept, which has garnered special attention from both the academic and industrial sectors. The main goal of a smart city is to promote economic growth while simultaneously improving quality of life using smart technologies at an urban level.

A mature and well-structured smart city exhibits key characteristics including safety, sustainability, efficiency, low-carbon emissions and resilience at several levels. Since the electrical energy system serves as the foundation for nearly all other urban infrastructures and services, ensuring the resilience of a city's operations begins with securing a resilient and strong power grid.

Resilience refers to the capacity of a system, community or society facing potential hazards to withstand, adjust to and recover from the impact of such hazards in a prompt and effective manner, preserving its fundamental structures and functions. When considering the urban power grid, resilience is characterised by four key aspects:

- Anticipation: ability to predict, warn, take preventive measures against risks before and during extreme events.
- Absorption: a resilient urban power grid should uphold essential functions and secure critical components by mitigating or isolating the effects of extreme events.
- Adaptation: the urban power grid should be able to deploy basic functional restoration before implementing recovery measures.
- Recovery: refers to the capacity to promptly regain partial functionality to an acceptable level following damage.

Therefore, resilience in the power systems context is a holistic concept that includes various demands and functionalities from traditional analytical methods and concepts. Its primary goal is to ensure the safety of the power grid even in the face of severe conditions. These conditions are checked using simulation models for different scenarios.

8.1 Requirements for simulation models

Complex and extensive networks need advanced and robust tools capable of rapidly and precisely optimising and enhancing the power systems [2–8]. Development of a mathematical model enables the assessment of several network scenarios and predictive operational strategies. Nowadays, a multitude of simulation tools are readily accessible and widely employed for power system analysis, research purposes and educational endeavours. There is a diverse range of options available in the market, with many developed by reputable and established companies.

8.1.1 Power system analysis tools

Power system analysis tools (PSATs) are software solutions that provide optimal performance while maintaining user-friendliness. Some of these tools come equipped with a graphical interface that offers features such as single-line diagrams for visual representation, data parameter input capabilities, presentation of simulation results and reports. The aim is to make the analysis and management of power systems both efficient and accessible for users.

Information systems such as supervisory control and data acquisition (SCADA) systems play a significant role in implementing novel solutions for monitoring and managing power systems, especially for non-conventional energy sources. Advanced technologies have enabled significant improvements in information exchange through more robust communication systems as an integral component of monitoring and control mechanisms.

Energy sources like solar and wind are inherently variable in nature, which poses significant challenges. The operational considerations associated with distributed generation (DG) from renewable sources of this kind necessitate online simulations that closely replicate the real behaviour of electrical systems.

Recent developments incorporate an internal convergence calculation engine that relies on measurements and mathematical models. This engine helps to establish control criteria for electric network automation, ensuring a more responsive and adaptive approach to managing the fluctuations and uncertainty that implies renewable energy generation.

Energy systems modelling and planning tools aim to provide quantitative analyses to support strategic development and decision-making processes. The main functions of these tools are addressed by the following, and sometimes overlapping, categories:

- Operational decision: involves making real-time decisions to maximise the efficiency and scheduling of energy technologies, often focusing on short-term and high temporal resolutions.
- Investment decision: entails optimising the energy capacity planning with a focus on long-term and low temporal resolutions. This involves strategic decisions on future infrastructure and resource allocation.
- Scenario planning: explores various long-term scenarios considering factors such as different policies, technologies, operational and environmental constraints. Scenarios may include variations in prices, technical learning curves and energy demand projections.

- Power systems analysis: assesses power flows, fault scenarios and system dynamics. This analysis typically occurs at a high spatial and temporal resolution, providing insights into the stability and reliability of the electrical grid.

Energy models demand a substantial quantity of input information and data for accurate modelling. The required data sets can vary in terms of temporal, spatial and resource details, depending on the scope of the energy system model analysis. These data sets encompass a wide range of information and may include:

Demand-side data

- Consumption profiles at annual or higher temporal resolutions.
- Data at citywide or higher spatial resolutions.
- Projections of future demand for long-term analysis.
- Information on demand elasticity for demand-side management (DSM) models.

Supply-side data

- Generation and fuel supply profiles.
- Segmentation by zoning sector and energy carrier.
- Details about supply technology, installed and planned capacities, technical specifications and operational performance.

Technical performance data

- Factors and costs related to technical performance by system size and application.
- Projection in future years for long-term models.

Technology installation potentials

- Estimates of renewable and non-renewable energy plants installation potentials.
- Consideration of restrictions, network capacities, availability and other factors impacting installation.

Energy resource potentials

- Estimates of renewable and non-renewable energy resource availability.
- Consideration of natural resources, import capacities and infrastructure limits.

Infrastructure data

- Infrastructure capacities, configurations and performance parameters.
- Covering networks, buildings and other relevant infrastructure valid for the models.

Emission data

- Data on emissions, including direct and indirect (e.g., lifecycle) emissions.
- Data for energy carriers, technologies and infrastructure to assess environmental impact.

Microclimate and weather data

- Local weather and climate information, including microclimate factors.
- Data used for renewable energy resources estimation and forecast, demand patterns and optimising smart energy management in urban areas.

8.1.2 New tendencies in PSAT applications

Recently, more complex and extensive networks have been implemented world-wide thanks to the emergence of more advanced simulation platforms that consider detailed requirements for network planning and operation with a wide range of technical, economic and environmental requirements. In this sense, simulations for power system analysis software tools have evolved significantly, now offering advanced modules and add-ons that have enabled remarkable capabilities to ease the integration of newer technologies.

For example, according to the variability of the generation and consumption systems in new systems like solar and wind generators, the simulation models must be adapted to the present conditions of the system. The above indicates that simulations must have a link between an equivalent mathematical model of the system and the real information of all or part of its measurements.

To enable the integration of PSATs with information systems in online simulations, modern platforms come equipped with features such as communication protocol applications, data synchronisation, information management and data storage capabilities. The most advanced PSATs should explore applications built in real-time integration, cloud computing and big data.

As mentioned, cloud computing stands out as a powerful technology and complement for smart grid (SG) applications, which implies innovative network operation solutions. It provides access to computational resources, including storage, servers and several applications, all without the need for desktop installations. Cloud computing leverages computing resources and services, facilitating access to information to enhance productivity, costs reduction, computational costs and minimises efforts.

The cloud computing concept is not only related to storage capability (one of the services with this system) but also uses informatics resources and services. It allows access to the information from any place, at any time. Different applications (infrastructure as a service IaaS, platform as a service PaaS and software as a service) allow multiple strategic implementations for information management systems, processing (calculation) and storage on the cloud. For the electrical infrastructure, different applications are integrated for network automation, asset management, big data, planning, analysis and power systems operation.

However, the cloud computing concept reduces hardware and software acquisitions because the licences can be services. Some advantages with cloud computing:

- Cost reduction
- Automated updating for simulation platforms
- Informatics resources optimisation
- High capability processing
- Updated technology

- Safety technology
- Resilience
- Accessibility to the information
- Digital systems integration

In the context of real-time simulations, calculations are executed based on a timestamp tied to a reference clock and hardware capabilities. These simulations operate using discrete time steps and by successively solving equation systems. This ensures an accurate representation of the dynamic system behaviour and real-time response of the system.

8.1.3 Advanced functionalities in PSATs

PSATs offer a wide range of specialised analyses to optimise network operation and planning. These analyses focus on various aspects of power system performance and efficiency. Some of these applications have been discussed and exemplified in previous chapters and are summarised as follows:

- Optimal capacitor placement: PSATs can determine the optimal locations for placing capacitor banks in the power system to reduce reactive power losses. This optimisation aims to maximise savings in megawatts (MW).
- Optimal topology: load flow simulations help identify the network topology that minimises total losses by considering different network configurations. This analysis considers voltage levels, overloading and other operational limits as the opening state of tie and line switches.
- Optimal power flow: determines the active power injection from generators based on objective functions such as minimising MW losses, MVAR losses, generation costs or other factors among others. It involves a nonlinear constrained optimisation with scalar nonlinear objective function.
- Contingency analysis: assesses the impact of removing specific elements (e.g., lines or generators) from the network and evaluates the system's response in steady state. This is crucial for planning and ensuring the network's ability to withstand disruptions.
- Hosting capacity: determines the amount of DG that can be installed on the grid without compromising power quality, reliability and security. It considers grid operational limits and the impact of the change in generation and load patterns.
- Probabilistic analysis: this analysis accounts for uncertainties in various network parameters such as network configuration, consumption, weather data and equipment failure rates. It uses probability distributions derived from historical data to describe system states stochastically.
- Phase balance optimisation: involves the automatic reconfiguration of loads, generators and branch elements to minimise power imbalances. It can be used to reduce unbalance at feeding points or within feeders considering various methodologies like heuristics, genetic algorithms and simulated annealing, among others.
- State estimation: is a tool used by electrical energy control centres for the construction, in real time, of the electrical model of the system and can use PSAT calculation engines.

The integration of real measurements into simulations constitutes a key tool in automation systems. Likewise, it represents new challenges in aspects of interoperability and cybersecurity.

Example 8.1

Figure 8.1 shows the measurements obtained from a SCADA system associated to a microgrid system of small power size.

A state estimator is used to implement an EMS. A model that represents the system is required to adjust the corresponding states to the real measurements from the SCADA system, based on power flow simulation. The information of the microgrid components is presented in Table 8.1.

In addition, Figure 8.2 presents the single line diagram of the system that allows to implement the model in the PSAT.

With this information, a power flow model is implemented as indicated in Figure 8.3.

The PSAT of this example uses Web Services to connect the simulation model with the real measurements from the SCADA system. Figure 8.4 illustrates the integration where the measurements are sent to the mathematical model and the optimal power flow determines the synchronous generation set point.

Figure 8.5 shows the equivalent state in a PSAT based on the power flow simulation using the real measurements from the SCADA. The synchronous generator is named 'SM' in the PSAT model, the Network corresponds to 'POI' and the Solar Panel corresponds to 'DG'.

From this initial state, an optimal power flow solution is performed to determine the dispatch from the synchronous generator. Generation costs are assumed, they consider that the cost from POI is greater than SM and that both are greater

Figure 8.1 Current measurements in the SCADA system

Table 8.1 Rating information of the microgrid components – Example 8.1

Component	Rating information	Component	Rating information
Synchronous generator	Name: SM Voltage: 120/208 V Current: 4.85/2.8 A Power: 0.8 kW RPM: 1800 Excitation system: 7.5–200 VDC	Three-phase transformer	Name: TRAFO Power: 2.5 kVA Type: Autotransformer Voltage: 400/208 V Current: 3.608/6.939 A
Circuit breaker	Name: PCC Remote with Power terminals, 52a and 52b contacts, trip and close coils.	Three-phase variable load	Name: LOAD 208 V/300 W (maximum) 1.2 H per phase 0–50 Ohms per phase
Three-phase inverter	Name: DG Adjustable power factor from 0.8 capacitive to 0.8 inductive. DC input voltage range: 250–1000 V MPP Voltage: 300–800 V DC current max: 11A Output voltage: 3 × 400V Power output: 3.2 kW	DC source	Name: SOLAR Closed-circuit current: 10 A Power output: 1500 W

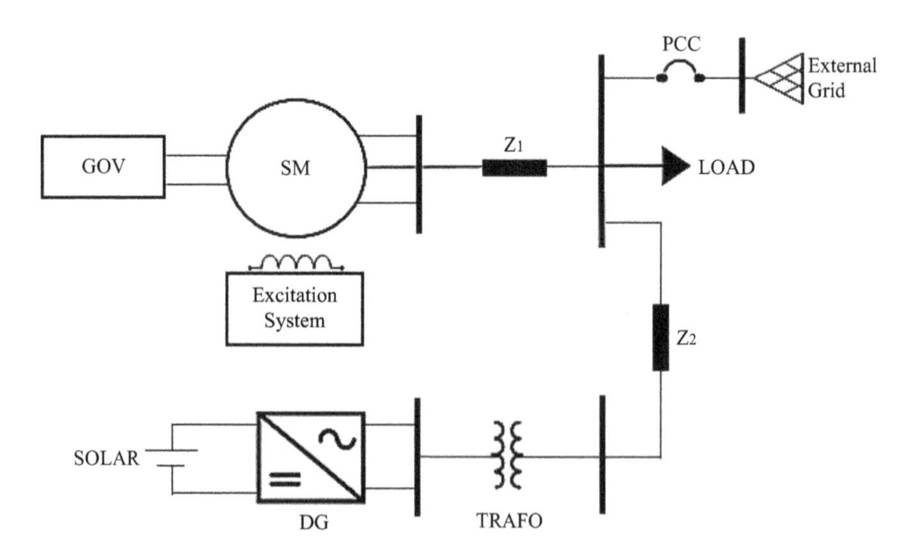

Figure 8.2 Single line diagram of the microgrid system – Example 8.1

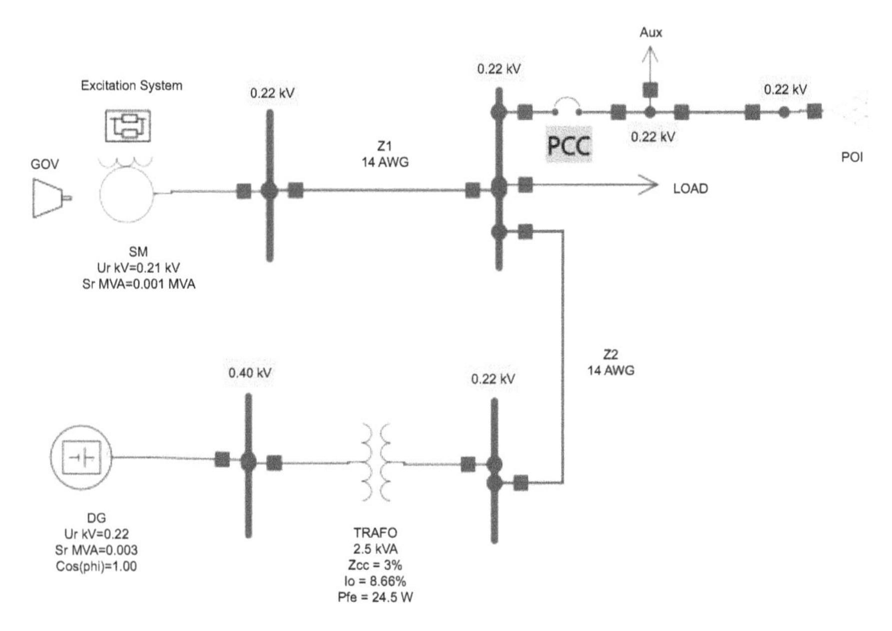

Figure 8.3 Power flow model of the microgrid system – Example 8.1

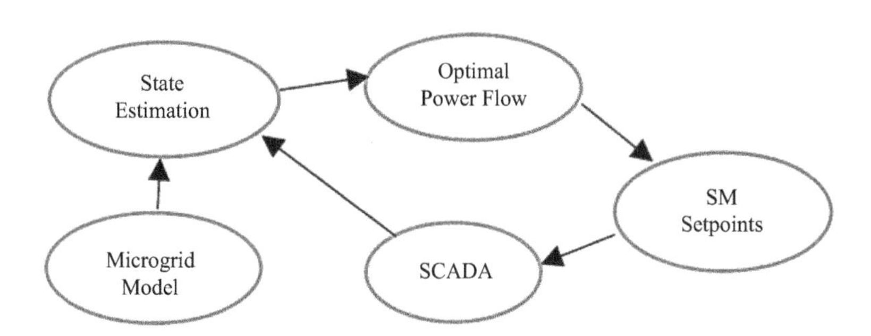

Figure 8.4 Sequence of the operation using Optimal Power flow – Example 8.1

than DG. Figure 8.6 presents the optimal power flow result when the reactive power compensation is assumed by SM.

For this example, it is shown in Figure 8.6 that the dispatchable resource is SM and that for this microgrid the local generation is preferred. The new set-point is sent to the SCADA and the corresponding controls will modify the generation dispatching.

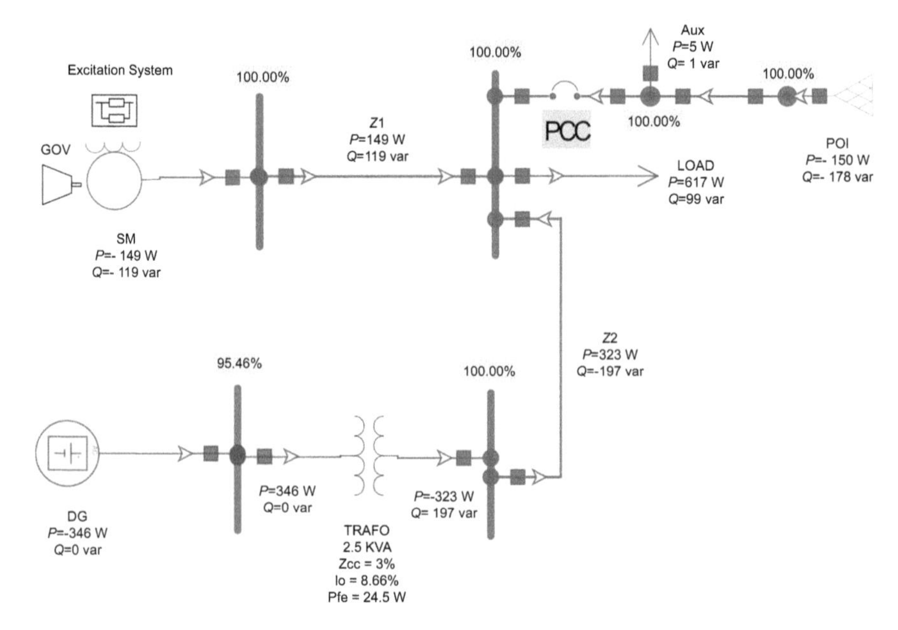

Figure 8.5 State estimation in PSAT using the SCADA measurements

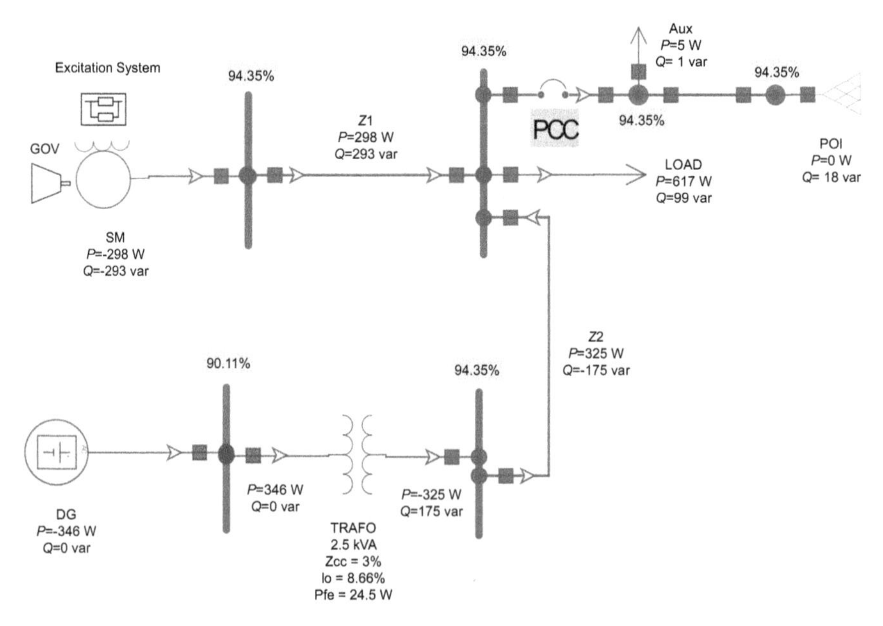

Figure 8.6 Optimal power flow from the state estimation in PSAT

8.2 Gaps and challenges

This section highlights the primary challenges in the field of urban and energy systems modelling (UESM) [9,10]. Not only technical and methodological challenges within current tools and approaches but also practical obstacles that hinder the wider adoption of UESM in local energy planning and decision-making processes. Each subsection not only addresses the gap but also delves into the factors responsible for its existence and the key barriers that impede its resolution. Figure 8.7 summarises the main challenges in the energy system model according to different characteristics.

8.2.1 Technical challenges

8.2.1.1 Availability and accessibility

Limited data availability and accessibility can be attributed to various factors. Challenges arise mostly from privacy-related concerns in terms of both data collection and distribution. Additionally, ownership of data across multiple sources can act as a barrier. Many data-issuing institutions have restrictive licences that prevent further sharing and publication, leading to redundancy in data collection and excessive efforts by researchers. These issues lead to inefficiencies in data collection and reproduction.

Furthermore, the absence of centralised regulatory standards and metrics creates inadequacies in data sharing platforms. Moreover, there is a lack of adequate ICT infrastructure for data collection. Enhancing ICT capabilities and data collection processes often implies substantial financial and time investments, adding to the challenges faced in this regard.

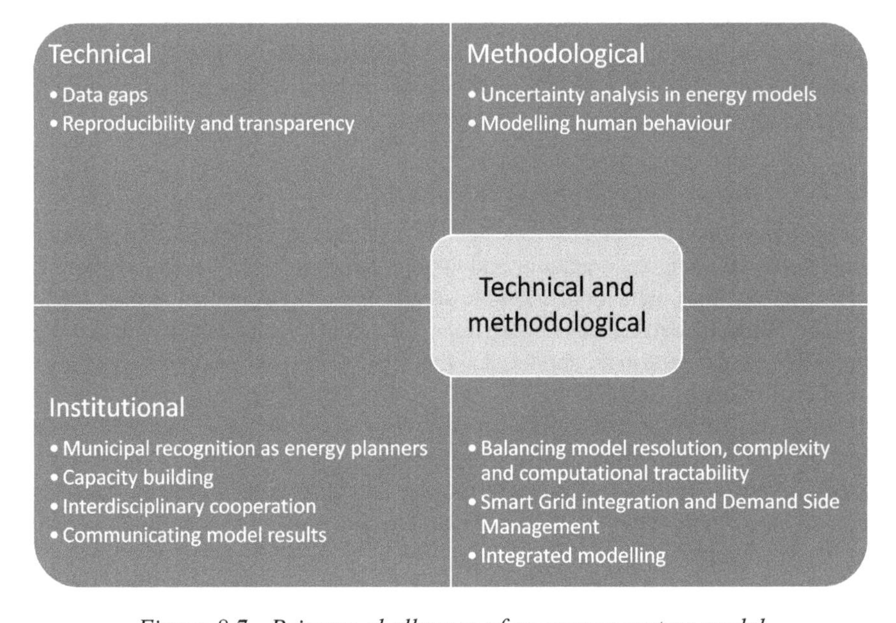

Figure 8.7 Primary challenges of an energy system model

8.2.1.2 Quality and consistency

As mentioned previously, the absence of centralised standards for the comprehensive collection and dissemination of energy-related data results in significant disparities across sources and sectors. This lack of uniformity leads to inconsistencies in the content, structure and quality of the data. Differences in physical and administrative scales contribute to the scarcity, dispersion and subpar quality of physical data, consequently diminishing the quality of models. Inadequate standards also create the potential for erroneous data, as there are limited quality control mechanisms in place.

8.2.1.3 Granularity

Many models resort to coarser resolutions when they lack the necessary spatial or temporal data. The significance of temporal resolution becomes evident and significant when dealing with relevant proportions of renewable energy sources. Models with inadequate temporal resolutions can lead to suboptimal investments in renewable technologies, an overestimation of the needed capacity and an underestimation of operating costs.

However, obtaining input data for models at the required temporal or spatial resolution remains a persistent challenge. Often, data is collected at lower temporal and spatial resolutions than what is ideal for modelling purposes. For instance, energy supply statistics are typically available on an annual basis for key sectors but may lack the higher temporal (e.g., hourly) or spatial (e.g., sector-specific) granularity needed for precise modelling.

8.2.1.4 Data management

As energy systems evolve towards greater decentralisation, reliance on renewable energy sources and intelligent systems, the volume of data exchanged experiences a significant surge. Additionally, several sources of big data are revolutionising the landscape of energy management and utilities.

8.2.2 Transparency and reproducibility

Transparency refers to the extent to which the underlying mechanisms of a modelling framework, including the equations and assumptions, are clearly elucidated through comprehensive documentation, access to source code and examination of input/output files. When modelling lacks transparency, it hinders comprehension, undermines credibility, erodes trust in the models and their results, and, consequently, diminishes the overall utility of UESMs.

The concept of reproducibility is closely linked with transparency and implies the extent to which models and their outcomes can be replicated or validated by external parties. Models transparency plays a crucial role as it defines the acceptance of sustainable planning recommendations derived from modelling outcomes within communities, for example.

Several factors contribute to the challenges regarding transparency and reproducibility in energy systems modelling; some of them are:

- Verification challenges: models are inherently challenging to verify against observable physical phenomena, making it difficult to ensure their accuracy.
- Stakeholder concerns: refers to concerns related to intellectual property rights, especially when considering a competitive edge in markets and research.
- Political considerations: implies sensitive model information and data as it prevents potential societal or economic impacts.

Addressing these challenges requires a multifaceted approach, including finding ways to balance intellectual property concerns with transparency, creating incentives to improve documentation and fostering a culture of openness and collaboration within academic and research communities.

8.2.3 Uncertainty in modelling

Energy models contend a substantial level of uncertainty. This uncertainty is particularly pronounced in long-term energy scenarios, related to assumptions and the model definition itself. These assumptions encompass factors like economic development, population and demand growth, future cost trends, resource availability over the long term, future policy directions and the effects of climate change, for example.

Scenarios are designed based on a combination of existing knowledge, historical patterns and assumptions for the sake of simplification, all of which carry uncertainty. This uncertainty in modelling can be broadly categorised into two types:

- Expected uncertainty: This refers to uncertainties that are anticipated and incorporated into the modelling process such as the range of possible future scenarios based on known factors and trends as well as data-related issues such as measurement errors, outliers and adjustments made for approximation purposes.
- Unexpected uncertainty: Encompasses uncertainties that arise from unanticipated events and may include external factors like sudden policy changes, geopolitical conflicts, unforeseen climate events or public health.

Unexpected uncertainty is often inadequately addressed in energy planning scenarios, whereas expected uncertainty is somewhat more manageable. Hence, navigating and mitigating these uncertainties is a significant challenge in energy system modelling. Strategies often involve sensitivity analyses, scenario planning and robust decision-making approaches to account for a range of potential futures and their associated uncertainties.

However, deterministic optimisation models, which are widely used in energy planning, often struggle to effectively represent uncertainty ranges and their impacts. This challenge may stem from the complexity of dealing with a large number of model parameters and uncertainties of unknown magnitude.

8.2.4 Bridging gaps

In this section, a variety of potential solutions are outlined to address the gaps identified in urban energy modelling. These solutions are aimed at enhancing the effectiveness of urban energy modelling approaches and the frameworks in which they are integrated, with the goal of increasing their utility and widespread adoption.

8.2.4.1 Data sharing platforms

Open data sharing platforms have the potential to make a significant impact on improving data availability, accessibility, transparency and reproducibility for urban and renewable energy system models. However, the majority of open energy system databases tend to focus on the national scale, which may not adequately meet the needs of urban and local-scale modelling.

Current data sharing initiatives would benefit from consolidation of a centralised resource. This would lead to more standardised data formats, structure and quality. Additionally, documented cases of other cities or areas using UESMs will benefit implementations and ease the scenario analysis for the deployment of solutions. Establishing experiential case study databases within transnational municipal networks provides an ideal platform for sharing such insights and best practices. This collaborative approach can foster mutual learning and enhance the effectiveness of urban energy planning and modelling worldwide.

8.2.4.2 Analytical methods

Analytical methods and mathematical analysis offer a practical means of addressing data gaps when collecting physical data is impractical due to factors like the physical scale, available resources or limited infrastructure. These approaches become especially critical in urban energy system modelling for cities in developing countries where alternative data sources may be scarce.

8.2.4.3 Privacy and security controls

Privacy and security safeguards in data collection are imperative for closing data gaps in modelling. The safeguarding of individual identities and corporate information is vital for promoting engagement in data-sharing initiatives. These privacy and security protocols also support the implementation of ICT programs by fostering public acceptance.

8.2.4.4 Communication architectures

The communication capabilities of current power systems are limited to basic functions primarily focused on system monitoring and control. These capabilities do not yet align with the communication demands of automated and intelligent power systems. Therefore, there is a pressing need for advancements in communication architectures and ICT to support district-level energy management within existing networks.

The development of robust and secure communication architectures (RCA) is crucial for the successful implementation of privacy and security solutions in data collection and transmission. These advancements collectively enable more efficient data collection, providing data with higher temporal and spatial resolutions.

8.2.4.5 Comprehensive scenario design

Scenarios serve as the foundation for a model as they define the assumed conditions and data inputs within the model, creating an internally consistent and coherent structure. Regarding urban energy models, comprehensive scenario design must be addressed based on the following key factors.

Socio-technical factors include representing human behaviour, social risks and opportunities, and factors particularly relevant in developing countries. For example, suppressed energy demand impacts can be seen as part of sensitivity analyses.

Scenarios can also be employed to model performance in the power sector, considering increased operation and maintenance costs, technical failures, load shedding and financial losses.

Climate impacts seen as urban microclimates and global climate changes exert considerable influence on local energy demand, economic growth, resource availability and demographic shifts. These impacts can be incorporated into energy scenarios as boundary conditions and input data.

Finally, urban landscape changes deal with shifts in demographics and land use, affecting energy demand profiles within models. Densification also alters the potential for installing renewable energy technologies on buildings, as well as the thermal performance of buildings.

8.3 Supervision and control challenges

The automation of power systems involves control centres that are interconnected with both generation and transmission systems, as well as distribution systems. These control centres can be seen as overlapping at the high-voltage/low-voltage (HV/LV) substations that also have their automation components [11].

Specific software applications are employed by these control centres to manage the assets and activate specific functionalities, with customisation for each application. For generation and transmission systems, applications are referred to as EMS and as distribution management systems (DMS) for distribution systems, which are illustrated in Figure 8.8.

There are two main types of architectures in control centres: centralised and decentralised systems. A decentralised or centralised architecture depends on the system's specific requirements and goals.

Centralised system: Centralised systems are controlled by a central unit, and communication is facilitated when they are located closely together. System communication is typically managed through a centralised point, making it efficient when devices are near one another.

Decentralised system: In these control systems, there is no single point of connection; rather, they share a common distribution infrastructure. The components or nodes in the system often communicate and coordinate with each other

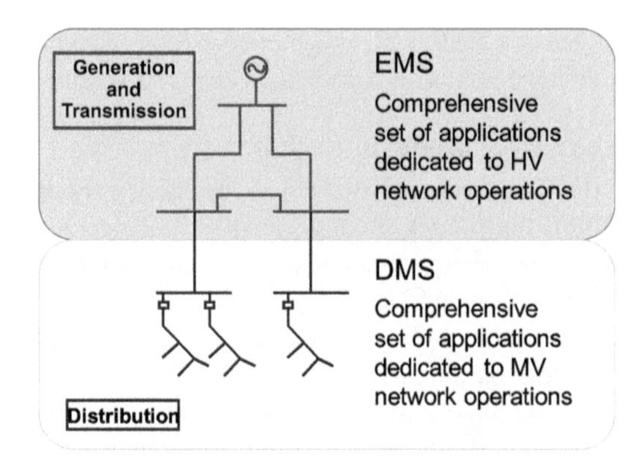

Figure 8.8 EMS and DMS functionalities

through a shared distribution infrastructure. This can provide advantages like increased fault tolerance and scalability, but it may introduce challenges related to coordination and consistency.

8.3.1 EMS functional scope

The EMS consists of two primary subsystems: SCADA and network analysis. Each of these subsystems includes various applications.

Within the SCADA subsystem, there are applications for multiple remote terminal unit protocols in a non-proprietary environment, load shedding, sequence-switching management and disturbance storage and analysis. These systems establish a form of communication between the user and the electrical network, which involves all monitoring and remote-control signals. SCADA systems are mechanisms that ease the implementation of automation schemes, since they are a way of establishing a digital vision of the magnitudes associated with the electrical network.

Network analysis applications provide the capability to perform power flow analysis, state estimation, contingency analysis, short-circuit analysis, security enhancement, optimal power flow analysis and Volt/VAR dispatch information, among others. These algorithms can be integrated using real measurements through communication protocols and interoperability standards like the Common Information Model.

8.3.2 DMS functional scope

A DMS is a comprehensive set of applications designed to efficiently and reliably monitor and control the distribution network in particular. It serves as a decision support system, aiding the control room and field personnel in overseeing and managing the electric distribution system. The primary goals of a DMS are to enhance the reliability and quality of service by reducing outages, minimising outage duration and maintaining acceptable quality performance indexes.

The SCADA subsystem within the DMS offers functionalities such as feeder topology, colour-coding, circuit tracing and mesh detection. The network analysis component provides capabilities for real-time power flow analysis, short-circuit analysis, Volt/VAR control and feeder reconfiguration, among others. These analyses are supported by features like colour-coding, circuit tracing and mesh detection.

8.3.3 Functionality of DMS

The functionality of DMS can be divided into steady-state performance improvement and dynamic performance improvement.

8.3.3.1 Steady-state performance improvement

In this section, the main functions associated with the analysis of the steady-state performance of the system are considered. These functions are characterised by fulfilling targets previously defined in planning studies.

- Voltage/VAR control: voltage/VAR control involves the detection and prediction of voltage violations, which necessitates close management of the relationship between voltage and reactive power equipment. Equipment includes capacitors, static VAR control (SVC) devices, load tap changers and voltage regulator controls.
- Feeder reconfiguration: involves the operation of switches, breakers or reclosers to modify the topology of feeders and enhance the system's operational condition. This functionality is activated under normal conditions to reduce losses, enhance reliability and minimise voltage drops.
- Demand-side management: is a method for controlling the load at the user's premises, based on an agreement previously established with the utility.
- Advanced metering infrastructure (AMI)/automatic meter reading (AMR): AMI/AMR plays an important role as it aids utilities in establishing competitive strategies that conduct the load profile. It identifies the major customers (aggregated load) and offers load information to customers as a special service.

8.3.3.2 Dynamic performance improvement

In this section, the main functions associated with the analysis of the dynamic performance of the system are considered. These functions are characterised by actions to be taken during faults, unpredicted events and emergency conditions.

8.3.4 Outage management systems

Another relevant functionality of DMS applications is the outage management system (OMS), which incorporates connectivity models and user interfaces equipped with features like handling trouble calls, outage analysis, prediction, crew management and reliability reporting. The system relies on connectivity maps of the distribution network to aid operators in managing outages, including partially restoring power and identifying nested outages. OMS is instrumental in pinpointing all affected customers and maintaining a complete, accurate and time-synchronised record of all data collected during an outage event.

OMS has become more integrated with other operational systems, including geographic information systems, customer information systems, work management

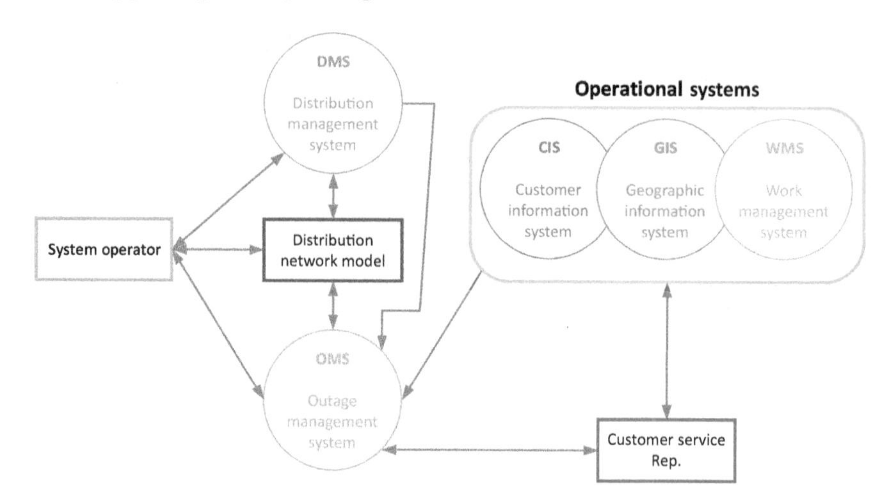

Figure 8.9 Integration of DMS applications

systems, mobile workforce management, SCADA, AMI and many others. These integrations enhance workflow efficiency and improves customer service.

Integration of DMS applications into OMS enhances outage performance. For example, a fault location algorithm uses the operated electric network model, incorporating the location of open switches, along with an electrical model of the distribution system, including line lengths and impedances, to estimate the fault location. The DMS fault location functionality utilises the electrical DMS model, ultimately enhancing the OMS process, as shown in Figure 8.9.

In a restoration switching analysis, the DMS/OMS system assesses potential isolation and restoration switching actions that can be taken when a permanent fault occurs. This application performs an unbalanced load flow analysis to identify overloaded lines for each switching action, providing the operator with a list of recommended switching actions. This functionality relies on the DMS model of the system but enhances the OMS process and its capabilities.

When DMS/OMS use the same operational model of the distribution system, a circuit analysis can be performed with consideration for time-based changes. This includes tasks like circuit tracing, trouble call and outage analysis, loop detection and parallel sources, fault location and load flow analysis. The result is a more comprehensive and accurate understanding of system conditions at any given moment.

8.3.5 Advanced distribution management system

The evolution of both the distribution system and consumer behaviour is characterised by a significant increase in data connectivity, interoperability and systems integration. These factors are essential for harnessing the benefits of new technologies on a large scale. Consequently, many utilities are exploring advanced distribution management systems (ADMS) as a solution to meet the demands of the modern, integrated distribution system [11–14].

ADMS offers a customisable approach where its components can be a combination of OMS, SCADA, DER management systems and various advanced applications from one or multiple vendors.

An ADMS platform provides operational and analytical tools that empower distribution utilities to efficiently manage assets, process near-real-time data and prepare for the increased levels of DER penetration and transactive loads.

Despite their well-known benefits, there are four major limitations to deploying and adopting ADMS functionalities:

- The complexity of integrating the various systems
- The availability and quality of the data model
- Benefits quantification
- The lack of a broad community of vendors to address the needs of utilities of all sizes

In an ADMS, various systems are seamlessly integrated to facilitate the sharing of network models, measurements, database values and control signals. A well-designed ADMS platform must strike a balance between the constraints of the past, the requirements of the present and the unforeseen challenges of the future. Utilities often deploy ADMS to support advanced applications like fault location, isolation

Table 8.2 Advanced DMS applications

	Energy efficiency	**Increased capacity utilisation**	**O&M improvement**	**Worker safety**	**Relatability improvement**	**Customer satisfaction**
State estimation	√	√	√	√	√	√
Switch order management			√	√	√	√
FLISR			√	√	√	√
Network recognition	√	√			√	
DER management	√	√	√	√	√	√
Load forecasting		√	√	√	√	√
Volt VAR optimisation	√	√				
Intelligent alarm processing		√	√	√	√	√
Relay study capability		√	√	√	√	√
Adaptive protection			√		√	√
Close loop			√		√	√

and service restoration; volt-var control/volt-var optimisation; the integration of DERs and other functionalities. A significant obstacle to interoperability arises from the fact that each software system frequently uses its proprietary data models and protocols, which are often not transferable between different vendors.

However, the process of connecting disparate systems from both single and multiple vendors introduces complex issues that are time-consuming and costly to resolve. Small- and medium-sized electric distribution utilities typically have limited IT resources and they require a scalable, cost-effective and cyber-secure solution for integrating ADMS with utility systems [15]. Table 8.2 summarises the main applications presented in ADMS.

8.4 Application of real-time digital simulation systems in network evaluation

Computational technologies have accelerated the evolution of simulation tools over the last decade. As computational technologies have improved in performance, simulation tools have also expanded in their ability to solve complex problems in a short period of time. Moreover, the cost of digital simulators has been decreasing steadily, making them more affordable and accessible for a wide range of applications.

When the time step of a clock is established as reference in a dynamic system, the possibility of simulating with a time stamp is identified and derives the processing and the data acquisition, in real time. A real-time simulator must perform calculations continuously in order to synchronise the simulation time with a reference clock.

Real-time processing has characteristics not only for the representation of dynamics in a power system but also for accurately emulating the response times of its physical counterpart. Real-time acquisition refers to the possibility for capturing online information from a dynamic system.

Based on the real-time simulation concept, some technologies can emulate dynamics associated with different subsystems in electrical networks. It allows the integration of external hardware to the simulation offering the possibility to test controllers and protection systems. These tests differ from conventional testing as the interaction is validated in closed loop over power system emulation, meaning full validation of automation schemes and control logics for pre-event, during the event and post-event conditions. Based on it, three main types of real-time simulation are known:

Model in the loop – MIL:
The power system and the protection/control system are simulated in real time.

Hardware in the loop – HIL:
The power system is simulated in real time and the protection/control system is connected physically. When amplifiers are used to generate power signals, the system adopts the power-hardware-in-the-loop (P-HIL) concept.

Rapid control prototype – RCP

The analyses are done over the actual power system, but the protection/control system is simulated in real time.

As modern power systems integrate distributed power generation, advanced load control, market operations and even more detailed functionalities, the operation of these systems becomes significantly intricate. This evolution has made real-time control and management in modern power systems a more complex task.

For this reason, smart control and EMS that differ from traditional approaches have become imperative for the efficient operation of modern power systems. Operators now require the capability to access real-time status information for equipment and system components, conduct thorough assessments of the system's state and predict emerging trends proactively. This allows them to respond swiftly even before issues begin to manifest and execute coordinated actions across the service area in real-time.

As mentioned in Chapter 2, real-time simulation is a highly dependable method that leverages electromagnetic transient (EMT) simulation to replicate real-world systems. It offers trustworthy, real-life insights into the consequences and advantages of new strategies or devices. This capability supports decision-making across various phases, from real-time operation and control to long-term planning.

The real-time simulation aims to emulate the actual behaviour of the power grid when subjected to different load, generation profiles or events while providing data on the electrical system's status. External signals control or influence the behaviour of modelled prosumers by updating photovoltaic (PV) generation outputs. During the simulation, the grid model requests the necessary values (active and reactive powers) to update the modelled PV generation outputs and receives the required data from the PV simulator through specific user datagram protocol blocks within the real-time model.

While there have been notable successes in hardware-in-the-loop (HIL) applications, the issue of simulation accuracy remains relatively unexplored and presents an ongoing research challenge that requires more robust solutions.

One proposed concept is the 'transparency performance index', which aims to assess the faithfulness of a HIL simulation. This method involves comparing the difference between the actual and the equivalent subsystems' impedance as perceived from the other side of the interface. A smaller difference suggests a higher level of transparency for the interface. However, this approach primarily focuses on the precision of the interface and doesn't fully consider the closed-loop nature of a HIL system. As a result, situations can arise where high transparency doesn't necessarily equate to low simulation errors.

To address this, two additional concepts are considered: 'performance mismatch (PMM)' and the 'probability of PMM'. These are intended to evaluate the simulation performance but come with certain challenges. The method relies on multiple approximations and assumptions, and even for simple scenarios, it can demand extensive computational resources, making it less practical for real-world applications.

With real-time simulation, the primary objective is to test the specific behaviour of a solution that is yet to be implemented, which also includes software solutions. Here, a method called software-in-the-loop (SIL) entails employing a precise computer simulator to evaluate the performance of newly designed software and serves to mitigate potential issues during the implementation.

8.4.1 Challenges of real-time simulation of SGs

The increasing complexity of modern electrical systems, including components such as SGs, and advanced power electronics, is driving the demand for more robust simulation tools. Real-time simulation of SGs faces several challenges:

- Real-time simulation of large power systems: Replicating large, distributed power systems with a significant number of components, modules and buses is a critical challenge. EMT simulations of power grids involve complex matrix computations, and the main challenge is to solve all the grid equations within a single simulation cycle.
- Accuracy of power electronics simulation: Simulation accuracy depends on the simulation step size, which should be small but not excessively so, as real-time processors must solve all equations within limited time cycles.

These challenges can be seen as a trade-off between the size of the simulated systems and the length of the simulation step size. A single simulator's versatility is limited by both processing time and the associated hardware needed to interface with the hardware under test. Real-time simulation of complex systems also involves incorporating communication networks into the simulation. This requires modelling the communication network and accounting for factors such as latency, bandwidth and packet losses, adding another layer of complexity.

Many research efforts are dedicated to developing novel integration schemes that leverage parallel platforms and promote modularity. However, to fully embrace real-time simulation, additional research is still required to ensure the simulation's fidelity to reality.

8.4.2 RT simulation towards innovation and research on SG

The SG approach offers fertile ground for innovation and research, as it doesn't rely on specific standardised technologies. However, testing and validating these innovations are challenging, especially when dealing with real installed systems.

Real-time simulation fills the gap by providing a platform for thorough testing of control strategies and software protection routines under a controlled environment. It allows for parallel testing and development alongside the actual physical system, reducing risks and enabling experimentation with risky or borderline conditions without the fear of damaging the real prototype.

Many researchers have introduced P-HIL as a reliable simulation method for testing SGs and microgrids. However, the emphasis often lies on power systems and power electronics rather than communication simulation. This is particularly important for designing AMIs, which can be prototyped using P-HIL simulations.

For the development of smart and microgrids, it's essential to establish a standard step-by-step process for conducting P-HIL tests. This procedure helps researchers and industries follow a common approach:

- Define the experiment or case study: List the electrical or technical requirements of the system to verify if the RT simulator and its components can accurately emulate the behaviour and dynamics of the system.
- Verify stability: Analyse the stability of the P-HIL simulator using classical methods such as the Nyquist criteria.
- Simulate the case study via software: Use tools like MATLAB®/Simulink® and LabVIEW to verify theoretical results, designs and controllers.
- Run the P-HIL simulation: Once all technical details have been considered, the system's stability has been verified, the results have been validated through offline simulations and the P-HIL test can be executed.

8.5 Grid codes and its relevance for the planning and operation of electrical systems

As more sectors of the economy rely on electricity, distribution grids must become more resilient, especially in the face of climate change and its potential for extreme operating conditions. Power grids need to evolve into inclusive platforms that enable electricity distributors to play a more flexible role. Their main goal is to have an energy system that includes various stakeholders, regulators, power producers, customers, prosumers, DER aggregators and electric vehicle charging operators. This transition will lead to a new era where all stakeholders collaborate on equal footing, optimising services, business solutions, technologies and processes through digital innovations.

Grid codes are set rules for power system and energy market operation. They enable network operators, generators, suppliers and consumers to function more effectively across the market. This ensures operational stability and supply security, and contributes to well-functioning wholesale markets. The grid code is distinct from a country's energy policy.

Rapid advancements in DER technology, accompanied by storage and control systems, have introduced new operational practices for power systems, increasing flexibility and allowing for a higher share of DERs. To ensure security while adopting these new technologies and operational practices, regulations and standards must be established. This typically involves the creation of a grid code, which is published by the grid operator or utility. The grid code sets the rules for all stakeholders involved in the operation of the electricity network, ensuring safe and reliable service delivery.

Developing a grid code, especially for smaller countries that are just embarking on the integration of DER technologies, can be a complex task. Striking the right balance between setting stringent requirements to ensure the operational security of the system while not creating barriers to investments can be challenging. For instance, grid operators have addressed this issue by directly adopting grid codes from other countries without considering the unique characteristics of their own systems and particular considerations. This approach often results in grid codes that are ill-suited for practical use.

Each grid code should be adapted to address the specific needs of its respective system, technological context as well as geographical, environmental and societal conditions. By learning from the experiences of other countries and collaborating internationally, grid codes can be more effectively tailored to accommodate the unique demands of DER integration and ensure the reliability and security of the power system. This approach allows for the development of grid codes that strike a balance between flexibility for renewable energy investment and stringent operational standards. Figure 8.10 shows this formulation based on the level of renewable energy integration and the size of the system.

The grid code defines the conditions under which entities can access and connect to the electricity grid. In some countries, DSOs establish separate rules for the low-voltage or distribution level in what is known as the distribution code.

The grid code governs various aspects, including how installations can connect to either the high and extra-high voltage grid or the low-voltage and distribution grid. It also addresses topics such as grid usage, system services, grid expansion, general grid operation and investment-related aspects. These rules are applicable to all systems that feed electricity into the grid network, regardless of their size to ensure the reliable operation of the grid.

While grid codes may vary from one country or operator to another, they typically encompass regulations on the following subjects:

- Connection requirements: define conditions related to voltage quality and the supply of active and reactive power by the connected system.

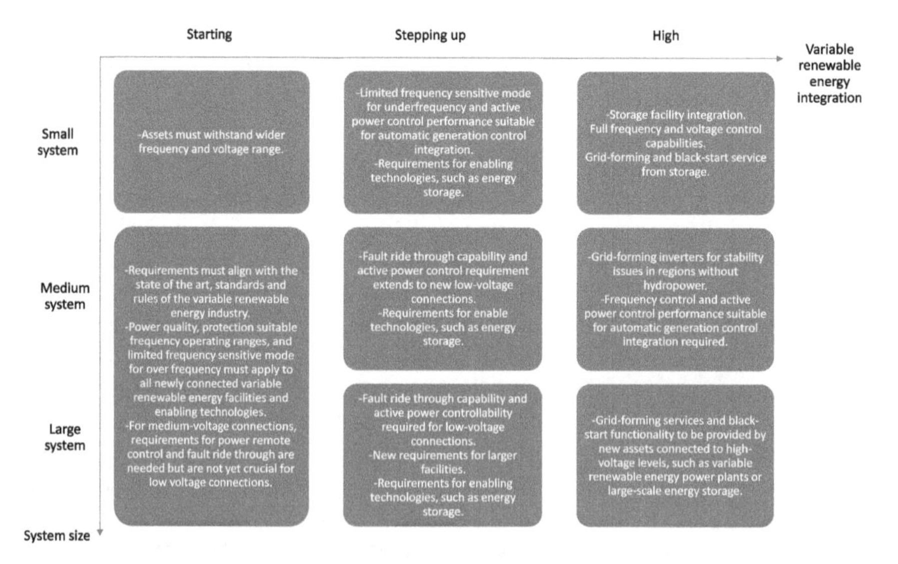

Figure 8.10 Grid code formulation guidance for DER integration

- System behaviour during grid failures: outline how the connected system should respond in the event of power grid failures, including provisions for islanding (isolating from the grid) and black start capability (restarting the system from a blacked-out state).
- Grid protection: detail measures to establish the proper protection systems.
- Remote control: may specify technical requirements allowing the grid operator to remotely control or limit power generation during grid disturbances.
- System services: define the provision of ancillary services, such as frequency regulation and voltage support, by the connected power generator.
- Network operator responsibilities: describe the roles of the network operator concerning grid capacity and architecture.

8.5.1 Grid code development

8.5.1.1 Relevance of requirements

The technical requirements outlined in a grid code should be tailored to the prevailing level of DER integration within the power system. In this regard, grid codes are closely intertwined with a country's energy policy, and coordination is needed to align technical requirements with anticipated increases in DER capacity [15,16]. It's essential to draft grid codes with a forward-looking perspective that considers the system security needs associated with future planned DER shares.

It is of great importance to strike a balance when evaluating the cost and benefit of each requirement specified in the grid code. Grid code requirements can influence the investment costs for DER generators, as they may need to install additional equipment to comply with the code. Requirements that are overly strict or unconventional can drive up the investment costs to the point where DER projects are no longer economically viable. On the contrary, lenient requirements can lower investment costs and lead to a rapid increase in DER shares.

8.5.1.2 Studies

The development of grid code requirements involves deep research into the needs of the power system itself, considering the capabilities of available generator systems and available infrastructure. Comprehensive planning and a deep understanding of grid code requirements are vital to support the successful integration of renewable energy sources and maintain the reliability of the power grid.

To ensure that grid code requirements don't hinder the adoption of DER, several studies and considerations are crucial:

- Load flow studies: are conducted to examine the required active/reactive power capabilities of generation units.
- Static and dynamic short-circuit studies: these studies are crucial for evaluating protection requirements and low-voltage ride-through (LVRT) requirements. They help ensure that the power system can withstand and respond effectively to critical events.
- Ramping study: this study focuses on reserve requirements and gradient limitations. It's essential to assess the ramping capabilities of generators, ideally including a frequency stability study.

8.5.1.3 Impact of technical requirements in small markets

Small power systems present a unique challenge when it comes to the development of grid codes. Several factors come into play:

- High development costs: The cost of developing grid codes can be disproportionately high when compared to the overall budget of the utility or system operator in a small power system.
- Rapid DER growth: Incentives for DER can lead to a rapid increase in DER shares within a very short time frame, which may outpace the development of grid codes.
- System stability: Small power systems are more vulnerable to issues affecting system stability.

The characteristic dynamics of small power systems mean that grid code development must strike a delicate balance. Here are some key considerations:

- Stricter requirements in large markets: If a significant player in DER development, such as a larger country with vast experience and a stable incentive system, decides to impose stricter or highly specific requirements, manufacturers are likely to adapt and offer the necessary technology to tap into the substantial market. Thus, a reasonable increase in prices would take place.
- Challenges in small markets: In contrast, small markets may face more difficulties if they attempt to introduce entirely new requirements. Manufacturers might find it economically unfeasible to develop new technology for such a limited market, potentially resulting in significantly higher investment costs.

Given these challenges, operators of small power systems are advised to perform a comprehensive analysis of the available technology options before establishing their grid code requirements.

8.5.1.4 Grid code revision process

It is essential to consider the anticipated and preferred future progress of the power system when formulating a grid code. A grid code tailored solely for the current situation can swiftly become outdated. This situation can potentially jeopardise the operational security and stability of the grid.

Regular revisions and modernisations of grid codes are crucial to keep pace with advancements in technology, changes in the economic landscape and the power system and load dynamics as they are. Establishing well-defined, routine procedures for assessing and amending the grid code as emerging challenges surface is highly advisable. Updates that occur too frequently may incur excessive costs. Conversely, waiting to update the grid code until issues have already become apparent may be too late and result in costly retrofits and significant damages in infrastructure and electrical equipment.

8.5.1.5 Grid code compliance

Establishing technical requirements for grid-connected generators would have limited effectiveness if there were no means to confirm compliance with these regulations. It is imperative to have well-defined processes that are tailored to the unique characteristics of the country, power system and grid code in question.

8.5.2 Grid codes overview

8.5.2.1 EU and UK framework

The framework, commonly referred to as the EU Network Codes, consists of eight primary documents [17–19]. Among these, only four are currently classified as actual network codes, while the remaining four have the status of guidelines (as indicated in Figure 8.8).

The entire framework was developed by ENTSO-E (European Network of Transmission System Operators for Electricity) in collaboration with the European Agency for the Coordination of Energy Regulators over the period of 2009–2015. It was officially incorporated into European law in 2016 and was put into effect in 2016 and 2017. National Transmission System Operators (TSOs) were required to align their regulations with the provisions outlined in the EU Network Codes indicated in Table 8.3.

Applying similar sets of requirements to all generators across the European Union faces a significant challenge due to the existence of five synchronous systems of varying sizes. The need for frequency stability functionalities may differ for smaller generators, as they can have a more pronounced impact on the stability of the entire system in smaller synchronous systems. To address this, the requirements for generators (RfG) specifies four generator size classes where each type is required to meet a specific set of requirements, as shown in Tables 8.4 and 8.5. Note that the units connected to a voltage level of 110 kV or higher are always type D, regardless of capacity.

The following aspects are excluded from the scope of these regulations:

- Evaluation of individual DER interconnections.
- Requirements specific to distribution systems such as voltage control, interface protection, islanding and safety.

Table 8.3 EU network codes

Connection	Operation	Market
Requirements for generators (RfG)	Emergency and restoration (ER)	Forward capacity allocation (FCA)
High-voltage direct-current connections (HVDCC)	Operation	Capacity allocation and congestion management (CACM)
Demand connection code (DCC)	No more than 20% of rated capacity	Electricity balancing (EB)

Table 8.4 Capacity threshold for power-generating modules of four types

Type	Continental Europe	Great Britain	Ireland	Nordic	Baltic
A	0.8 kW	0.8 kW	0.8 kW	0.8 kW	0.8 kW
B	1 MW	1 MW	0.1 MW	1.5 MW	0.5 MW
C	50 MW	50 MW	5 MW	10 MW	10 MW
D	75 MW	75 MW	10 MW	30 MW	15 MW

Table 8.5 Main requirements in the EU network code RfG

	Type A	Type B	Type C	Type D	Comments
Frequency range	√	√	√	√	
LFSM-O	√	√	√	√	
LFSM-U			√	√	
LVRT		√	√	√	
Dynamic fault current		√	√	√	
LVRT to 0 voltage				√	
Protection coordination		√	√	√	
FSM			√	√	
Black start			√	√	Non-mandatory
Island operation			√	√	Non-mandatory
Fault recorder			√	√	
Simulation models			√	√	
Voltage ranges				√	
Reactive power		√	√	√	Type B: synchronous only

- Compliance testing.
- Implementation of communication and control measures.

These aspects are still governed by local laws, standards, codes or regulations at the national or utility level.

8.5.2.2 North America

In Canada and the United States, transmission systems are overseen and operated by various independent system operators (ISO) [1]. These entities typically perform both power system operation and market operation roles, but they may not own the grid infrastructure, as separate transmission companies often focus on maintenance and asset management. However, the sub transmission and distribution grids are managed by a range of different grid operators and utility companies.

Even though the North American power system involves only two countries (though they are highly federalised), the landscape of power system structures and grid codes in North America is as fragmented and varied as it is in Europe.

The North American Electric Reliability Corporation (NERC) plays a crucial role in this context. NERC is a non-profit association of network operators, serving a role analogous to ENTSO-E in Europe. NERC is responsible for ensuring the reliability of the bulk power system at the transmission level as it operates through eight regional entities and focuses on maintaining the proper functioning of the electricity market and operation of the system. Currently, there are 102 NERC standards that apply to North American power systems, as detailed in Table 8.6.

When it comes to technical requirements for generators, the NERC Reliability Standards in North America cover a significantly smaller number of items compared to the European Network Codes for Requirements for Generators (EU NC RfG). The NERC standards primarily focus on requirements that are essential for ensuring system

Table 8.6 Applicable NERC standards for TOs/ISOs

Section	**Acronym**	**Number of standards**
Resource and demand balancing	BAL	8
Critical infrastructure protection	CIP	16
Communications	COM	2
Emergency preparedness and operations	EOP	6
Facilities, design, connection and maintenance	FAC	9
Interchange scheduling and coordination	INT	2
Interconnection reliability operation and coordination	IRO	12
Modelling, data and analysis	MOD	12
Nuclear	NUC	1
Personal performance, training and qualifications	PER	4
Protection and control	PRC	20
Transmission operations	TOP	4
Transmission planning	TPL	3
Voltage and reactive	VAR	3

Table 8.7 Frequency range requirements and minimum operation times according to NERC

	WECC	**ERCOT**	**Quebec**
>66.0			Instantaneous trip
≥63.0			5 s
≥61.8		Instantaneous trip	
≥61.7	Instantaneous trip		
≥61.6	30 s	30 s	
≥61.5			90 s
≥60.6	180 s	540 s	660 s
59.4–60.6	Continuous operation		
≤59.4	180 s	540 s	660 s
≤58.5			90 s
≤58.4	30 s	30 s	
≤58.0		2 s	
≤57.8	7.5 s		
≤57.5		Instantaneous trip	10 s
≤57.3	0.75 s		
≤57.0	Instantaneous trip		2 s
≤56.5			0.35 s
<55.5			Instantaneous trip

stability at the synchronous system level. These requirements are generally technology-neutral, meaning they do not prescribe specific technological solutions but rather specify what needs to be achieved in terms of system stability. One example is the frequency requirements that can change according to one of the regional entities in which the generator is located: Western Electricity Coordinating Council (WECC), Electric Reliability Council of Texas (ERCOT) or Quebec, as shown in Table 8.7.

However, other technical requirements for synchronous generators and inverter-based resources are often determined by individual ISOs. This approach allows ISOs to tailor specific technical requirements to their regions and systems, considering their unique characteristics and needs.

References

[1] North American Electric Reliability Corporation (NERC), 'Distributed Energy Resources, Connection, Modeling and Reliability Considerations – Report', 2017.

[2] DIgSILENT, 'Distribution Network Tools – DIgSILENT' [Online]. Available: https://www.digsilent.de/en/distribution-network-tools.html#top [accessed 25 March 2024].

[3] F. Mišurović and S. Mujović, 'Numerical probabilistic load flow analysis in modern power systems with intermittent energy sources', *Energies (Basel)*, vol. 15, no. 6, 2038, 2022, doi:10.3390/en15062038.

[4] S. M. Mirbagheri, D. Falabretti, V. Ilea, and M. Merlo, 'Hosting capacity analysis: a review and a new evaluation method in case of parameters uncertainty and multi-generator', *IEEE International Conference on Environment and Electrical Engineering and 2018 IEEE Industrial and Commercial Power Systems Europe (EEEIC / I&CPS Europe)*, Palermo, Italy, 2018, pp. 1–6, doi:10.1109/EEEIC.2018.8494572.

[5] L. Ibarra, A. Rosales, P. Ponce, A. Molina, and R. Ayyanar, 'Overview of real-time simulation as a supporting effort to smart-grid attainment', *Energies (Basel)*, vol. 10, no. 6, 817, 2017, doi:10.3390/en10060817.

[6] O. Nzimako and R. Wierckx, 'Modeling and simulation of a grid-integrated photovoltaic system using a real-time digital simulator', *IEEE Trans. Ind. Appl.*, vol. 53, no. 2, pp. 1326–1336, 2017, doi:10.1109/TIA.2016.2631120.

[7] W. Ren and Others, 'Interfacing issues in real-time digital simulators', *IEEE Trans. Power Deliv.*, vol. 26, no. 2, pp. 1221–1230, 2011, doi: 10.1109/TPWRD.2010.2072792.

[8] T. Logenthiran, D. Srinivasan, A. M. Khambadkone, and H. N. Aung, 'Multiagent system for real-time operation of a microgrid in real-time digital simulator', *IEEE Trans Smart Grid*, vol. 3, no. 2, pp. 925–933, 2012, doi: 10.1109/TSG.2012.2189028.

[9] Y. Song, C. Wan, X. Hu, H. Qin, and K. Lao, 'Resilient power grid for smart city', *iEnergy*, vol. 1, no. 3, pp. 325–340, 2022, doi:10.23919/ien.2022.0043.

[10] M. Yazdanie and K. Orehounig, 'Advancing urban energy system planning and modeling approaches: Gaps and solutions in perspective', *Renewable and Sustainable Energy Reviews*, vol. 137, 110607,2021, doi:10.1016/j.rser.2020.110607.

[11] J. M. Gers, *Distribution System Analysis and Automation*, 2nd ed. London: The Institution of Engineering and Technology, 2020.

[12] L. Bottaccioli, A. Estebsari, E. Patti, E. Pons, and A. Acquaviva, 'Planning and real-time management of smart grids with high PV penetration in Italy', *Proceedings of the Institution of Civil Engineers – Engineering Sustainability*, vol. 172, no. 6, pp. 272–282, 2019, doi:10.1680/jensu.17.00066.
[13] Capgemini, 'Advanced Distribution Management Systems', 2012. Available: https://www.capgemini.com/wp-content/uploads/2017/07/Advanced_Distribution_Management_Systems.pdf.
[14] A. Razon, T. Thomas, and V. Banunarayanan, 'Advanced distribution management systems: connectivity through standardized interoperability protocols', *IEEE Power and Energy Magazine*, vol. 18, no. 1, pp. 26–33, 2020, doi:10.1109/MPE.2019.2947816.
[15] International Renewable Energy Agency, 'Grid Codes for Renewable Powered Systems', 2022. Available: https://www.irena.org/publications/2022/Apr/Grid-codes-for-renewable-powered-systems.
[16] Enel Group, 'The Future of Electric Grid', 2022. [Online]. Available: https://www.enel.com/company/services-and-products/enel-grids [accessed 24 March 2024].
[17] Austrian Institute of Technology, 'Grid Codes in Europe. Overview on the current requirements in European codes and national interconnection standards', in *NEDO/IEA PVPS Task 14 Workshop*, 2019 [Online]. Available: https://www.researchgate.net/publication/338800967_Grid_Codes_in_Europe_-_Overview_on_the_current_requirements_in_European_codes_and_national_interconnection_standards [accessed 24 March 2024].
[18] Next Kraftwerke GmbH, 'The Grid Code: Rules for the Power Network', 2019. [Online]. Available: https://www.next-kraftwerke.com/knowledge/what-is-the-grid-code [accessed 24 March 2024].
[19] T. Ackermann, P.-P. Schierhorn, and D.-I. N. Martensen, 'The role of grid codes for VRE integration into power systems development of power system specific requirements', in *6th International Workshop Integration of Solar Power into Power Systems*, Vienna, 2016.

Index

Printed in the USA
CPSIA information can be obtained
at www.ICGtesting.com
JSHW010041061024
71105JS00003B/19